THERMOPHYSICAL PROPERTIES OF MATERIALS

GÖRAN GRIMVALL

The Royal Institute of Technology
Stockholm, Sweden

1986

NORTH-HOLLAND
AMSTERDAM · OXFORD · NEW YORK · TOKYO

ISBN: 0 444 86985 9

Published by:
North-Holland Physics Publishing
a division of
Elsevier Science Publishers B.V.
P.O. Box 103
1000 AC Amsterdam
The Netherlands

Sole distributors for the U.S.A. and Canada:
Elsevier Science Publishing Company, Inc.
52 Vanderbilt Avenue
New York, N.Y. 10017
U.S.A.

Library of Congress Cataloging in Publication Data

Grimvall, Göran.
Thermophysical properties of materials.

Bibliography: p.
Includes index.
1. Materials--Thermal properties. I. Title.
TA418.52.G75 1986 620.1'121 85-28394
ISBN 0-444-86985-9 (U.S.)

Printed in The Netherlands

THERMOPHYSICAL PROPERTIES OF MATERIALS

SERIES OF MONOGRAPHS ON SELECTED TOPICS IN SOLID STATE PHYSICS

Editor: E.P. WOHLFARTH

1 Magnetostatic principles in ferromagnetism, *W.F. Brown, Jr.*
2 The physics of magnetic recording, *C.D. Mee*
3 Symmetry and magnetism, *R.R. Birss*
4 Ferromagnetism and ferromagnetic domains, *D.J. Craik* and *R.S. Tebble*
5 The growth of crystals from the melt, *J.C. Brice*
6 X-ray determination of electron distributions, *R.J. Weiss*
7 Ferroelectricity, *E. Fatuzzo* and *W.J. Merz*
8 Experimental magnetochemistry, *M.M. Schieber*
9 Experimental methods in magnetism, *H. Zijlstra*
10 Group theory and electronic energy bands in solids, *J.F. Cornwell*
11 The electrodynamics of magneto-electric media, *T.H. O'Dell*
12 The growth of crystals from liquids, *J.C. Brice*
13 Soft modes in ferroelectrics and antiferroelectrics, *R. Blinc* and *B. Zeks*
14 Magnetic bubbles, *A.H. Bobeck* and *E. Della Torre*
15 Magnetism and metallurgy of soft magnetic materials, *C.W. Chen*
16 The electron–phonon interaction in metals, *G. Grimvall*
17 Amorphous magnetism and metallic magnetic materials – digest, *A.R. Ferchmin* and *S. Kobe*
18 Thermophysical properties of materials, *G. Grimvall*

PREFACE

Materials science is a very broad field, dealing with fundamental solid state theory as well as practical aspects of engineering materials. It is often assumed that these two extremes are linked, in that a good knowledge of the basic concepts is necessary, or at least of much help, in order to tackle applied problems. I had my first scientific training in solid state theory, covering Green functions and similar concepts. I also spent eight years, part-time, at a department of engineering materials, after I gained my Ph.D. It was then very frustrating to realise the large gap between my knowledge and that which was required for an understanding of "real" materials. This book has been written in an attempt partly to bridge that gap, in a specific area of materials science. Most textbooks in solid state physics make assumptions such as that of a single crystal having a monatomic cubic lattice, without lattice defects, anharmonicity in the lattice vibrations, etc. This book gives information about the effect of non-cubic structures, non-perfect lattices, non-harmonic vibrations and similar complications.

It is always a problem to decide what to include and what to leave out of a book covering a broad field. The following criteria have been used in this case. Low-temperature phenomena, which are studied in basic research at helium temperatures, are almost entirely neglected. Instead, emphasis is on properties at ambient and high temperatures. Thermophysical phenomena which clearly belong to a specific research field, for instance superconductivity, semiconductor physics and magnetism, are also left out. This leaves room for a more detailed treatment of aspects which are of general importance, such as elasticity, vibrational amplitudes and imhomogeneous materials. What might be called chemical thermodynamics, which is not based on a description in terms of electrons and phonons, is not treated. Finally, I have not dealt with phenomena or materials for which our understanding has not yet reached the maturity of traditional solid state physics. That means neglect of, e.g., atomic mass transport and polymers. Still, there is a wide field to cover. It is hoped that the reader will find the text itself, or the references to more detailed accounts, helpful on the route from basic to applied solid state physics, or *vice versa*.

Since this book is written when we face the age of "electronic publication" it may be of interest to future generations of scientists to know how it came into being. I have not made use of databases to search for information, but have instead read the contents pages of very many journals. The manuscript was typed by myself, on a terminal in my office, using a word-processing program and the main computer of our university. A paper printout was then sent to the publisher for ordinary typesetting.

As alluded to above, I spent the first thirteen years of my scientific life at the Department of Theoretical Physics and the Department of Metallic Engineering Materials, both at the Chalmers Institute of Technology, Gothenburg, Sweden. I want to thank, particularly, Professor Stig Lundqvist who introduced me to solid state theory and Professor Hellmut Fischmeister (now in Stuttgart) for letting me enter his field of engineering materials. It is this mixed background that has lead me to write the book. Finally I also want to deeply thank my wife, Siv, who has relieved me of the household work during the last hectic months before the deadline, even though she also has full-time employment.

Göran Grimvall
Stockholm, April 1985

LIST OF MOST IMPORTANT SYMBOLS*

C	heat capacity
C_g	phonon group velocity (42)
C_{har}	harmonic phonon heat capacity (86)
$C(\boldsymbol{q}, \lambda)$	sound velocity of mode $(\boldsymbol{q}, \lambda)$ (37)
C_p	heat capacity at constant pressure (182)
C_V	heat capacity at constant volume (182)
c	concentration of impurities, etc.
c_{ij}	elastic stiffnesses (26)
D	dynamical matrix (308)
E	energy
E	Young's modulus (26)
E_F	Fermi energy (303)
e	electronic charge
F	Helmholtz free energy
$F(\omega)$	phonon density of states (67)
f	Fermi–Dirac distribution function (158)
f_i	volume fraction of phase i (260)
G	Gibbs free energy
G	shear modulus (26)
H	enthalpy
K	bulk modulus (26)
K_S	isentropic (adiabatic) bulk modulus (182)
K_T	isothermal bulk modulus (182)
k_B	Boltzmann's constant
$\boldsymbol{k}$	electron wave vector
k_F	Fermi wave number (303)
M	ion (atom) mass
m	free electron mass
m_b	electronic band mass (307)
m_{th}	electronic thermal mass (168)

* (Page where the quantity is introduced or defined within parentheses.)

N	total number of ions (atoms)
$N(E)$	electronic density of states (306)
$N(E_F)$	density of states at Fermi level (157, 303)
n	number of electrons per unit volume (303)
n	Bose–Einstein distribution function (79)
p	pressure
$\boldsymbol{q}$	phonon wave vector
q_D	Debye wave number (69)
$(\boldsymbol{q}, \lambda)$	label on phonon state (66)
R	position of ion (atom)
r_s	electron density parameter (303)
S	entropy
s_{ij}	elastic compliances (26)
T	temperature
T_F	Fermi temperature (304)
T_m	melting temperature
U	lattice energy (1)
$\boldsymbol{u}$	displacement vector of ion (atom) (89, 308)
V	crystal volume
$\boldsymbol{v}, \boldsymbol{v_k}$	electron velocity (208)
v_F	Fermi velocity (303)
Z	ionic charge
α	linear expansion coefficient (177)
$\alpha_{tr}^2 F(\omega)$	transport coupling function (215)
β	cubic expansion coefficient (177)
γ	electronic heat capacity coefficient (157)
γ_G	Grüneisen parameter (183)
$\gamma(n)$	generalised Grüneisen parameter (121)
$\gamma(\boldsymbol{q}, \lambda)$	Grüneisen parameter of phonon mode $(\boldsymbol{q}, \lambda)$ (119)
$\delta_{2(3,4)}$	anharmonic phonon frequency shifts (125)
$\varepsilon, \varepsilon_{\boldsymbol{k}}$	electron energy
ε	elastic strain (27)
$\boldsymbol{\varepsilon}(\boldsymbol{q}, \lambda)$	phonon eigenvector (66)
Θ_D	Debye temperature (69)
$\Theta_D(n)$	generalised Debye temperature (76)
Θ_E	Einstein temperature (100)
κ	compressibility (26)
κ_S	isentropic (adiabatic) compressibility (182)
κ_T	isothermal compressibility (182)
κ	label on atom in unit cell (308)

κ	thermal conductivity (227)
κ_{ph}	phonon part of thermal conductivity (231)
κ_e	electron part of thermal conductivity (249)
λ	phonon branch index (38, 66)
λ	electron–phonon interaction parameter (157)
ν	Poisson's ratio (26)
ρ	electrical resistivity (206)
ρ	mass density of a solid
σ	electrical conductivity (206)
σ	elastic stress (27)
τ	scattering time (206)
τ	shear stress (29)
Ω_a	atomic volume (303)
ω	phonon frequency
ω_D	Debye frequency (69)
$\omega_D(n)$	generalised Debye frequency (76)
$\omega(\boldsymbol{q}, \lambda)$	phonon frequency of mode $(\boldsymbol{q}, \lambda)$ (66)

CONTENTS

CHAPTER 1

ENERGIES AND STRUCTURES OF LATTICES

1. Introduction

The crystal structure plays a major role in determining many properties of materials, in particular their mechanical behaviour. Usually, the structure and the lattice parameters are assumed to be known and a theory is built on that framework. In this introductory chapter, however, we shall deal with the prediction of lattice structures and atomic volumes.

Consider a certain crystal structure, with a characteristic linear dimension λ (λ can be a lattice parameter, the distance between the centers of nearest neighbours, the cube root of the molar volume, etc.). The total energy $E(\lambda)$ often varies with λ as shown schematically in fig. 1.1. This curve determines three parameters which have a direct physical interpretation. The position of the minimum, $\lambda_{\min}$, gives the equilibrium lattice parameter (or atomic volume, etc.). The curvature at the minimum is related to the bulk modulus K,

$$K = -V(\partial p/\partial V) = V(\partial^2 E/\partial V^2) = V(\partial^2 E/\partial \lambda^2)(\partial \lambda/\partial V)^2, \tag{1.1}$$

where V is the volume of the sample and p is the pressure. The derivatives are evaluated at $\lambda_{\min}$. The depth of the minimum gives the crystal binding energy U_b;

$$U_b = E(\lambda \to \infty) - E(\lambda_{\min}). \tag{1.2}$$

The binding energy may be different from the cohesive energy U_c. In an ionic solid, like NaCl, U_b is most naturally measured relative to a state with separated Na^+ and Cl^- ions. In our free-electron description of a metal (§ 2.2), $E(\lambda \to \infty)$ corresponds to separated ions. The cohesive energy of a solid, however, is defined relative to a state with all atoms neutral and infinitely separated. The reader is warned that the

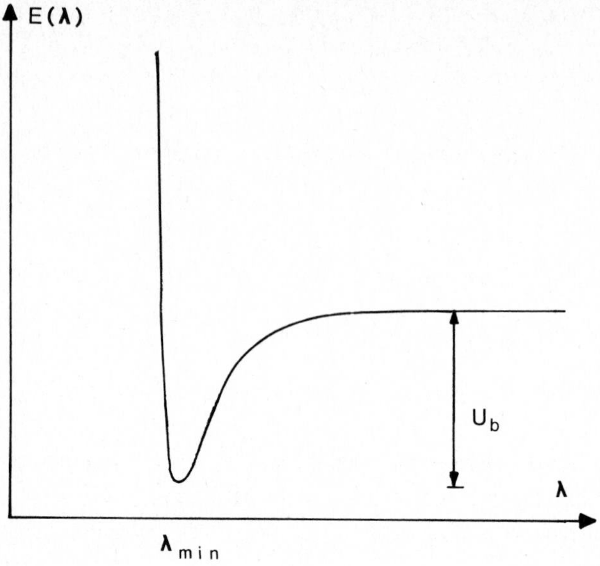

Fig. 1.1. The total crystal energy $E(\lambda)$ as a function of a lattice parameter λ. U_b is the binding energy.

distinction between the binding energy and the cohesive energy is not always made in the literature.

To predict the actual crystal structure of a solid, one has to compare binding energies U_b (or cohesive energies U_c) for all conceivable lattice structures and find the lowest U_b. In practice the comparison is limited to the most likely structures, such as fcc, bcc and hcp lattices in the case of metals. Naturally, this procedure would fail to predict, for example, the unusual crystal structures of gallium. An additional complication, which is often neglected in calculations, is that of dynamical instability. For instance, a cubic lattice may have a minimum in $E(\lambda)$ when λ corresponds to a certain value of the lattice parameter a, but a further lowering of E may occur if the lattice is sheared. Therefore one should consider $E(\lambda_1, \lambda_2, \ldots, \lambda_n)$, where the parameters λ_i describe all possible deformations of a lattice. Figure 1.1 corresponds to a minimum when E is a function of only one λ_i, but it does not say if this is a true minimum or, say, a saddle point in the complete λ space.

It is instructive to express some characteristic energies in the unit $k_B T_m$ (per atom), where T_m is the melting temperature. Table 1.1 gives typical values for U_c, the energy difference ΔE_{struct} between the most stable and

Table 1.1
Typical energy scales in solids. Energy per atom, in units of $k_B T_m$.

U_c	ΔE_{struct}	ΔH_{AB}	$E_{vib}\,(T = 0)$
25–40	0.1–1	~ 1	0.1–0.2

the second most stable crystal structure, the heat of formation ΔH_{AB} of an intermetallic compound formed by the elements A and B and the zero-point vibrational energy $E_{vib}(T = 0)$.

2. Simple models of cohesive properties

2.1. Introduction

Our goal is to account for structural properties such as the atomic volume, the binding energy and the bulk modulus. There are a number of simple models with this aim. Although some of them are outdated in their details, they give a qualitative understanding of important features and provide a background to our subsequent discussion of ab initio calculations. Quite different approaches will be taken to describe non-transition metals (also called free-electron-like or simple metals), transition metals, ionic and covalent solids. For details, see, for example, the monograph by Harrison (1980).

2.2. Free-electron-like metals

The simplest representation of a metal is the jellium model. The ion charges are "smeared out" into a uniform positive background. The distribution of the conduction electrons is also spatially uniform. Since there is charge neutrality everywhere, this system has no electrostatic energy. Let there be N atoms in a volume V. The only energy which varies with the atomic volume $\Omega_a = V/N$ is the kinetic energy of the electron

$$Z\langle E_{kin}\rangle = 2.210 Z r_s^{-2}\,[\mathrm{Ry}]. \tag{1.3}$$

Here and in the rest of this chapter, energies refer to an average per atom (or stoichiometric unit in compounds). The dimensionless parameter r_s is a measure of the electron number density, $r_s a_0$ being the radius of a sphere with volume Ω_a/Z, where Z is the number of valence electrons contributed by each ion and a_0 is the Bohr radius. Typically, r_s lies between 2 and 4 (cf. fig. 1.2). The energy is expressed in Rydberg units ($1\,\mathrm{Ry} = me^4/2\hbar^2$). See appendix A for details.

Since the energy (1.3) can always be lowered if the system is allowed to expand, i.e. if r_s increases, this model does not lead to cohesion. We therefore modify the model and concentrate the positive charges in point ions Ze, but let the electron gas remain uniformly distributed. Then the local charge neutrality is lost, and we get a Coulomb energy (appendix A)

$$\langle E_C \rangle = -\alpha_C Z^{5/3} r_s^{-1}\,[\mathrm{Ry}]. \tag{1.4}$$

Here α_C is a Madelung constant which depends on the spatial arrangement of the positive point charges (table 1.2).

Table 1.2
The Madelung constant α_C

Lattice structure		α_C
Single point ion		1.8000 [a]
Simple cubic		1.760 [b]
Body-centered-cubic		1.79186 [c]
Face-centered-cubic		1.79172 [c]
Hexagonal close-packed	($c/a = 1.633$)	1.79168 [b]
	($c/a = 1.5$)	1.78998 [d]
	($c/a = 1.8$)	1.78909 [d]
Diamond		1.671 [d]
White tin	($c/a = 0.554$)	1.77302 [e]

[a] Appendix A. [b] Carr (1961). [c] Fuchs (1935).
[d] Harrison (1966). [e] Ihm and Cohen (1980).

The energy E_{tot}, the sum of the kinetic energy (1.3) and the Coulomb energy (1.4), has a minimum when $\partial E_{tot}/\partial r_s = 0$, i.e. at

$$r_s = (4.42/\alpha_C) Z^{-2/3}. \tag{1.5}$$

This gives the binding energy

$$U_b = 0.113 \alpha_C^2 Z^{7/3}\,[\mathrm{Ry}]. \tag{1.6}$$

As an improvement, but still within the jellium model, we add an exchange energy $-0.916\, r_s^{-1}$ and an approximation for the correlation energy (appendix A) to get the exchange and correlation contribution

$$\langle E_{x,c}\rangle = -0.916 r_s^{-1} + 0.031 \ln(r_s) - 0.115\,[\text{Ry}]. \tag{1.7}$$

Finally, we also add a structure independent energy E_{ec} calculated from the empty core (or Ashcroft) pseudopotential when the electron density is still assumed to be uniform (and hence screening is neglected). One has

$$E_{ec} = 2\pi Z e^2 r_c^2 n = 3(r_c/a_0)^2 Z r_s^{-3}\,[\text{Ry}]. \tag{1.8}$$

Here r_c is a parameter (the core radius) in the pseudopotential and a_0 is the Bohr radius. Typically, $r_c/a_0 \approx 1$. E_{ec} is a repulsive term which arises because the conduction electrons tend to stay away from the core electrons.

Figure 1.2 exemplifies (for the ion charge $Z = 2$) how the energy per

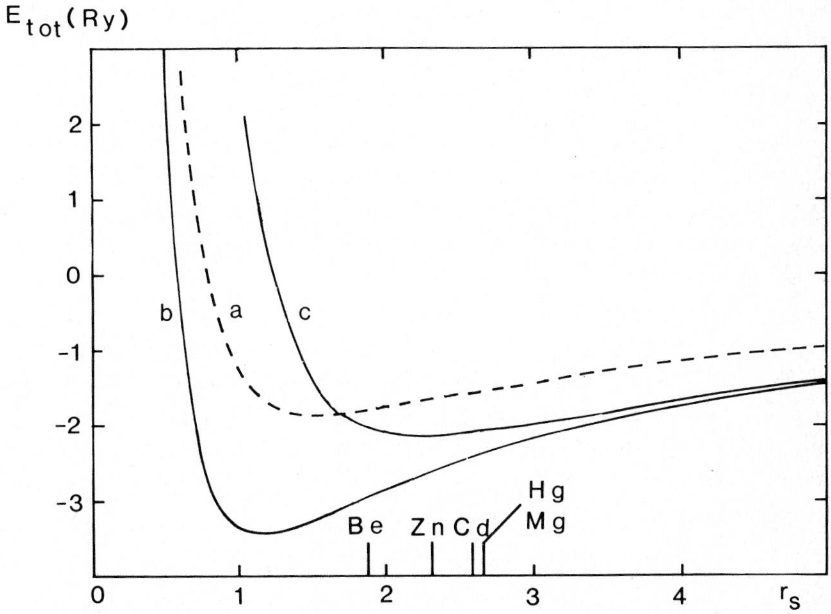

Fig. 1.2. The energy $E_{tot}(r_s)$ of a jellium metal with $Z = 2$. Curve a includes kinetic and Coulomb energies, curve b has an exchange and correlation energy added while curve c includes an "empty core" pseudopotential term. The actual values of r_s for divalent simple metals are marked.

atom varies with the atomic volume (represented by r_s) when E_{tot} includes first only (1.3) and (1.4), then (1.7) and finally also (1.8). We note that the exchange and correlation energies and the "empty-core" term have a strong influence on the total binding energy. Since α_C does not vary much with the lattice type, the structural dependence of U_b is too small to be shown in fig. 1.2. It is obvious that a theoretical account of the relative crystal structure stability requires a much more refined theory than presented here. However, the low value of α_C for the diamond lattice makes that structure improbable for a free-electron-like metal.

The atomic volume is obtained from (1.5) as

$$\Omega_a = (4.42/\alpha_C)^3(4\pi a_0^3/3)Z^{-1} = 9.17Z^{-1}10^{-30}\,[\mathrm{m}^3]. \tag{1.9}$$

In the last step, we have taken $\alpha_C = 1.8$ for all structures. Since the energy expression which yields Ω_a in (1.9) is not very accurate, we cannot expect Ω_a to agree well with experiment. However, it is significant that Ω_a depends very weakly on the crystal structure. Ashcroft and Langreth (1967) have calculated binding energies and bulk moduli of simple metals, using an approach similar to the one outlined here. Theory and experiment typically agree to better than 10%.

Example: c/a in hcp Cd–Mg alloys. Our simple theory should apply also to substitutional alloys of non-transition metals, at least when the components have the same valence. The Cd–Mg system is an interesting case. Mg and Cd have $Z = 2$, crystallise in the hcp structure and form a solid solution Cd_xMg_{1-x} for all x, at high temperatures. The axis ratio c/a has the anomalous value 1.90 for pure Cd, while an ideal close-packing of rigid spheres gives $c/a = \sqrt{(8/3)} \approx 1.633$. On alloying with Mg, c/a changes considerably but there is a simultaneous change in a such that the atomic volume $\Omega_a = (\sqrt{3/4})ca^2$ varies less than c and a (fig. 1.3). This provides further support for the idea that the crystal structure is of little importance in determining the atomic volume of simple metals.

2.3. *Transition metals*

The d-electrons play a major role in the transition metals. We shall use a simple model due to Friedel (1969), which neglects the s- and p-electrons altogether. Let the d-state of an isolated atom have the energy E_d^0

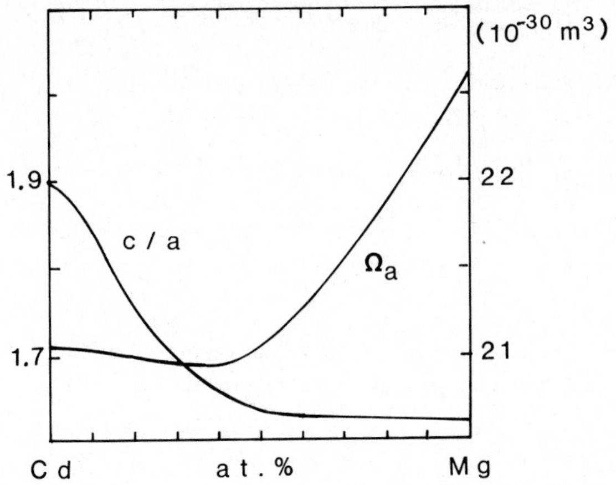

Fig. 1.3. Variation of the ratio c/a of the hcp lattice parameters (left-hand scale) and of the atomic volume Ω_a (right-hand scale), of hcp solid solutions of Cd–Mg.

relative to some reference level. When the atoms are brought together in a solid, the d-level broadens into a band described by an electron density of states $N_d(E)$. If a metal has n_d d-electrons, the cohesive energy (per atom) becomes

$$E_c = n_d E_d^0 - 2\int_{E'}^{E_F} N_d(E) E \mathrm{d}E, \tag{1.10}$$

where E' is the bottom of the d-band and E_F is the Fermi level. Friedel assumed that $N_d(E)$ is rectangular in shape, with a width W_d and a "center of gravity" shifted from the atomic level E_d^0 to E_d (fig. 1.4). The total number of d-states is 10, which fixes the height of $N_d(E)$ to $5/W_d$, if $N_d(E)$ refers to one spin. Then,

$$E_c = (E_d^0 - E_d)n_d + (1/20)W_d n_d(10 - n_d). \tag{1.11}$$

Neglecting the fact that W_d, E_d and E_d^0 vary with n_d, which is a crude but a reasonable approximation in the context, this model predicts that the cohesive energy varies parabolically with n_d. Typically, $W_d = 0.5$ [Ry] and $E_d^0 - E_d < 0.1$ [Ry]. Our model neglects any structural dependence of $N_d(E)$ but this is not too serious since E_c is an integrated quantity of $N_d(E)$ and the most important factor is the width W_d of $N_d(E)$. As seen in fig. 1.5 the cohesive energy calculated in this approximate manner is

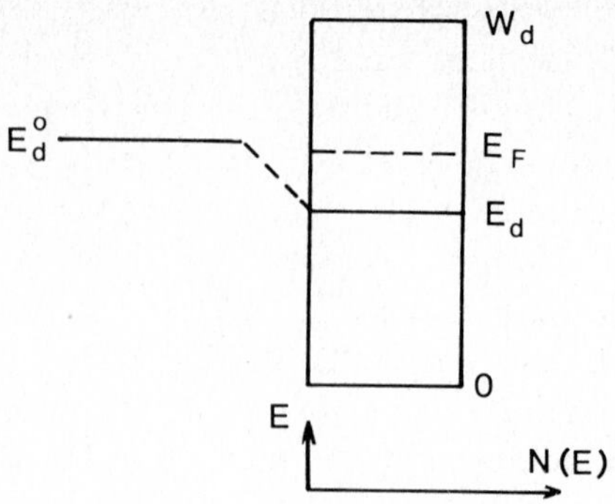

Fig. 1.4. A schematic picture of how an electron d-level shifts from its value E_d^0 in an atom and broadens into a band of width W_d in the solid.

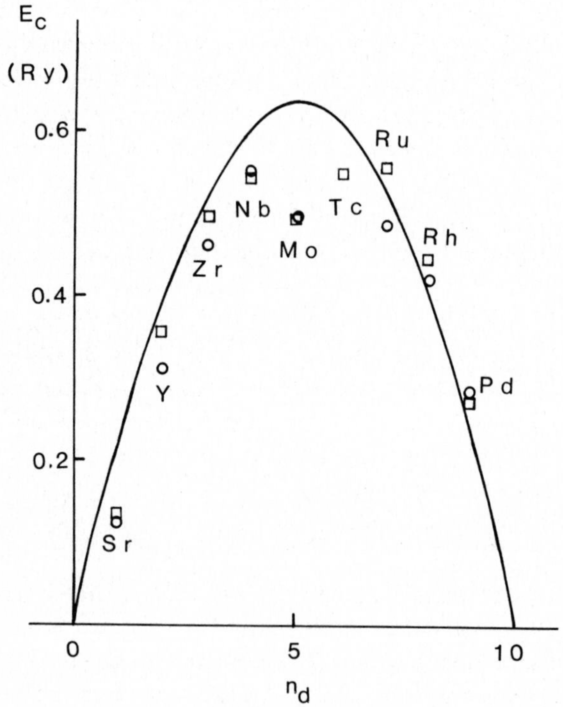

Fig. 1.5. The cohesive energy E_c (solid line), calculated from (1.11) when $E_d^0 = E_d$ and $W_d = 0.5$ Ry. Circles are experimental values and squares are theoretical results from a first-principles calculation (Moruzzi et al. 1977, 1978).

in qualitative agreement with experiments. Since the model makes no reference to how E_c changes with volume, we cannot estimate the atomic volume. Such considerations should use the fact that W_d increases with decreasing volume (ch. 10, §8) which leads to an attractive force between the atoms. This is balanced by the repulsive force arising when the conduction electrons (s- and p-electrons) are forced into the ion cores on compression. Thus the s- and p-electrons are important in determining the atomic volume and the bulk modulus, but not for the cohesive energy. This picture, which is supported by work of Gelatt et al. (1977), is contrary to a suggestion by Engel (1967) and Brewer (1967), that the lattice structure of transition metals is determined by the number of s- and p-electrons.

Example: Relative stability of fcc, hcp and bcc structures. Pettifor (1972, 1977, 1983) has given a qualitative account of how the crystal structure changes as one goes across the 3d, 4d and 5d rows of transition metals in the Periodic Table. We shall very briefly indicate the main idea, which is to use Friedel's expression (1.10) but with a realistic density of states $N_d(E)$ taken from band-structure calculations. Figure 1.6 shows characteristic $N_d(E)$ for bcc and fcc lattices, and the energy difference $U_c(\text{bcc}) - U_c(\text{fcc})$, as a function of the number of d-electrons, n_d. The bcc structure has the lower energy near $n_d = 4$. In reality the bcc structure occurs for $n_d = 5$ and 6, a result which can be obtained in a more refined treatment. Pettifor also compared the fcc and hcp lattices,

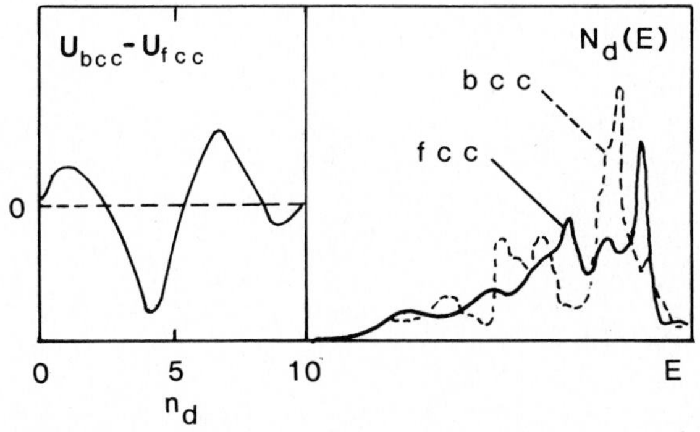

Fig. 1.6. The relative energy (arbitrary scale–left frame) of bcc and fcc structures in transition metals, as a function of the number of d-electrons, n_d. The right frame shows the characteristic shapes of the electron density of states $N_d(E)$ of bcc and fcc structures. Based on calculations by Pettifor (1972, 1977, 1983).

and discussed the effect of s- and p-electrons. Similar arguments can account for the structural trends among metal–nonmetal (pd-bonded) binary compounds (Pettifor and Podloucky 1984). Of course there are energy terms which have not been included in this simple picture, but they are insensitive to the crystal structure.

Example: The atomic volume of transition metals. In the preceding example, different structures were considered at the same atomic volume. This is a reasonable assumption, as is shown by table 1.3. It gives experimental results for the relative change in the atomic volume, $\Delta\Omega_a/\Omega_a$.

Table 1.3
The relative change in atomic volume when the crystal structure is changed

Transition	$\Delta\Omega_a/\Omega_a$ (%)
Fe(bcc) → Fe(fcc), 1183 K [a]	−1.0
Fe(fcc) → Fe(bcc), 1663 K [a]	+0.3
Fe(bcc) → Fe(hcp), 130 kbar [a]	−6
$Ni_{0.76}P_{0.24}$ (amorphous → cryst.) [b]	−0.8
$Co_{0.75}P_{0.25}$ (amorphous → cryst.) [b]	−1.3
$(FeNi)_{0.8}(BP)_{0.2}$ (amorphous → cryst.) [c]	−0.3 to −0.8 [d]

[a] Donohue (1974). [b] Logan and Ashby (1974). [c] Van den Beukel and Radelaar (1983). [d] Cooling rate to produce the amorphous structure varying from 10^3 to 10^6 K/s.

2.4. *Ionic compounds*

The cohesive properties of ionic compounds can, in their main features, be explained in terms of classical physics. This is in contrast to the metals considered in §2.2 and §2.3, where quantum mechanics played a major role. We assume that two ions, i and j, interact through a potential (e.g., Born and Huang 1954)

$$V(r) = e^2 Z_i Z_j / r + B/r^n. \tag{1.12}$$

The first term is the Coulomb interaction between charges $Z_i e$ and $Z_j e$. When summed over the lattice, it gives the Madelung energy $\langle E_M \rangle$. The next term represents a repulsive interaction which prevents the ions from coming too close to each other. We assume that it acts only between the nearest neighbours of unlike ions. As an example, consider diatomic compounds (e.g., NaCl, MgO). Let R be the shortest distance between anions and cations. The total energy of a lattice with ν nearest unlike neighbours is (per stoichiometric unit)

$$E_{\text{tot}} = -\alpha_R(Ze)^2/R + \nu B/R^n. \tag{1.13}$$

The equilibrium distance R_0 is obtained from $\partial E_{\text{tot}}/\partial R = 0$,

$$(R_0)^{n-1} = \nu Bn/\alpha_R(Ze)^2. \tag{1.14}$$

The binding energy $U_b = -E_{\text{tot}}(R_0)$ depends on three parameters: R_0, α_R and νB. If eq. (1.14) is used to eliminate νB, we obtain

$$U_b = [\alpha_R(Ze)^2/R_0](1-1/n). \tag{1.15}$$

If we only want to account for U_b, with no calculation of the molar volume, R_0 can be taken from experiments. Born related the remaining parameter, n, to the bulk modulus, i.e. to $\partial^2 E_{\text{tot}}/\partial R^2$. He found that n lies between 8 and 10 for many ionic compounds. That uncertainty in n alters U_b in (1.15) by only a few percent. The electrostatic energy, taken between point charges, is therefore by far the most important contribution to the binding energy of ionic crystals.

If (1.14) is used to eliminate R_0 in U_b we obtain

$$U_b = \alpha_R(Ze)^2[\alpha_R(Ze)^2/B\nu n]^{1/(n-1)}[1-1/n]. \tag{1.16}$$

For a given compound, we now let B and n be fixed. The relative stability of different crystal structures is determined by the relative magnitude of

$$(\alpha_R)^{n/(n-1)}\nu^{-1/(n-1)}. \tag{1.17}$$

A comparison of the ZnS, NaCl and CsCl structures on this basis shows that the ZnS structure is the most stable (has the largest U_b) when $n <$ 6.3. The stability ranges for the NaCl and CsCl lattices are $6.3 < n < 33$ and $n > 33$ respectively. This, together with the finding that typically $n \sim 9$, is in qualitative agreement with the fact that the NaCl structure is the most common among diatomic ionic compounds. However, a value $n > 33$ has no physical meaning and we need some additional argument to account for the observed stability of the CsCl structure. Born and

Huang (1954) discuss the effect of contact between like ions when the positive and negative ions have very different ionic radii. It is possible that this is a qualitatively correct way to explain the stability of the CsCl structure, but there is still uncertainty on this point (Harrison 1980).

Quantum mechanics provides a more fundamental understanding of the repulsive interaction. Since the outer parts of the electron-core wave functions decay exponentially with the distance to the nucleus, Born and Mayer (1932) suggested a repulsive interaction, between unlike nearest neighbour ions, of the form

$$V(r) = B_{BM}\exp(-r/\rho). \tag{1.18}$$

Again, there are two unknown parameters, B_{BM} and ρ. Proceeding as above, one obtains for the binding energy

$$U_b = [\alpha_R(Ze)^2/R_0][1-\rho/R_0]. \tag{1.19}$$

R_0 is the distance between two neighbouring ions and the parameter ρ approximately equals the radius of an ion core, $\rho/R_0 \sim 0.1$. Also this estimation of U_b shows that its major contribution comes from the interaction between point charges. The Madelung constant α_R is approximately the same for several crystal structures (table 1.4). An account of the relative stability of lattices therefore requires a detailed calculation, such as is discussed in § 3.

Table 1.4
Madelung constant α_R and the number ν of nearest unlike neighbours

Structure	ν	α_R
Sodium chloride	6	1.74756
Cesium chloride	8	1.76267
Sphalerite (ZnS, cubic)	4	1.63806

Example: Pressure-induced polymorphism in alkali halides. From the simple model result (1.14) and table 1.4 it follows that the CsCl structure has a smaller molar volume (i.e. a smaller R_0) but a higher energy (at R_0) than the NaCl structure. This implies the possibility of a structural transformation from the NaCl to the CsCl lattice when a pressure p is applied. The relative lattice stability is determined by the enthalpy $H = E + pV$. At the transition pressure p_t,

$$E_1(V_1) + p_t V_1 = E_2(V_2) + p_t V_2. \tag{1.20}$$

Thus, a transformation occurs at a pressure

$$p_t = [E_2(V_2) - E_1(V_1)]/(V_1 - V_2). \tag{1.21}$$

The analysis above suggests that p_t lies within easy experimental reach. This is experimentally born out by the fact that at least eight of the twelve alkali halides which crystallise in the NaCl structure at zero pressure transform to the CsCl structure at pressures below 100 GPa (Kim and Gordon 1974). Figure 1.7 shows $E(V)$ for sodium chloride in the B1 (i.e. NaCl) and B2 (i.e. CsCl) structures. The transformation

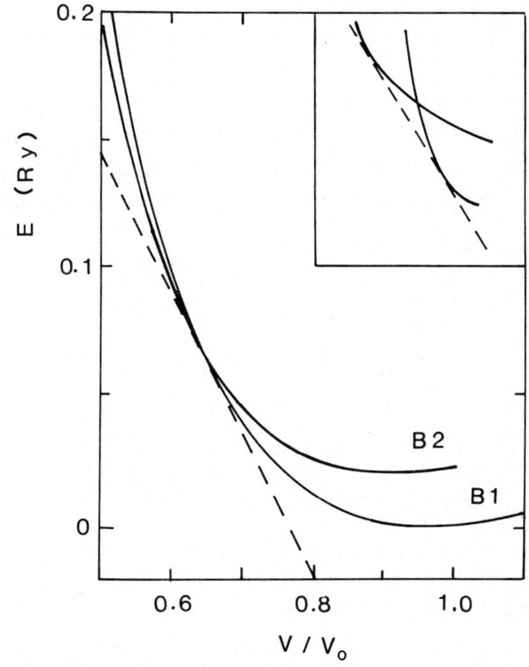

Fig. 1.7. The total crystal energy, E, of sodium chloride in the NaCl (B1) and CsCl (B2) structures, from calculations by Froyen and Cohen (1984). V_0 is the observed volume at $p = 0$. The slope of the tangent line (dashed) gives the transformation pressure p_t. The inset shows schematically how a tangent line is drawn.

pressure p_t is given by the slope of the common tangent to the $E(V)$ curves. The volumes are normalised to the observed molar volume of NaCl (B1) at ambient pressure and temperature. Since our approach only considers the static energies, we cannot say anything about the kinetic aspects, such as the nucleation and growth of a phase (Yamada et al. 1984).

2.5. Covalent solids

Covalent solids are similar to metals in that the (valence) electrons outside the filled cores to a large extent occupy regions in between the ions. This is contrary to ionic solids, where the electron density is low at the boundaries of the Jones cells. (A Jones cell is a polyhedron assigned to each atom such that the polyhedra fill the space of the solid.) Harrison (1980) writes the cohesive energy of a covalent solid as the sum of three terms:

$$U_{\mathrm{coh}} = -E_{\mathrm{pro}} - V_0(a) + E_{\mathrm{bond}}(a). \tag{1.22}$$

E_{pro} is the "promotion" energy. It arises as an increase of the energy of an atom when its two 2s and two 2p electron states form four sp^3 hybridised states. When Jones cells containing atoms in these hybridised states are stacked to form a solid, there will be an electron redistribution leading to an energy-lowering E_{bond}. Finally there is a repulsive interaction due to overlap between the sp^3 states of neighbouring atoms. This gives rise to the energy $V_0(a)$. V_0 and E_{bond} depend on the lattice parameter a. In principle, we can minimise U_{coh} with respect to a and obtain the equilibrium atomic volume and the corresponding cohesive energy. However, the terms in (1.22) cannot be expressed in only simple atomic parameters like the valency (as was the case for simple metals and ionic solids). Therefore we cannot give simple closed forms for a and U_{coh}.

3. First-principles calculations of structural energies

It would be desirable to calculate structural properties in a scheme which only uses as input parameters the atomic numbers of the atoms and the crystal structure, including the lattice parameters. Then one could scan through a number of crystal structures, varying the lattice parameters, and obtain the most stable structure, its atomic volume and compressibility etc. We shall discuss two such methods.

In the ab initio pseudopotential method (e.g., Cohen 1985, 1986, Yin and Cohen 1982, Martin 1985) one assumes that the closed electron shells around each atom (i.e. the core electrons) do not depend on the position or kind of neighbouring atoms. The outer electrons (the valence electrons) interact with the nucleus and the core electrons through a pseudopotential. That potential depends on the charge distribution in the solid and it is calculated self-consistently by iterating solutions to the Schrödinger equation. The total lattice energy is (Yin and Cohen 1982)

$$E_{\text{tot}} = E_{\text{kin}} + E_{\text{el-core}} + E_{\text{Hartree}} + E_{\text{x, c}} + E_{\text{cc}}. \tag{1.23}$$

Here we recognise several of the energy terms considered in § 2.2. E_{kin} is the kinetic energy of the electrons, $E_{\text{el-core}}$ is the valence electron–core electron interaction (described by a pseudopotential), E_{Hartree} is the direct electron–electron Coulomb interaction, $E_{\text{x, c}}$ is the exchange and correlation energy (calculated in the local density scheme) and E_{cc} is an Ewald-type Coulomb interaction between the ions formed by a nucleus and its core electrons. In a real calculation, the valence-electron wave function is expanded in a plane-wave basis set. The wave function obeys the Schrödinger equation, but that equation contains a potential which depends on the charge density in the solid, i.e. on the wave function itself. Through iteration, one arrives at a self-consistent solution. The ab initio method has been applied to a large number of solids, e.g. Al (Lam and Cohen 1983), Be (Chou et al. 1983), Si, Ge (Yin and Cohen 1982, Hybertsen and Louie 1984), III–V semiconductors (Froyen and Cohen 1983) and NaCl (Froyen and Cohen 1984). As an example, in the calculation of Hybertsen and Louie, for Si and Ge, the lattice parameter is within 1% and the bulk modulus within 10% of the experimental values.

The ab initio pseudopotential method assumes that the core electrons are non-overlapping and the valence electrons are predominantly outside the core region. These prerequisites are not fulfilled for transition metals, which have large cores and no clear distinction between core electrons and valence electrons. Still, one can attempt an ab initio calculation, in which all electrons are treated on the same footing and with their energies described within the local density scheme. Moruzzi et al. (1977, 1978) have used that approach to calculate the energy as a function of the atomic volume for 3d and 4d transition metals. Figure 1.5 shows the cohesive energy thus obtained.

4. Empirical rules related to atomic volumes

In an alloy, or a mixed crystal, the crystal volume varies with the composition c_i. Although this variation is often quite small, we shall encounter examples where the small volume change is of importance. In a solid solution of two elements (1 and 2) with the same crystal structure, it is often assumed that the lattice parameter a varies as

$$a = c_1 a_1 + c_2 a_2. \tag{1.24}$$

This relation is known as Vegard's law. It was originally applied to solid solutions of ionic crystals (Vegard 1921), but has been frequently used in alloys. However, the "law" is often strongly violated and must be amended to be useful (Gschneidner and Vineyard 1962). In the light of our previous discussion, it may seem better to use the corresponding rule for atomic volumes,

$$\Omega = c_1 \Omega_1 + c_2 \Omega_2. \tag{1.25}$$

This is sometimes called Zen's rule (Zen 1956). The solids 1 and 2 need not have the same crystal structure. Note that (1.24) and (1.25) cannot hold simultaneously. Zen's rule may be of some value, but it should only be used to establish trends (cf. fig. 1.3). Details on lattice parameters in metals and alloys are found in the extensive work by Pearson (1972).

In chemistry and metallurgy one frequently uses the concept of partial molar quantities (Swalin 1962). The partial molar volume $\bar{V}_i$ of component i in a system containing other components j, is defined by

$$\bar{V}_i \equiv \left(\frac{\partial V}{\partial n_i}\right)_{T,\, p,\, j}, \tag{1.26}$$

where V is the volume of the system and n_i is the number of moles of component i. If Zen's rule holds and with L = Avogadro's number

$$\bar{V}_1/L = \left(\frac{\partial[n_1\Omega_1 + n_2\Omega_2]}{\partial n_1}\right)_{p,\, T,\, n_2} = \Omega_1. \tag{1.27}$$

Example: Interstitial carbon in austenite (fcc Fe). When the bcc and fcc lattices are represented by hard spheres, the largest "holes" where one can inscribe a sphere in an interstitial position are found in the more densely packed fcc lattice. This explains why carbon is almost insoluble in bcc Fe but forms a solid solution in fcc Fe (up to 9 at. % at 1426 K). If carbon atoms could be viewed as rigid spheres which precisely fitted the "holes" in the lattice, alloying with carbon would not change the lattice parameter of fcc Fe. Then the partial molar volume of C in Fe would be zero. In reality, the iron lattice expands. Ridley and Stuart (1970) measured the lattice parameter as a function of the carbon concentration. At 900 K, they obtained the partial molar volumes

$$\bar{V}_{\mathrm{Fe(fcc)}} = 7.20\,\mathrm{cm}^3, \qquad \bar{V}_{\mathrm{C\ in\ Fe(fcc)}} = 3.76\,\mathrm{cm}^3.$$

This can be compared with the molar volume $\bar{V}_{\mathrm{C}} = 5.34\,\mathrm{cm}^3$ of pure carbon (graphite).

CHAPTER 2

CRYSTAL DEFECTS

1. Introduction

Vacancies and some other point-like defects may be thermally generated and therefore present in thermal equilibrium. Other defects have such high defect energies that their presence depends on the detailed pre-history (cold-working, annealing, etc.) of the sample. Since we are interested in the temperature dependence of phenomena related to the defects, it is illuminating to express the magnitude of some characteristic energies in units of $k_B T_m$, where k_B is Boltzmann's constant and T_m is the melting temperature; see table 2.1, based on Grimvall and Sjödin (1974).

Henderson (1972) has written an excellent elementary introduction to various defects in crystalline solids, and Watts (1977) gives a somewhat more advanced account, with emphasis on non-metals. Grain boundaries are dealt with in the monograph by Chadwick and Smith (1976). There are a large number of review articles, for instance in the Festschrift for Sir Nevil Mott (Hirsch 1975). March and Rousseau (1971), Seeger (1973) and Kovács and El Sayed (1976) cover point defects, mainly in metals. Extended defects in materials have been reviewed by Friedel (1980), and various aspects of defects are covered in a Les Houches summer school (Balian et al. 1981).

Table 2.1
Some characteristic energies in elements, expressed in units of $k_B T_m$

Vacancy formation energy	8–13
Activation energy of self-diffusion	15–19
Heat of fusion [a]	0.9–1.3
Surface energy [b]	5–10
Grain boundary energy [b]	3–6

[a] Per atom. [b] Per area of one atom (in a monolayer).

2. General thermodynamic relations

2.1. Formation energy, formation volume, enthalpy, entropy and free energy

The formation energy, E_d, of a defect is the energy difference between a defect and a perfect crystal containing the same number of atoms. The formation volume V_d is the difference in volume between the two crystals. Similarly, one can define the thermodynamic quantities H_d, F_d, G_d and S_d. The usual thermodynamic relations for the enthalpy, the free energy, etc. are also valid for quantities like H_d and F_d, if they involve derivatives with respect to intensive variables, such as p and T. For instance, the formation volume can be written

$$V_d = \left(\frac{\partial G_d}{\partial p}\right)_T. \tag{2.1}$$

One has to be careful in the use of relations involving derivatives with respect to extensive variables, such as the volume V (Howard and Lidiard 1964, Levinson and Nabarro 1967).

2.2. Defect concentration in thermodynamic equilibrium

The defects present in a state of complete thermodynamic equilibrium must have low formation energies E_d. This limits us to point imperfections and small aggregates of them. Let N be the number of lattice sites. The number of defects in equilibrium is

$$N_d = N \exp(S'_d/k_B) \exp(-H_d/k_B T). \tag{2.2}$$

This expression assumes that $N_d \ll N$ and that interactions between defects can be neglected. At ambient pressure, we can put $p = 0$ so that $H_d = E_d$. The entropy S'_d has two parts:

$$S'_d = S_g + S_d. \tag{2.3}$$

$S_g = k_B \ln(z_d)$ is a temperature-independent geometrical term, z_d being the number of configurations of a defect associated with a particular site ı the lattice, and with proper consideration of double counting. We write

$$\exp(S_g/k_B) = g_d. \tag{2.4}$$

If there are several atoms per primitive cell, one may have to assign different values H_d, S_g and S_d to different configurations or lattice sites.

2.3. Defect parameters from an Arrhenius plot

H_d (E_d) is determined from a plot of $\ln(N_d/N)$ versus $1/k_B T$ giving a straight line with the slope $-H_d$ (fig. 2.1). One has

$$-\frac{\partial \ln(N_d/N)}{\partial(1/k_B T)} = T^2\left(\frac{\partial S_d}{\partial T}\right) - T\left(\frac{\partial H_d}{\partial T}\right) + H_d = H_d(T). \tag{2.5}$$

Here we have used the general thermodynamic relation $S = -(\partial G/\partial T)_p$, from which $T(\partial S/\partial T)_p = (\partial H/\partial T)_p$. Thus, a meaningful formation enthalpy $H_d(T)$, or formation energy $E_d(T)$, can be obtained even when these quantities are temperature dependent, so that the Arrhenius plot is curved.

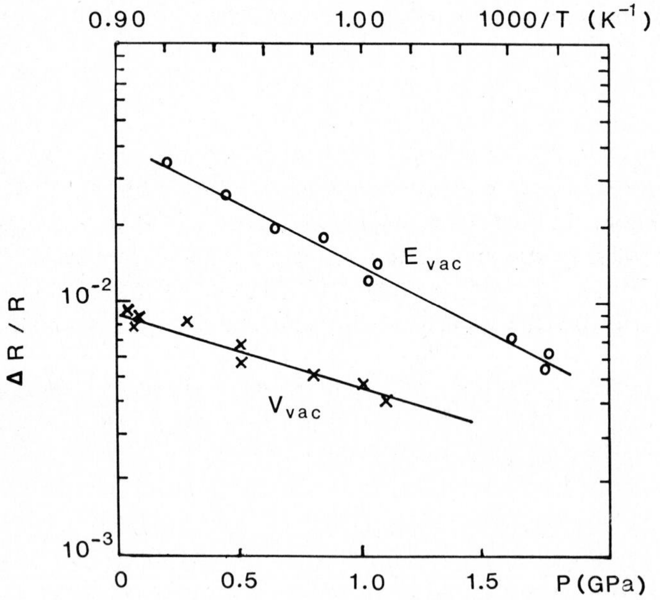

Fig. 2.1. The formation energy E_{vac} and the formation volume V_{vac} of a vacancy in gold, derived from the slope of the quenched-in electrical resistance $\Delta R/R$ plotted versus $1/T$ (upper scale, slope $-E_{vac}$) and p (lower scale, slope $-V_{vac}/k_B T$). Data from Huebener and Homan (1963).

From (2.1) and (2.2) we get the formation volume

$$V_\mathrm{d} = \left(\frac{\partial G_\mathrm{d}}{\partial p}\right)_T = \left(\frac{\partial (H_\mathrm{d} - TS_\mathrm{d})}{\partial p}\right)_T = -k_\mathrm{B}T\left(\frac{\partial \ln(N_\mathrm{d}/N)}{\partial p}\right)_T. \tag{2.6}$$

A plot of $\ln(N_\mathrm{d}/N)$ versus p thus gives the formation volume V_d (Levinson and Nabarro 1967).

Example: Temperature-dependent $H_\mathrm{d}(T)$ for vacancies. We shall estimate the variation of $H_\mathrm{d}(T)$ with T and make an expansion in T near the melting temperature T_m:

$$\begin{aligned} H_\mathrm{d}(T) &= H_\mathrm{d}(T_\mathrm{m}) + (T - T_\mathrm{m})\left(\frac{\partial H_\mathrm{d}}{\partial T}\right)_{p;\, T = T_\mathrm{m}} \\ &= H_\mathrm{d}(T_\mathrm{m}) + (T - T_\mathrm{m})T_\mathrm{m}\left(\frac{\partial S_\mathrm{d}}{\partial T}\right)_{p;\, T = T_\mathrm{m}}. \end{aligned} \tag{2.7}$$

For harmonic lattice vibrations, and in the high-temperature limit, we have (cf. eq. (4.55))

$$S_\mathrm{d} = 3Nk_\mathrm{B}\ln[\Theta(0)/\Theta_\mathrm{d}(0)]. \tag{2.8}$$

$\Theta_\mathrm{d}(0)$ and $\Theta(0)$ are "entropy Debye temperatures" for the crystal with and without a vacancy. From (2.8), $(\partial S_\mathrm{d}/\partial T)_p = 0$. To obtain a temperature-dependent H_d, anharmonic effects must be included. If this is done within the quasiharmonic model, and with equal Grüneisen parameters for the perfect and the defect state, we still obtain $(\partial S_\mathrm{d}/\partial T)_p = 0$. One has to go beyond such a simple description to get a temperature-dependent H_d (Girifalco 1967, Levinson and Nabarro 1967).

2.4. Constant pressure and constant volume

Sometimes it is essential to distinguish between quantities considered at constant volume (subscript V) and at constant pressure (subscript p). Experiments are usually performed at constant pressure ($p = 0$) but model calculations may be more conveniently carried out at constant volume. If $(V_\mathrm{d})_p$ is the increase in the specimen volume when a certain defect is introduced at constant pressure, and $(p_\mathrm{d})_V$ is the increase in pressure if the defect is added at constant volume, they are related by

$$(V_d)_p K_T = (p_d)_V V, \tag{2.9}$$

with similar relations for other thermodynamic properties of defects (Catlow et al. 1981).

3. Vacancies

3.1. Thermodynamic relations

Here we shall mainly have in mind thermally generated vacancies in elements and alloys. The formation energy $E_d = E_{vac}$ of a monovacancy is the energy difference between a perfect lattice with N occupied lattice sites and a similar lattice with $N+1$ sites, one of which is void. The formation volume V_{vac} of a monovacancy is the difference in volume between these two crystals. The concentration of monovacancies in a monatomic solid is, in thermal equilibrium and at $p = 0$,

$$c_{vac} = \exp(S_{vac}/k_B)\exp(-E_{vac}/k_B T). \tag{2.10}$$

S_{vac} is an entropy term related to changes in the vibrational spectrum of the lattice (ch. 6 § 5). The configurational entropy is zero, i.e. $g_d = 1$. Under a pressure p, E_{vac} in (2.10) is replaced by $E_{vac} + pV_{vac}$.

Example: Vacancies in Au. If a specimen is quenched from a (high) temperature T, the residual resistivity $R_0 + \Delta R$ of the quenched sample has a part R_0 caused by impurities and a part ΔR arising from the N_d quenched-in defects. Since ΔR is proportional to N_d, a plot of $\ln(N_d/N)$ can be replaced by a plot of $\ln(\Delta R/R_0)$. Huebener and Homan (1963) measured ΔR for vacancies in gold at varying temperatures and pressures. Figure 2.1 shows their results. From the slope of the straight lines one obtains $H_{vac} = 0.98$ eV and $V_{vac} = 9.16 \times 10^{-30}$ m^3 $= 0.53\,\Omega_a$, where Ω_a is the atomic volume. H_{vac} refers to $p = 1.8$ kbar. Then $pV_{vac} = 0.01$ eV is negligible.

3.2. Theoretical calculations of vacancy properties

The most elaborate theoretical calculations of the vacancy formation energy E_{vac} may be in error by a factor of two or more, while simple empirical relations, like $E_{vac}/k_B T_m = 10$, give a much better estimation of

E_{vac}. We now briefly review some theoretical approaches. If there are only pairwise central interactions, described by a potential $V(r)$, E_{vac} equals the binding energy U_b (per atom). In metals, $E_{vac} \sim U_b/3$. Both volume-dependent energy terms (not accounted for by $V(r)$) and atomic relaxation must be included in a realistic calculation. In the jellium model, a vacancy is represented by a spherical cavity. For aluminium, the resulting E_{vac} is negative (Manninen et al. 1975). This means that the jellium is unstable against the formation of voids. The reason is that the r_s-value ($= 2.07$) of Al falls well below the r_s at which the energy of a jellium has its minimum. Calculations based on an empirical pseudopotential, which have proved to give reasonable results for properties like phonon energies and the electrical conductivity, may overestimate E_{vac} of Al by a factor of five (So and Woo 1981, Manninen et al. 1981). It is obvious from these examples that although many simple models have yielded results in fair agreement with experiment, this is largely fortuitous. Calculations of formation volumes (e.g., Finnis and Sachdev 1976) are less uncertain.

4. Divacancies and vacancy clusters

Divacancies and vacancy clusters can be treated by a direct generalisation of the approach given in §3. The concentration of divacancies at thermal equilibrium, in a monatomic solid, may be written

$$c_{divac} = (z/2)(c_{vac})^2 \exp(S_{divac}/k_B)\exp(E^b_{divac}/k_BT). \tag{2.11}$$

Here c_{vac} is the monovacancy concentration, E^b_{divac} is the binding energy between two vacancies (the formation energy of a divacancy is $2E_{vac} - E^b_{divac}$) and S_{divac} is the change in the vibrational entropy when two separate vacancies form a divacancy. If a certain vacant site has z equivalent neighbouring lattice sites, one can form z differently oriented divacancies. After a correction by a factor 1/2 to avoid double counting, we get the prefactor $g_d = g_{divac} = z/2$ in (2.4). There may be configurations with different E^b_{divac} and S_{divac}. Then the total concentration of divacancies is a sum of terms such as the right-hand side of (2.11). For instance, in an hcp lattice one may have to distinguish between divacancies lying in, or perpendicular to, the basal lattice plane. In fcc lattices, E^b_{divac} may be so large ($> k_BT$) for a pair of vacancies occupying next-nearest lattice positions, that a term corresponding to this configuration must also be included in (2.11). Typically, $E^{\mathrm{b}}_{\mathrm{divac}}/k_{\mathrm{B}}T_{\mathrm{m}} \sim 1$ to 2. Then, about 10% of the vacant sites are associated with divacancies.

A liquid is stabilised relative to the solid phase by the large entropy of spatial disorder. There is evidence for a similar stabilisation of a liquid-like structure surrounding vacancies or interstitials, and involving nearest neighbours to the original point defect (Gösele et al. 1983, Pokorny and Grimvall 1984).

5. Interactions between point defects and other defects

Consider a dilute substitutional solid solution, with concentration c_{sol} of solute atoms. The coordination number of the lattice is z. The equilibrium number of vacancies may be higher in the alloy than in the pure host because vacancies bind to impurities and thus are associated with a lowered formation energy. The equilibrium concentration of vacancies in the alloy is given by Lomer's equation:

$$c_{\mathrm{vac}} = (c_{\mathrm{vac}})^0[1 - zc_{\mathrm{sol}}] + (c_{\mathrm{vac}})^0 zc_{\mathrm{sol}} \exp[G_{\mathrm{vac-sol}}/k_{\mathrm{B}}T]. \tag{2.12}$$

$G_{\mathrm{vac-sol}}$ is the Gibbs free energy of binding between a vacancy and a solute atom. Often, the entropy term in G is neglected and G in eq. (2.12) is replaced by the binding energy $E_{\mathrm{vac-sol}}$. We may justify (2.12) as follows: The fraction $1 - zc_{\mathrm{sol}}$ of all lattice sites is not adjacent to a solute atom. The probability for a vacant site at these lattice positions is the same as in the pure host (if interactions beyond nearest neighbours are neglected). This gives the first term on the right-hand side of (2.12). The remaining sites have a vacancy bound to a solute atom, giving an extra $\exp(G_{\mathrm{vac-sol}}/k_{\mathrm{B}}T)$ in the Boltzmann factor. Thus we obtain the second term on the right-hand side of (2.12). Lomer (1958), Lidiard (1960) and Burke (1972) have given more stringent arguments for eq. (2.12). It fails if the solute atoms show clustering, and the validity is often limited to $c_{\mathrm{sol}} < 0.01$. March and Rousseau (1971) and Seeger (1973) have reviewed various aspects of vacancy–solute interactions.

Example: Vacancy–solute interactions in Al–Ag alloys. Vacancies increase the length L of a specimen but they do not affect the lattice parameter a, as measured by X-ray methods. Simmons and Balluffi (1960) used this fact to derive the vacancy formation energy of Al from the relation

$$c_{\text{vac}} = 3\left(\frac{\Delta L}{L} - \frac{\Delta a}{a}\right). \tag{2.13}$$

Here ΔL includes the effect of vacancies as well as ordinary thermal expansion, while Δa only includes the thermal expansion. If the experiment is done on a pure element (unprimed quantities) and a dilute alloy of the same element (primed quantities), one has from (2.12)

$$(c_{\text{vac}})^0 c_{\text{sol}} z[\exp(E_{\text{vac-sol}}/k_B T) - 1] = 3\left\{\frac{\Delta L'}{L'} - \frac{\Delta L}{L}\right\} - 3\left\{\frac{\Delta a'}{a'} - \frac{\Delta a}{a}\right\}. \tag{2.14}$$

Using this method for Al with 0.94 at.% Ag, Beaman et al. (1964) obtained $E_{\text{vac-sol}}/E_{\text{vac}} = 0.10$. One should note that eq. (2.13) holds even when there are relaxations around the defects (Simmons and Balluffi 1960, Seeger 1973).

CHAPTER 3

ELASTIC PROPERTIES

1. Introduction

The elastic properties of materials can be described in different ways. The engineer usually deals with polycrystalline one- or multi-phase systems which are characterised by macroscopic elastic parameters such as the bulk modulus K and the shear modulus G. The physicist is more interested in the properties of a single crystal, described by the elastic stiffness coefficients c_{ij}, and often he considers elasticity as the special case of long-wavelength lattice vibrations. The terminology in the field is not quite clear. In this book we shall adopt the following notation and terminology (Ledbetter and Reed 1973):

K = bulk modulus	elastic moduli	(polycrystal) engineering elastic constants	elastic constants
$\kappa = 1/K$ = compressibility			
G = shear modulus			
E = Young modulus			
ν = Poisson ratio			
c_{ij} = elastic stiffnesses		(single-crystal) elastic coefficients	
s_{ij} = elastic compliances			

Some authors use B for the bulk modulus and K for the compressibility, call our K the modulus of (hydrostatic) compression and our κ the coefficient of (hydrostatic) compression, call G the rigidity modulus and denote it μ, call s_{ij} elastic moduli and c_{ij} elastic constants or even call s_{ij} elastic constants while c_{ij} are called elastic coefficients.

Much of the following account of elasticity is covered in monographs and articles by Nye (1957), Huntington (1958), Hearmon (1961), Fedorov (1968), Musgrave (1970) and Schreiber et al. (1973). We shall mainly consider that which forms a necessary background for our discussion of thermophysical properties in other parts of this book.

2. General considerations

The elastic strain ε_{kl} is related to the stress σ_{ij} by Hooke's law,

$$\sigma_{ij} = \sum_{k,l=1}^{3} c_{ijkl}\,\varepsilon_{kl}, \tag{3.1}$$

where i, j, k and l are indices running from 1 to 3. The elastic properties of a material are described by a fourth-rank elasticity tensor with $3^4 = 81$ elements ε_{ijkl}. Because $c_{ijkl} = c_{klij} = c_{jikl} = c_{ijlk}$, there are at most 21 different elements ε_{ijkl}. They can be arranged in a 6×6 matrix which is symmetric, i.e. with elements $c_{\alpha\beta} = c_{\beta\alpha}$. The relations between $c_{\alpha\beta}$ and c_{ijkl} are summarised in table 3.1.

Table 3.1
Voigt's (1928) contraction scheme for indices in $c_{\alpha\beta}$ and c_{ijkl}

i, j or k, l	11	22	33	23 or 32	13 or 31	12 or 21
α or β	1	2	3	4	5	6

We can write (3.1) as

$$\sigma_\alpha = \sum_{\beta=1}^{6} c_{\alpha\beta}\,\varepsilon_\beta, \tag{3.2}$$

where

$$\sigma_\alpha = \sigma_{ij}, \tag{3.3}$$

$$\varepsilon_\beta = \varepsilon_{kl} \quad \text{if } \beta = 1, 2 \text{ or } 3, \tag{3.4}$$

$$\varepsilon_\beta = 2\varepsilon_{kl} \quad \text{if } \beta = 4, 5 \text{ or } 6. \tag{3.5}$$

The elastic stiffness tensor $\boldsymbol{c}$ is related to the elastic compliance tensor $\boldsymbol{s}$ by

$$\boldsymbol{cs} = \boldsymbol{I}_6, \tag{3.6}$$

where $\boldsymbol{I}_6$ is a 6×6 identity matrix. The relation (3.2) for the stresses σ

expressed in the strains ε can be inverted to give the strains in terms of the stresses:

$$\varepsilon_\alpha = \sum_{\beta=1}^{6} s_{\alpha\beta}\sigma_\beta. \tag{3.7}$$

Hooke's law, (3.1), holds for both polycrystalline and single-crystal specimens. The number of independent parameters required to specify the tensor ε_{ijkl} depends on the symmetry of the system; cf. table 3.2. Some tetragonal and trigonal symmetry classes have additional $c_{\alpha\beta} \neq 0$ with certain natural choices of coordinate axes; see, for example, Fedorov (1968).

Table 3.2
Parameters describing the elastic properties

Polycrystalline[a]	Two of K, G, E and ν
Cubic	c_{11}, c_{12}, c_{44}
Hexagonal	$c_{11}, c_{12}, c_{13}, c_{33}, c_{44}$
Tetragonal	$c_{11}, c_{12}, c_{13}, c_{33}, c_{44}, c_{66}$
Trigonal	$c_{11}, c_{12}, c_{13}, c_{14}, c_{33}, c_{44}$
Orthorhombic	$c_{11}, c_{12}, c_{13}, c_{23}, c_{22}, c_{33}, c_{44}, c_{55}, c_{66}$

[a] Statistically isotropic and homogeneous.

If a material has spherically symmetric elastic properties, two parameters suffice for their description. It follows that for isotropic and statistically homogeneous materials, there are only two linearly independent engineering elastic constants. Then there are 12 relations between K, G, E and ν:

$$K = \frac{GE}{3(3G-E)} = \frac{E}{3(1-2\nu)} = \frac{2G(1+\nu)}{3(1-2\nu)}, \tag{3.8}$$

$$G = \frac{3KE}{9K-E} = \frac{E}{2(1+\nu)} = \frac{3K(1-2\nu)}{2(1+\nu)}, \tag{3.9}$$

$$E = \frac{9KG}{3K+G} = 2G(1+\nu) = 3K(1-2\nu), \tag{3.10}$$

$$\nu = \frac{3K-2G}{2(3K+G)} = E/2G-1 = 1/2-E/6K. \tag{3.11}$$

Sometimes one prefers to take as independent elastic parameters the

Lamé constants λ and μ, with $\lambda = K - 2G/3$ and $\mu = G$. Cubic lattices have three independent elastic coefficients. Usually one chooses them to be c_{11}, c_{12} and c_{44}. An alternative set of constants was introduced by Zener (1948); $C = c_{44}$, $C' = (c_{11} - c_{12})/2$, $K = (c_{11} + 2c_{12})/3$. It can be shown that K is the bulk modulus. C and C' are shear moduli for shear in the (100) and (110) planes; see eqs. (3.32) and (3.33). If the cubic lattice is elastically isotropic, $C' = C = c_{44}$ equals the shear modulus G.

3. Hooke's law applied to certain systems

3.1. Statistically isotropic and homogeneous polycrystalline materials

Expressed in K and G, Hooke's law, (3.1), takes the form

$$\begin{pmatrix} \sigma_1 \\ \sigma_2 \\ \sigma_3 \\ \sigma_4 \\ \sigma_5 \\ \sigma_6 \end{pmatrix} = \begin{pmatrix} K+4G/3 & K-2G/3 & K-2G/3 & 0 & 0 & 0 \\ K-2G/3 & K+4G/3 & K-2G/3 & 0 & 0 & 0 \\ K-2G/3 & K-2G/3 & K+4G/3 & 0 & 0 & 0 \\ 0 & 0 & 0 & G & 0 & 0 \\ 0 & 0 & 0 & 0 & G & 0 \\ 0 & 0 & 0 & 0 & 0 & G \end{pmatrix} \begin{pmatrix} \varepsilon_1 \\ \varepsilon_2 \\ \varepsilon_3 \\ \varepsilon_4 \\ \varepsilon_5 \\ \varepsilon_6 \end{pmatrix} \tag{3.12}$$

We illustrate the application of (3.12) by three simple cases; uniaxial stress, hydrostatic pressure and shear.

Uniaxial stress. A uniaxial stress $\sigma_{xx} = \sigma$ corresponds to $\sigma_1 = \sigma$ and all other $\sigma_i = 0$. Further, only ε_1 and $\varepsilon_2 (= \varepsilon_3)$ are non-zero. From eq. (3.12) we then get $\sigma_1 (= \sigma) = (K + 4G/3)\varepsilon_1 + (K - 2G/3)\varepsilon_2 + (K - 2G/3)\varepsilon_3$, with a similar equation for $\sigma_2 (= 0)$. Elimination of $\varepsilon_2 (= \varepsilon_3)$ leads to

$$\varepsilon_1 = \sigma/E, \quad \varepsilon_2 = \varepsilon_3 = -\nu\sigma/E. \tag{3.13}$$

Hydrostatic pressure. A hydrostatic pressure p corresponds to $\sigma_1 = \sigma_2 = \sigma_3 = -p$ and $\sigma_4 = \sigma_5 = \sigma_6 = 0$. From (3.12) we obtain $\sigma_1 (= -p) = (K + 4G/3)\varepsilon_1 + (K - 2G/3)\varepsilon_2 + (K - 2G/3)\varepsilon_3$, with similar equations for σ_2 and σ_3. Then

$$\varepsilon_1 = \varepsilon_2 = \varepsilon_3 = -p/3K. \tag{3.14}$$

The relative volume change, $\Delta V/V$, is

$$\Delta V/V = \varepsilon_1 + \varepsilon_2 + \varepsilon_3 = -p/K. \tag{3.15}$$

Shear. Shear, specified by $\tau_{yz} = \tau_{zy} = \tau$, has $\sigma_4 = \tau$ and all other $\sigma_i = 0$. From (3.12) one obtains

$$\varepsilon_4 = \tau/G, \qquad \text{all other } \varepsilon_i = 0. \tag{3.16}$$

3.2. Single crystal with cubic lattice symmetry

The elastic properties of a cubic lattice may be described by the elastic stiffness parameters c_{11}, c_{12} and c_{44}, or by the elastic compliance parameters s_{11}, s_{12} and s_{44}. They are connected by eq. (3.6). The explicit relations are

$$s_{11} = \frac{c_{11}+c_{12}}{(c_{11}-c_{12})(c_{11}+2c_{12})}, \tag{3.17}$$

$$s_{12} = -\frac{c_{12}}{(c_{11}-c_{12})(c_{11}+2c_{12})}, \tag{3.18}$$

$$s_{44} = \frac{1}{c_{44}}. \tag{3.19}$$

The corresponding relations for $c_{\alpha\beta}$ expressed in $s_{\alpha\beta}$ are obtained if the symbols s and c are interchanged in (3.17)–(3.19).

Hooke's law, in the case of cubic elastic symmetry, has the form

$$\begin{pmatrix} \sigma_1 \\ \sigma_2 \\ \sigma_3 \\ \sigma_4 \\ \sigma_5 \\ \sigma_6 \end{pmatrix} = \begin{pmatrix} c_{11} & c_{12} & c_{12} & 0 & 0 & 0 \\ c_{12} & c_{11} & c_{12} & 0 & 0 & 0 \\ c_{12} & c_{12} & c_{11} & 0 & 0 & 0 \\ 0 & 0 & 0 & c_{44} & 0 & 0 \\ 0 & 0 & 0 & 0 & c_{44} & 0 \\ 0 & 0 & 0 & 0 & 0 & c_{44} \end{pmatrix} \begin{pmatrix} \varepsilon_1 \\ \varepsilon_2 \\ \varepsilon_3 \\ \varepsilon_4 \\ \varepsilon_5 \\ \varepsilon_6 \end{pmatrix} \tag{3.20}$$

tensor of compliance coefficients has all non-zero $c_{\alpha\beta}$ in (3.20) ced by $s_{\alpha\beta}$ (c_{11} replaced by s_{11}, etc.). A particularly simple case arises when the crystal is elastically isotropic. Then

$$c_{11}-c_{12}-2c_{44} = 0, \tag{3.21}$$

or, equivalently,

$$2s_{11}-2s_{12}-s_{44} = 0. \tag{3.22}$$

Zener (1948) introduced, as a measure of anisotropy,

$$A_Z = 2c_{44}/(c_{11}-c_{12}). \tag{3.23}$$

There is no unique measure of elastic anisotropy, but in the equation for sound propagation, a natural choice is (Every 1980)

$$A_{\mathrm{E}} = \frac{c_{11}-c_{12}-2c_{44}}{c_{11}-c_{44}}. \tag{3.24}$$

For isotropic systems, $A_{\mathrm{Z}} = 1$ and $A_{\mathrm{E}} = 0$. Note that A_{Z} and A_{E} are not trivially related, i.e. A_{Z} cannot be expressed in terms of A_{E}. Fedorov (1963, 1968) has defined the anisotropy by the mean-square deviation of the matrix $c_{\alpha\beta}$ from a certain average $\bar{c}_{\alpha\beta}$. That makes it possible to assign an anisotropy measure in a lattice of arbitrary symmetry. However, Fedorov's parameter (denoted Λ'_{m}) for a cubic lattice is more complicated than A_{E}.

In anisotropic crystals, a specimen twists and bends at the same time, even if only a bending (or twisting) moment is applied. The relations below, for E, G and ν, refer to this situation. See Schreiber et al. (1973) or Hearmon (1946) for a discussion of other load situations.

Young's modulus. E is defined as the ratio of uniaxial stress to strain measured along the same axis, when the body is unconstrained perpendicular to that axes. Let σ_1 be along the [100] axis, and $\sigma_i = 0$ $(i \neq 1)$. Further, $\varepsilon_2 = \varepsilon_3$ and $\varepsilon_4 = \varepsilon_5 = \varepsilon_6 = 0$. Then, from eq. (3.20), we get Young's modulus in the [100] direction:

$$E[100] = \sigma_1/\varepsilon_1 = (c_{11}-c_{12})(c_{11}+2c_{12})/(c_{11}+c_{12}) = 1/s_{11}. \tag{3.25}$$

In an arbitrary direction $[hkl]$, Young's modulus is best given in terms of s_{ij}. One has (Hearmon 1946, Schreiber et al. 1973)

$$\{E[hkl]\}^{-1} = s_{11}-(2s_{11}-2s_{12}-s_{44})N^4, \tag{3.26}$$

where

$$N^4 = n_1^2n_2^2+n_1^2n_3^2+n_2^2n_3^2, \tag{3.27}$$

and n_1, n_2, n_3 are direction cosines for the direction $[hkl]$, i.e. $n_1 = h/(h^2+k^2+l^2)^{1/2}$. Uniaxial tension in the [110] direction gives $n_1 = n_2 = 1/\sqrt{2}$ and, by (3.26),

$$\{E[110]\}^{-1} = (2s_{11}+2s_{12}+s_{44})/4. \tag{3.28}$$

In an elastically isotropic medium, $2s_{11}-2s_{12}-s_{44}=0$ and the term containing N^4 in (3.26) vanishes. Then

$$E = 1/s_{11} = (c_{11}-c_{12})(c_{11}+2c_{12})/(c_{11}+c_{12}). \tag{3.29}$$

Bulk modulus. K refers to hydrostatic pressure, and is an isotropic quantity

$$K = (c_{11}+2c_{12})/3 = [3(s_{11}+2s_{12})]^{-1}. \tag{3.30}$$

The relative volume change $\Delta V/V$, under pressure p, is

$$\Delta V/V = -p/K = -3p/(c_{11}+2c_{12}) = -3p(s_{11}+2s_{12}). \tag{3.31}$$

Shear modulus. A shear stress τ_{yx} gives a shear in the (100) plane and the [010] direction. The corresponding shear modulus is

$$G(100)[010] = c_{44} = 1/s_{44}. \tag{3.32}$$

Similarly, for shear in the (110) plane and the $[1\bar{1}0]$ direction,

$$\{G(110)[1\bar{1}0]\}^{-1} = 2(s_{11}-s_{12}) = 2/(c_{11}-c_{12}) = 1/C', \tag{3.33}$$

where C' is Zener's shear constant.

If we let $G(hkl)$ refer to torsion around $[hkl]$, shear occurs in all directions in the plane (hkl). The general expression for that shear modulus is (Hearmon 1946)

$$\{G(hkl)\}^{-1} = s_{44}+2(2s_{11}-2s_{12}-s_{44})N^4. \tag{3.34}$$

In an isotropic system, the term containing N^4 in (3.34) vanishes and

$$G = c_{44} = 1/s_{44}. \tag{3.35}$$

A general expression $G(hkl)[h'k'l']$ is found, for example, in Schreiber et al. (1973) and Turley and Sines (1971).

Poisson ratio. ν is defined as the negative ratio of transverse strain to longitudinal strain, for the case of uniaxial stress. If the stress is along [100] one gets

$$\nu[100] = -\varepsilon_2/\varepsilon_1 = c_{12}/(c_{11}+c_{12}) = -s_{12}/s_{11}. \tag{3.36}$$

More generally, ν must be specified both with respect to the direction $[hkl]$ of the longitudinal strain and the direction $[h'k'l']$ of the transverse strain. Then (Schreiber et al. 1973),

$$\nu[h'k'l'][hkl] = -\frac{s_{12}+(s_{11}-s_{12}-s_{44}/2)M^4}{s_{11}-(2s_{11}-2s_{12}-s_{44})N^4}. \tag{3.37}$$

N^4 was defined in (3.27) and

$$M^4 = n_1^2m_1^2+n_2^2m_2^2+n_3^2m_3^2, \tag{3.38}$$

where m_1, m_2, m_3 are direction cosines for $[h'k'l']$. In an elastically isotropic crystal

$$\nu = -s_{12}/s_{11}. \tag{3.39}$$

The result (3.36) is obtained from (3.37) with $n_1 = 1$, $n_2 = n_3 = 0$ and $m_1 = m_2 = 0$, $m_3 = 1$.

Example: Direction of largest E. N^4 varies between 0 (in $\langle 100\rangle$ directions) and 1/3 (in $\langle 111\rangle$ directions). Usually, $2s_{11}-2s_{12}-s_{44} > 0$ (i.e. $A_E < 0$ or $A_Z > 1$). Then E has its largest value in $\langle 111\rangle$ directions. In a few elements of cubic structure (V, Nb, Cr, Mo) and in, for example, alkali halides, $A_E > 0$ and E has its largest value in $\langle 111\rangle$ directions.

Example: Anisotropic Poisson ratio with $\nu > 0.5$ and $\nu < 0$. Consider a single crystal of cubic elastic symmetry, under tension in the $[110]$ direction. We want to calculate ν referring to contractions in the $[1\bar{1}0]$ and $[001]$ directions, respectively. Equation (3.37), with $n_1 = n_2 = 1/\sqrt{2}$, $n_3 = 0$ and $m_1 = -m_2 = 1/\sqrt{2}$, $m_3 = 0$ or $m_1 = m_2 = 0$, $m_3 = 1$, gives

$$\nu[1\bar{1}0][110] = -\frac{2s_{11}+2s_{12}-s_{44}}{2s_{11}+2s_{12}+s_{44}}, \tag{3.40}$$

$$\nu[001][110] = -\frac{4s_{12}}{2s_{11}+2s_{12}+s_{44}}. \tag{3.41}$$

In an elastically isotropic material, $\nu < 0.5$. Negative ν are thermodynamically allowed, but have not been observed with certainty. However, the anisotropic Poisson ratio often exceeds these limits. Bcc iron has $s_{11} = 7.67$, $s_{12} = -2.83$ and $s_{44} = 8.57$ (in $(\mathrm{TPa})^{-1}$; from Hearmon (1979)). This yields $\nu[1\bar{1}0][110] = -0.06$ and $\nu[001][110] = 0.62$. Extreme values of ν arise for elastically very anisotropic materials (Kitagawa et al. 1980, Date and Andrews 1969).

3.3. *Single crystal of arbitrary symmetry*

The matrix c_{ij} for a hexagonal structure is

$$\begin{pmatrix} c_{11} & c_{12} & c_{13} & 0 & 0 & 0 \\ c_{12} & c_{11} & c_{13} & 0 & 0 & 0 \\ c_{13} & c_{13} & c_{33} & 0 & 0 & 0 \\ 0 & 0 & 0 & c_{44} & 0 & 0 \\ 0 & 0 & 0 & 0 & c_{44} & 0 \\ 0 & 0 & 0 & 0 & 0 & (c_{11}-c_{12})/2 \end{pmatrix} \tag{3.42}$$

Examples of solids with this symmetry are Mg, Zn, Ti, Zr, ice, graphite and β-quartz (SiO_2). The matrix of compliances for hexagonal structures have all c_{ij} in (3.42) replaced by s_{ij} (c_{12} replaced by s_{12}, etc.) except that $(c_{11}-c_{12})/2$ is replaced by $2(s_{11}-s_{12})$. The corresponding matrix for tetragonal structures has $(c_{11}-c_{12})/2$ in the lower right corner of (3.42) replaced by c_{66}. Examples of solids with that symmetry are In, β-Sn and $BaTiO_3$. The matrix of compliance coefficients has all $c_{\alpha\beta}$ replaced by $s_{\alpha\beta}$. Matrices of stiffness and compliance coefficients for other lattice structures are given, for example, by Nye (1957), Fedorov (1968) and Schrieber et al. (1973).

The compressibility κ, expressed in s_{ij} for a solid of arbitrary elastic symmetry, is (e.g., Nye 1957)

$$\kappa = K^{-1} = s_{11}+s_{22}+s_{33}+2(s_{12}+s_{13}+s_{23}). \tag{3.43}$$

In a cubic structure, $s_{11} = s_{22} = s_{33}$ and $s_{12} = s_{13} = s_{23}$, and we recover (3.30). Crystals of hexagonal and trigonal symmetry have $s_{11} = s_{22} \neq s_{33}$ and $s_{13} = s_{23} \neq s_{13}$. One then obtains (Thurston 1965)

$$K = \frac{(c_{11}+c_{12})c_{33}-2c_{13}^2}{c_{11}+c_{12}+2c_{33}-4c_{13}}. \tag{3.44}$$

The volume change under a hydrostatic pressure is $\Delta V = -pV/K$, but the strains ε_i are not equal. In an hcp lattice, $\varepsilon_1(=\varepsilon_2)$ and ε_3 are given by (cf. Nye 1957, Musgrave 1970)

$$\varepsilon_{1(2)} = -p\frac{c_{33}-c_{13}}{(c_{11}+c_{12})c_{33}-2c_{13}^2} = -p(s_{11}+s_{12}+s_{13}), \tag{3.45}$$

$$\varepsilon_3 = -p\frac{c_{11}+c_{12}-2c_{13}}{(c_{11}+c_{12})c_{33}-2c_{13}^2} = -p(2s_{13}+s_{33}). \tag{3.46}$$

Analogous results for other crystal symmetries are found in Nye (1957). As a partial check of (3.45) and (3.46) we restrict the discussion to elastic isotropy. Then $c_{11} = c_{33}$ and $c_{12} = c_{13}$, and we recover the result in a cubic structure.

Expressions for $E[hkl]$, referring to non-cubic lattices, are found in Hearmon (1946), Boas and Mackenzie (1950) and Nye (1957). They also give explicit relations between c_{ij} and s_{ij}. The anisotropic Poisson ratio has been derived for hexagonal (Li 1976), tetragonal (Chung et al. 1975) and trigonal (Gunton and Saunders 1972) lattices. In analogy to the example on p. 33 a negative ν is found for α-SiO_2 (Kittinger et al. 1981).

Example: Nearly isotropic tetragonal and hexagonal structures. A crystal may have elastic coefficients c_{ij} which, accidentally, yield an isotropic elastic behaviour. For an hcp lattice, the isotropy conditions are $c_{11} = c_{33}$, $c_{12} = c_{13}$ and $(c_{11}-c_{12})/2 = c_{44}$. The last requirement is the same as for a cubic lattice. The tetragonal lattice is elastically isotropic if $c_{11} = c_{33}$, $c_{12} = c_{13}$ and $c_{44} = c_{66}$. Table 3.3 shows that the isotropy conditions are approximately fulfilled for Mg and ice, less well for In, and not for Zn. The latter fact may be compared with the anomalous

Table 3.3
Elastic stiffnesses in In, Mg, Zn, ice and graphite, in units of GPa

	c_{11}	c_{33}	c_{12}	c_{13}	c_{44}	c_{66} [a]
Indium (tetragonal)	45.2	44.9	40.0	41.2	6.52	12.0
Magnesium (hcp)	59.3	61.5	25.7	21.4	16.4	16.8
Zinc (hcp)	165	61.8	31.1	50.0	39.6	67.0
Ice (hexagonal)	13.7	14.7	7.0	5.6	3.0	3.4
Graphite (hexagonal)	1060	36.5	180	15	4	440

[a] For hexagonal symmetry, $c_{66} = (c_{11}-c_{12})/2$.

ratio $c/a = 1.86$ in Zn (cf. ch. 1 §2.2). Graphite, which has a layered hexagonal structure, is very anisotropic. Data from the Landolt–Börnstein tables (Hearmon 1979, 1984).

Example: Nearly isotropic orthorhombic structures. An orthorhombic structure is elastically isotropic if its nine independent elastic coefficients (table 3.2) reduce to two, through the relations $c_{11} = c_{22} = c_{33}$, $c_{44} = c_{55} = c_{66}$, $c_{12} = c_{13} = c_{23}$ and the additional requirement that $c_{11} - c_{12} = 2c_{44}$. Table 3.4, with c_{ij} in GPa from the Landolt–Börnstein tables (Hearmon 1979), shows that Rb_2SO_4 is quite isotropic, in contrast to Na_2SO_4. Gallium and α-uranium are moderately anisotropic. This behaviour of the elastic properties can be compared with the anisotropy of transport properties (Touloukian et al. 1970), where gallium is highly anisotropic.

Table 3.4
Elastic stiffnesses in orthorhombic structures, in units of GPa

	c_{11}	c_{22}	c_{33}	c_{44}	c_{55}	c_{66}	c_{12}	c_{13}	c_{23}
Rb_2SO_4	50	51	48	16	16	14	20	20	19
Na_2SO_4	80	105	67	15	18	24	30	26	17
Ga	100	90	135	35	42	40	37	33	31
α-U	215	199	267	124	73	74	46	22	108

Example: Hydrostatic compression of hexagonal structures. The strains $\varepsilon_1(=\varepsilon_2)$ and ε_3 in eqs. (3.45) and (3.46) are equal only if

$$s_{11} + s_{12} = s_{13} + s_{33}. \tag{3.47}$$

In a previous example, we noted that Mg and ice are elastically quite isotropic while Zn and graphite are very anisotropic. With s_{ij} from the Landolt–Börnstein tables (Hearmon 1979, 1984), and the formulae (3.45) and (3.46), we get $\varepsilon_{1(2)}/p$ and ε_3/p as in table 3.5.

Table 3.5
Anisotropic hydrostatic compression in hcp structures

	Mg	Ice	Zn	Graphite
$-\varepsilon_{1(2)}/p$ $[\text{TPa}]^{-1}$	9.2	38	1.8	0.49
$-\varepsilon_3/p$ $[\text{TPa}]^{-1}$	9.7	40	13.7	26.9

Example: Expansion of tellurium under pressure. Tellurium has a trigonal chain-like structure. Under a hydrostatic pressure p, the strains are (Boas and Mackenzie 1950)

$$\varepsilon_j/p = (s_{11}+s_{12}-s_{13}-s_{33})n^2-(s_{11}+s_{12}+s_{13}), \tag{3.48}$$

where $n^2 = 1$ for strain along the chain axis ($j = 3$) and $n^2 = 0$ for an axis perpendicular to the chain. Thus

$$\varepsilon_3/p = -(2s_{13}+s_{33}). \tag{3.49}$$

With data from the Landolt–Börnstein tables (Hearmon 1984), we get for tellurium $\varepsilon_3/p = -[2(-14.2)+24.6]$ $[\text{TPa}]^{-1} = 3.8$ $[\text{TPa}]^{-1}$, i.e. a positive value. Hence, the crystal *expands* along the chain direction when a hydrostatic pressure is applied.

4. Sound waves

4.1. Introduction

In an isotropic engineering material, e.g. a texture-free piece of polycrystalline iron, there are two kinds of sound waves, corresponding to longitudinal and transverse vibrations. The longitudinal mode has the velocity $C_\text{L} = \{[K+(4/3)G]/\rho\}^{1/2}$ and the transverse mode the velocity $C_\text{T} = \{G/\rho\}^{1/2}$. Here ρ is the mass density of the medium. In this case there is no distinction between phase velocity and group velocity. The transverse mode is degenerate; its vibrations can be along any of two (arbitrary) directions perpendicular to the velocity vector.

In a single crystal there are still three modes for sound waves, with velocities expressed in the elastic constants of the medium and the crystallographic directions of the wave vector. However, they can no longer be classified as pure longitudinal or transverse waves, and the phase velocity is not parallel to the group velocity except in certain symmetry directions of the lattice. Below we give expressions for the phase velocities. The approach is related to the treatment of lattice vibrations in ch. 4, and sound waves may be viewed as the long-wavelength limit of phonons.

4.2. Formulation of the secular equation

Let the displacement vector $\boldsymbol{u} = (u_x, u_y, u_z)$ of a sound wave be

$$u_j = A_j \exp\left[\mathrm{i}(q_1 x + q_2 y + q_3 z - \omega t)\right]. \tag{3.50}$$

The index j refers to Cartesian coordinates. We shall write $j = x, y, z$ or $j = 1, 2, 3$ alternatively. The displacement in (3.50) is a complex quantity, while the actual displacement is real. However, the real and imaginary parts of $\boldsymbol{u}$ both are solutions to the equation of motion. It is convenient to lump the two solutions together, as in (3.50). The sound wave has the frequency ω and the wave vector $\boldsymbol{q} = (q_1, q_2, q_3)$. The (phase) velocity of the mode λ is

$$C_\lambda(\boldsymbol{q}) = \omega(\boldsymbol{q}, \lambda)/|\boldsymbol{q}|. \tag{3.51}$$

When there is no risk of confusion with the Cartesian index j, we let the mode label λ have any of the values 1, 2 or 3. The wave properties are obtained from the equation

$$\begin{pmatrix} \Gamma_{11} - \rho\omega^2 & \Gamma_{12} & \Gamma_{13} \\ \Gamma_{21} & \Gamma_{22} - \rho\omega^2 & \Gamma_{23} \\ \Gamma_{31} & \Gamma_{32} & \Gamma_{33} - \rho\omega^2 \end{pmatrix} \begin{pmatrix} A_1 \\ A_2 \\ A_3 \end{pmatrix} = 0. \tag{3.52}$$

This is known as the Christoffel (1877) (also Christoffel–Kelvin or Green–Christoffel) equation. The quantities Γ_{jk} are related to $\boldsymbol{q}$ and the elastic constants by

$$\Gamma_{jk} = \sum_{i,l=1}^{3} (1/2)(c_{ijkl} + c_{ijlk}) q_i q_l. \tag{3.53}$$

There are non-trivial solutions (A_1, A_2, A_3) to eq. (3.52) only if ω is a solution to the secular equation

$$|\Gamma_{ij} - \delta_{ij}\rho\omega^2| = 0. \tag{3.54}$$

Here $|\cdots|$ is a 3×3 determinant and $\delta_{ij} = 0$ for $i \neq j$ and 1 for $i = j$. Often, (3.54) is written as an equation for the sound velocities C, i.e. with ω^2 replaced by $C^2 q^2$.

4.3. General solution of the secular equation

Since (3.54) is a cubic equation in C^2 (or ω^2), it has solutions in a closed mathematical form, for any lattice symmetry. A convenient expression for C_λ is (Every 1979, 1980)

$$3\rho[C_\lambda]^2 = T + 2\sqrt{G}\cos\left[\Psi + (2\pi/3)(\lambda - 1)\right]; \qquad \lambda = 1, 2, 3. \tag{3.55}$$

The quantities T, G and Ψ contain the elastic constants and the direction cosines of $\boldsymbol{q}$. If $\Psi = 0$, the cosine terms in the last part of (3.55) are equal for $\lambda = 2$ and $\lambda = 3$. Then these modes have equal velocities.

From a well-known relation, (C.12), between $\sum \omega_\lambda^2$ and the trace of the determinant in (3.54), we get

$$3\langle C_\lambda^2\rangle = \sum_{\lambda=1}^{3} C_\lambda^2 = (\Gamma_{11} + \Gamma_{22} + \Gamma_{33})/\rho q^2, \tag{3.56}$$

for any lattice symmetry. In a cubic lattice (cf. (3.58)),

$$\langle C_\lambda^2\rangle = (c_{11} + 2c_{44})/3\rho. \tag{3.57}$$

Explicit expressions for C_λ in other lattice symmetries are given by Every (1980).

4.4. The secular equation for cubic symmetry

Among all c_{ijkl}, there are only three independent parameters; c_{11}, c_{12}, c_{44}. The explicit form of (3.54) is

$$\begin{vmatrix} (c_{11}-c_{44})q_1^2 + c_{44}q^2 - \rho\omega^2 & (c_{12}+c_{44})q_1q_2 & (c_{12}+c_{44})q_1q_2 \\ (c_{12}+c_{44})q_1q_2 & (c_{11}-c_{44})q_2^2 + c_{44}q^2 - \rho\omega^2 & (c_{12}+c_{44})q_2q_3 \\ (c_{12}+c_{44})q_1q_3 & (c_{12}+c_{44})q_2q_3 & (c_{11}-c_{44})q_3^2 + c_{44}q^2 - \rho q^2 \end{vmatrix} = 0. \tag{3.58}$$

When the solution to this equation is expressed in the general form (3.55), the angle Ψ only depends on the direction cosines of $\boldsymbol{q}$ and on the combination $A_\mathrm{E} = (c_{11} - c_{12} - 2c_{44})/(c_{11} - c_{44})$ of the elastic parameters. For this reason, A_E is a natural parameter to measure the anisotropy in cubic lattices; cf. §3.2. The solutions to (3.58) have very simple forms when $\boldsymbol{q}$ is along the principal crystallographic directions. Table 3.6 gives the quantity $\rho C^2 = \rho\omega^2/q^2$ for these cases.

Table 3.6
$\rho C^2 = \rho\omega^2/q^2$ in three directions $[hkl]$ in a cubic lattice.

Mode	[100]	[110]	[111]
Longitudinal	c_{11}	$(c_{11}+c_{12}+2c_{44})/2$	$c_{11}+2c_{12}+4c_{44})/3$
Transverse	c_{44}	c_{44} [a]	$(c_{11}-c_{12}+c_{44})/3$
		$(c_{11}-c_{12})/2$ [b]	

[a] Polarised along [001]. [b] Polarised along [1$\bar{1}$0].

4.5. The secular equation for hexagonal symmetry

The Christoffel equation for an hcp lattice separates into a linear and a second-order equation in ω^2. The sound velocities (phase velocities) are (Hearmon 1961, Musgrave 1970)

$$\rho[C_1]^2 = c_{44}+(1/2)(1-n^2)(c_{11}-c_{12}-2c_{44}), \quad (3.59)$$

$$\rho[C_{2,3}]^2 = c_{44}+(1/2)[n^2P+(1-n^2)Q] \pm(1/2)\{[n^2P+(1-n^2)Q]^2+4n^2(1-n^2)(R^2-PQ)\}^{1/2}, \quad (3.60)$$

with n related to the angle θ between $\boldsymbol{q}$ and the crystallographic c-axes; $n = q_3/|\boldsymbol{q}| = \cos\theta$. Further,

$$P = c_{33}-c_{44}; \qquad Q = c_{11}-c_{44}; \qquad R = c_{13}+c_{44}. \quad (3.61)$$

In the basal plane of the hcp lattice, $n = \cos\theta = 0$. Then

$$\rho[C_1]^2 = (c_{11}-c_{12})/2; \quad (3.62)$$

$$\rho[C_3]^2 = c_{11}; \qquad \rho[C_2]^2 = c_{44}. \quad (3.63)$$

The sound velocities C_1 and $C_{2(3)}$ in (3.59) and (3.60) are isotropic if

$$c_{11}-c_{12}-2c_{44} = 0; \qquad P = Q; \qquad PQ = R^2. \quad (3.64)$$

These relations are equivalent to the isotropy conditions for c_{ij} on p. 35. Finally, with $\boldsymbol{q}$ parallel to the crystallographic c-axis, $n = 1$ and

$$\rho[C_1]^2 = \rho[C_2]^2 = c_{44}; \qquad \rho[C_3]^2 = c_{33}. \quad (3.65)$$

4.6. The secular equation for isotropic polycrystalline materials

In an isotropic material we can identify K and G with $(c_{11}+2c_{12})/3$ and c_{44}; cf. eq. (3.12) with (3.20). Then,

$$\rho C_L^2 = K+\tfrac{4}{3}G = \frac{3K(1-\nu)}{1+\nu} = \frac{G(4G-E)}{3G-E}$$

$$= \frac{E(1-\nu)}{(1+\nu)(1-2\nu)} = \frac{2G(1-\nu)}{1-2\nu} = \frac{3K(3K+E)}{9K-E}. \qquad (3.66)$$

The velocity C_T of transverse sound waves is given by

$$\rho C_T^2 = G. \qquad (3.67)$$

Since K and $G > 0$,

$$C_L > \sqrt{(4/3)}C_T. \qquad (3.68)$$

This inequality may be sharpened because, almost universally, $\nu > 0$ and then $C_L > \sqrt{2}\,C_T$. Also, when $G < K$, $C_L > \sqrt{(7/3)}\,C_T$.

Example: An important average of sound velocities. In the Debye theory of the lattice heat capacity, one encounters an average sound velocity $\bar{C}$, defined by

$$\frac{3}{\bar{C}^3} = \frac{1}{C_L^3} + \frac{2}{C_T^3}. \qquad (3.69)$$

By eqs. (3.66) and (3.67), the average velocity $\bar{C}$ can be written

$$\bar{C} = C_T\left\{\frac{2}{3}+\frac{1}{3}\left[\frac{1-2\nu}{2(1-\nu)}\right]^{3/2}\right\}^{-1/3}. \qquad (3.70)$$

The factor in the parenthesis varies slowly with the Poisson ratio ν, e.g., $\bar{C} = (1.12\pm 0.02)C_T$ if $\nu = 0.31\pm 0.14$. Since $C_T = \sqrt{(G/\rho)}$, one may in this way connect G with the Debye temperature, see ch. 4 §4 and Schreiber et al. (1973).

4.7. Pure and non-pure modes. Phase and group velocity

We define phase velocities

$$\boldsymbol{C}_{\text{ph}}(\boldsymbol{q}, \lambda) = \hat{\boldsymbol{q}}\omega(\boldsymbol{q}, \lambda)/|\boldsymbol{q}| \tag{3.71}$$

and group velocities

$$\boldsymbol{C}_{\text{g}}(\boldsymbol{q}, \lambda) = \nabla_{\boldsymbol{q}}\omega(\boldsymbol{q}, \lambda). \tag{3.72}$$

When $\boldsymbol{C}_{\text{ph}}$ and $\boldsymbol{C}_{\text{g}}$ are isotropic, the label $\boldsymbol{q}$ can be dropped. Then, the angle ϕ between $\boldsymbol{C}_{\text{ph}}(\boldsymbol{q}, \lambda)$ and the displacement vector $\boldsymbol{u}(\boldsymbol{q}, \lambda) = (u_x, u_y, u_z)$ for the atomic vibrations is either zero (the longitudinal mode) or 90° (the two transverse modes). In this case, one speaks of "pure" modes. The longitudinal mode has the phase velocity $C_{\text{L}} = \sqrt{(c_{11}/\rho)}$ and the two degenerate modes have velocities $C_{\text{T}} = \sqrt{(c_{44}/\rho)}$.

A single crystal often has quite anisotropic elastic properties. Then, $\boldsymbol{C}_{\text{ph}}(\boldsymbol{q}, \text{L})$, $\boldsymbol{C}_{\text{g}}(\boldsymbol{q}, \text{L})$ and $\boldsymbol{u}(\boldsymbol{q}, \text{L})$ are not parallel (except for $\boldsymbol{q}$-values in certain symmetry directions) and $\boldsymbol{C}_{\text{ph}}$ and $\boldsymbol{C}_{\text{g}}$ differ in magnitude. Still it is customary to call the modes "longitudinal" (or quasi-longitudinal) and "transverse" (quasi-transverse). The label of a branch is then retained as one moves with the $\boldsymbol{q}$-vector away from a symmetry direction where the modes are pure. Since eigenvectors referring to different eigenvalues are normal to each other, the vectors $\boldsymbol{u}(\boldsymbol{q}, \lambda)$ for $\lambda = 1$, 2 and 3 are orthogonal. Brugger (1965a) has listed the crystallographic directions of pure modes, and the corresponding phase velocities, for cubic, hexagonal, orthorhombic, tetragonal and rhombohedral lattice symmetries.

The energy transport by an elastic wave can be described by a ray velocity, analogous to the Poynting vector in electromagnetism. In a non-dissipative medium, the ray velocity equals the group velocity (cf. Every 1980). The group velocity $\boldsymbol{C}_{\text{g}}$ can be derived from (3.72) and the explicit relations for the phase velocity $\boldsymbol{C}_{\text{ph}}$ given by Every (1980). However, $\boldsymbol{C}_{\text{g}}(\boldsymbol{q}, \lambda)$ does not have as simple a form as $\boldsymbol{C}_{\text{ph}}$. The wave vector $\boldsymbol{q}$, the phase velocity $\boldsymbol{C}_{\text{ph}}$ and the group velocity $\boldsymbol{C}_{\text{g}}$ are, for small $\boldsymbol{q}$, related by

$$\boldsymbol{C}_{\text{g}} \cdot \boldsymbol{q} = |\boldsymbol{q}||\boldsymbol{C}_{\text{ph}}|. \tag{3.73}$$

Hence, $|\boldsymbol{C}_{\text{g}}| \geqslant |\boldsymbol{C}_{\text{ph}}|$, with equality only in pure modes.

Example: Displacement vectors in anisotropic cubic lattices. Every (1979, 1980) has shown that the angle ϕ between the displacement vectors and $\boldsymbol{q}$ is a function only of A_E and the direction cosines $q_i/|\boldsymbol{q}|$. Figure 3.1 gives the maximum angle ϕ_{max} between $\boldsymbol{q}$ and $\boldsymbol{u}(\boldsymbol{q}, \lambda)$ of the quasi-longitudinal mode (ϕ_{max} from Grimvall (1969a)) as a function of $A_E = (c_{11} - c_{12} - 2c_{44})/(c_{11} - c_{44})$. Note that the anisotropy is not correlated to the lattice type (bcc or fcc).

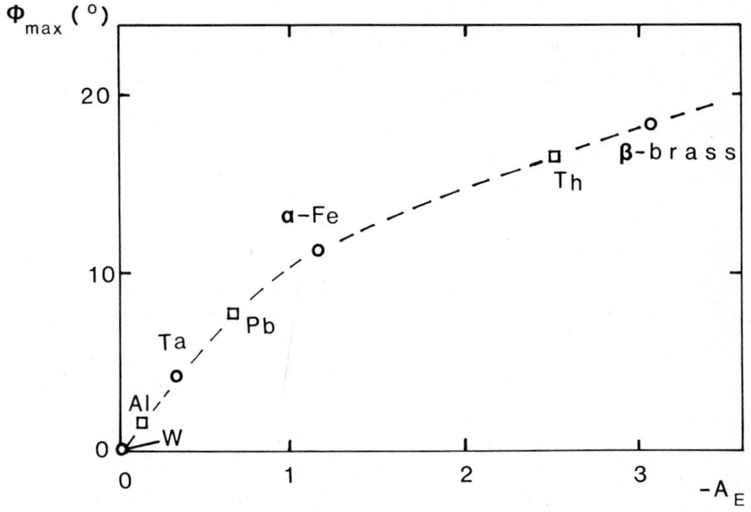

Fig. 3.1. The maximum angle, ϕ_{max}, between the wave vector $\boldsymbol{q}$ and the displacement vector $\boldsymbol{u}$, for the elastic limit of the "longitudinal" mode, plotted versus the index A_E of elastic anisotropy. Circles denote bcc lattices and squares fcc lattices. Values of ϕ from Grimvall (1969a).

Example: Anisotropic group velocities and phonon focusing. The group velocity $\boldsymbol{C}_g(\boldsymbol{q}, \lambda) = \nabla_{\boldsymbol{q}}\omega(\boldsymbol{q}, \lambda)$ is a vector normal to the surface $\omega(\boldsymbol{q}, \lambda) =$ constant. Figure 3.2 shows the shape of $\omega(\boldsymbol{q}, T_2) =$ constant, and the direction of some $\boldsymbol{C}_g(\boldsymbol{q}, T_2)$, for the low-velocity transverse mode (T_2) in germanium, when $\boldsymbol{q}$ is in the (001) plane. $\boldsymbol{C}_g$ tends to be directed along "channels", when $\boldsymbol{q}$ lies near a point where $\omega(\boldsymbol{q}, \lambda) =$ constant has an inflection point in $\boldsymbol{q}$-space. This is the physical basis for experiments on "phonon focusing" in crystals (Wolfe 1980). One may define an enhancement factor A by

$$A = \frac{\Delta\Omega(\boldsymbol{q})}{\Delta\Omega(\text{group velocity})}, \tag{3.74}$$

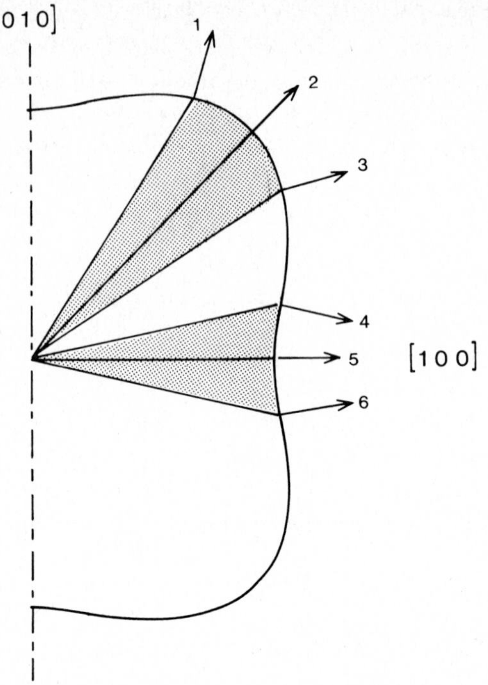

Fig. 3.2. The shape of the curve $\omega(\boldsymbol{q}, T_2)$ = constant for the low-velocity transverse mode (T_2) in the (001) $\boldsymbol{q}$-plane of Ge, and the direction $\boldsymbol{C}_g(\boldsymbol{q}, T_2)$ (arrows 1–6) of group velocities. The shaded areas correspond to the same angle $\Delta\Omega(\boldsymbol{q})$. The group velocities of the modes 1–3 and 4–6 span angles with $\Delta\Omega$(group velocity) $\neq \Delta\Omega(\boldsymbol{q})$. After Wolfe (1980).

where $\Delta\Omega(\boldsymbol{q})$ is the solid angle spanned by a certain set of $\boldsymbol{q}$-vectors and $\Delta\Omega$(group velocity) is the solid angle spanned by the directions of $\boldsymbol{C}_g(\boldsymbol{q}, \lambda)$ when $\boldsymbol{q}$ lies in $\Delta\Omega(\boldsymbol{q})$. The factor A measures the enhancement of energy flow over what would result from an isotropic distribution of propagating phonons. Calculations of group velocities have been performed for LiF, KCl and Al_2O_3 (Taylor et al. 1971), α-SiO_2 and Al_2O_3 (Farnell 1961, Rösch and Weis 1976a), and diamond, Si and Ge (Rösch and Weis 1976b). Among the experiments, emphasis has been on germanium (e.g., Hensel and Dynes 1979, Dietsche et al. 1981).

5. What values can the elastic constants take?

5.1. Stability requirements

The elastic energy U associated with arbitrary but small deformations ε_α is

$$U = (1/2)\, V_0 \sum_{\alpha,\beta=1}^{6} c_{\alpha\beta}\, \varepsilon_\alpha \varepsilon_\beta. \tag{3.75}$$

In this chapter, we denote the energy by U, to distinguish it from Young's modulus E. V_0 is the volume of the unstrained sample. Lattice stability requires that U is positive for any small deformation. This poses restrictions on $c_{\alpha\beta}$ which mathematically are expressed by the requirement that the principal minors of the determinant with elements $c_{\alpha\beta}$ are all positive. For the engineering elastic constants, one has

$$K,\ G,\ E > 0, \tag{3.76}$$

$$0.5 > \nu > -1. \tag{3.77}$$

It follows that

$$3G > E. \tag{3.78}$$

The stability criteria for c_{ij} are (e.g., Born and Huang 1954, Alers and Neighbours 1957)

$$c_{11} > |c_{12}|; \qquad c_{11}+2c_{12} > 0; \qquad c_{44} > 0. \tag{3.79}$$

For an isotropic system, i.e. when $c_{11}-c_{12} = 2c_{44}$ and with the identification $K = (c_{11}+2c_{12})/3$ and $G = c_{44}$, (3.79) is equivalent with (3.76). Note that $c_{11}+2c_{12} > 0$ together with the isotropy condition $c_{11}-c_{12} = 2c_{44}$ implies that $3c_{11} > 4c_{44}$. Then the sound velocities obey $(C_L/C_T)^2 > \frac{4}{3}$ (cf. eq. (3.68)). In appendix C we introduced a very simple model, with central nearest neighbour interactions, for the lattice vibrations in an fcc lattice. That model is dynamically stable (i.e. $\omega^2 > 0$ for all phonon frequencies). Yet it is statically unstable since $c_{11} = c_{44}$, which would imply that $C_L = C_T$ and thus violate (3.68). In a system of hexagonal elastic symmetry, the stability requirements are

$$c_{11} > |c_{12}|; \qquad c_{33}(c_{11}+c_{12}) > 2(c_{13})^2; \tag{3.80a}$$

$$c_{11}c_{33} > (c_{13})^2; \qquad c_{44} > 0. \tag{3.80b}$$

5.2. Cauchy relations and central interatomic forces

If the interatomic forces can be described by a potential $\phi(r)$ which only depends on the distance r between atoms, and if all atoms occupy centres of inversion symmetry in the lattice, $c_{\alpha\beta}$ obey the Cauchy relations:

$$c_{23} = c_{44}; \qquad c_{13} = c_{55}; \qquad c_{12} = c_{66}; \tag{3.81a, b, c}$$

$$c_{14} = c_{56}; \qquad c_{25} = c_{46}; \qquad c_{36} = c_{45}. \tag{3.81d, e, f}$$

This form of the potential ϕ excludes torsional, or bending, forces (present in covalent crystals) and interatomic forces which vary with the atomic volume (present in metals). The Cauchy relations need not be valid in non-Bravais lattices, or in a specimen under pressure. For instance, the atomic sites in the diamond structure do not have inversion symmetry and hence the accidental fulfillment of (3.81) does not imply central forces. Another example is provided by numerically simulated glasses, in which the atoms interact through a central potential (Weaire et al. 1971). The material is macroscopically isotropic. If we neglect internal displacements under stresses and shears, the Cauchy relations plus the isotropy condition gives $K = (5/3)G$. However, there *are* internal displacements and K and G must be calculated separately.

When eqs. (3.81) are fulfilled, the maximum number of linearly independent elements $c_{\alpha\beta}$ is reduced from 21 to 15. For cubic lattices, the three independent elastic coefficients c_{11}, c_{12} and c_{44} are reduced to two since the Cauchy relations in that case can be summarised by c_{12} $(= c_{66}) = c_{44}$.

In this context we note that a central potential $\phi(r)$ may lead to instabilities. Misra (1940) (see also Born and Huang 1954) showed that with $\phi(r) = ar^{-m} - br^{-n}$, simple cubic lattices are never stable, fcc lattices are always stable and bcc lattices are stable for certain small m and n. As another example, consider a bcc lattice with only nearest-neighbour and central interactions. Then $c_{11} = c_{12}$ and the lattice is unstable against shear (Zener 1948). In §5.1 we noted that an fcc lattice with central nearest-neighbour interactions is statically unstable.

5.3. *Ranges for elastic constants in real systems*

In cubic lattices, there does not seem to be a case with $c_{11} < c_{44}$, although this is physically possible. It is not unusual that $c_{12} < c_{44}$, but negative c_{12} are very rare. (Some of the negative c_{12} reported in the Landolt–Börnstein tables (Hearmon 1979, 1984) are uncertain and may be due to indirect effects, e.g., twinning). Table 3.7 lists c_{ij}, from Hearmon (1979, 1984), for some cubic materials with noteworthy properties. The extreme anisotropy of Pu is further discussed by Ledbetter and Moment (1976). With $0.5 > \nu > 0$ one has $3G > E > 2G$ and $K > 2G/3$. These inequalities are shown in figs. 3.3 and 3.4 for polycrystalline elements, with data from Gschneidner (1964).

Table 3.7
Elastic stiffnesses c_{ij} in some cubic lattices, in units of GPa

Material	c_{11}	c_{12}	c_{44}	A_E	Comment
Al	108	62	28	− 0.12	almost isotropic ($A_E \simeq 0$)
Li	13.4	11.3	9.6	− 4.5	very anisotropic
Pu (fcc)	36	27	34	−21	extremely anisotropic
W	523	203	160	0.00	high c_{ij}; isotropic
Fe (bcc)	230	135	117	− 1.2	rather anisotropic
Mo	465	163	109	0.24	high c_{11}; $A_E > 0$
Ir	600	260	270	− 0.60	largest c_{11}, c_{44} (in metals), c_{12}
Cr	347	66	100	0.32	$A_E > 0$; $c_{12} < c_{44}$
β-brass	130	102	74	− 2	very anisotropic
Diamond	1040	170	550	− 0.47	largest c_{11}, c_{44}; $c_{12} \ll c_{44}$
TiC	418	89	217	− 0.52	high c_{11}, c_{44}; $c_{12} \ll c_{44}$
UCd_{11}	101	36	32	0.01	almost isotropic
NaBr	41.1	9.9	10.0	− 0.36	Cauchy relation obeyed
BaF_2	91	41	25	0.0	isotropic
^{3}He	0.024	0.020	0.012	~ −5	very low c_{ij} very anisotropic

6. Fundamental definitions of elastic constants

6.1. *Adiabatic and isothermal elastic constants*

Hooke's law is a phenomenological expression of how a solid responds to applied stresses. Before we discuss the calculation of elastic properties

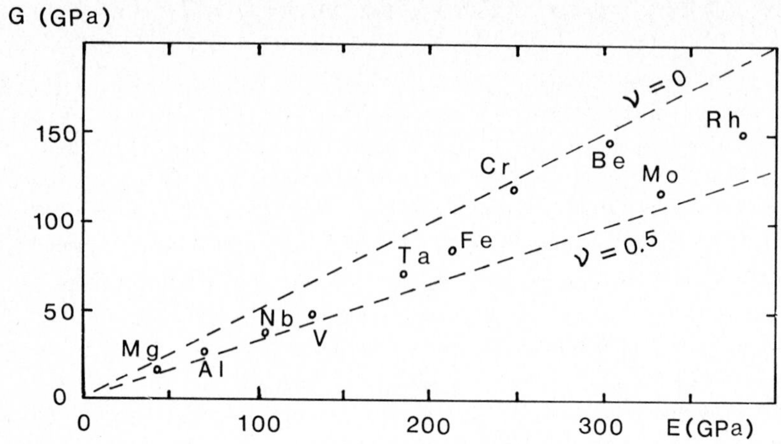

Fig. 3.3. The shear modulus G of polycrystalline metals, versus Young's modulus E. The dashed lines give the upper and lower bounds, corresponding to Poisson ratios $\nu = 0$ and $\nu = 0.5$.

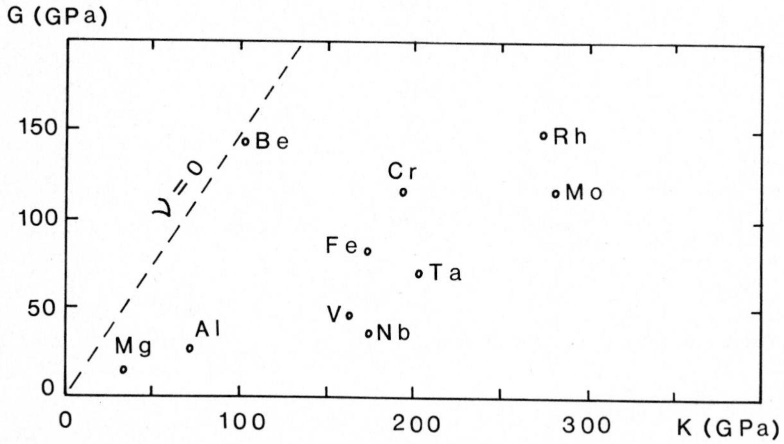

Fig. 3.4. The shear modulus G versus the bulk modulus K. The dashed line is an upper bound, corresponding to the Poisson ratio $\nu = 0$. The lower bound (for $\nu = 0.5$) is $G = 0$.

from first principles (quantum mechanics) or the influence of temperature, pressure, etc. on the elastic constants, we should express $c_{\alpha\beta}$ in fundamental terms. This means a definition of $c_{\alpha\beta}$ in terms of appropriate derivatives of thermodynamic functions. The first law of thermodynamics, in most cases just written $dU = TdS - pdV$, has a generalised form when one resolves the forces and deformations into

Cartesian components;

$$dU = TdS + V_0 \sum_{i=1}^{6} \sigma_i d\varepsilon_i. \tag{3.82}$$

For an adiabatic deformation ($dS = 0$, i.e. no heat flows in or out of the volume V_0), we can take σ_i from Hooke's law (3.2) and integrate (3.82), which gives the strain energy of (3.75). The components of the stress tensor can be defined by

$$\sigma_i = (1/V_0)(\partial U/\partial \varepsilon_i)_{S,\varepsilon'}. \tag{3.83}$$

The subscripts S, ε' mean that the derivatives are taken at constant entropy S and with all strains $\varepsilon_j \neq \varepsilon_i$ held constant. The prefactor $1/V_0$ ensures that σ_i is independent of the size of the specimen. Alternatively, we could have defined U as an energy density, i.e. energy per unit volume, and dropped V_0 in (3.83).

The components $c_{\alpha\beta}$ of the stiffness tensor are defined by

$$(c_{\alpha\beta})_S = (\partial \sigma_\alpha/\partial \varepsilon_\beta)_{S,\varepsilon'} = (1/V_0)(\partial^2 U/\partial \varepsilon_\alpha \partial \varepsilon_\beta)_{S,\varepsilon'}. \tag{3.84}$$

Now, the subscripts S, ε' mean that S and all ε except ε_α and ε_β are kept constant. The $c_{\alpha\beta}$ defined by (3.84) are called adiabatic elastic constants since they are taken at constant entropy. We can also define elastic coefficients at constant temperature T, if we start from the Helmholtz free energy. One has

$$dF = -SdT + V_0 \sum_{i=1}^{6} \sigma_i d\varepsilon_i, \tag{3.85}$$

which leads to (Brugger 1964)

$$(c_{\alpha\beta})_T = (\partial \sigma_\alpha/\partial \varepsilon_\beta)_{T,\varepsilon'} = (1/V_0)(\partial^2 F/\partial \varepsilon_\alpha \partial \varepsilon_\beta)_{T,\varepsilon'}. \tag{3.86}$$

The adiabatic compliances $(s_{\alpha\beta})_S$ are defined from the enthalpy,

$$(s_{\alpha\beta})_S = (\partial \varepsilon_\alpha/\partial \sigma_\beta)_{S,\sigma'} = -(1/V_0)(\partial^2 H/\partial \sigma_\alpha \partial \sigma_\beta)_{S,\sigma'}, \tag{3.87}$$

and the isothermal compliances follow from the Gibbs free energy,

$$(s_{\alpha\beta})_T = (\partial \varepsilon_\alpha/\partial \sigma_\beta)_{T,\sigma'} = -(1/V_0)(\partial^2 G/\partial \sigma_\alpha \partial \sigma_\beta)_{T,\sigma'}, \tag{3.88}$$

see Brugger (1964, 1965b) for details on the definitions of $c_{\alpha\beta}$ and $s_{\alpha\beta}$ and how they are related to thermodynamics and measured elastic constants.

The adiabatic and isothermal compressibilities κ_S and κ_T, and bulk moduli K_S and K_T, are of the form

$$\kappa_S = (K_S)^{-1} = \sum_{\alpha,\beta=1}^{3} (s_{\alpha\beta})_S, \tag{3.89}$$

$$\kappa_T = (K_T)^{-1} = \sum_{\alpha,\beta=1}^{3} (s_{\alpha\beta})_T. \tag{3.90}$$

These relations, valid for any crystal structure, are discussed further in ch. 10, in connection with thermal expansion. Here we just note that $K_S/K_T = C_p/C_V$, (10.35). There is no significant difference between adiabatic and isothermal elastic constants when $T < \theta_D$ (the Debye temperature), but close to the melting temperature they may differ by 15% or more.

Example: The bulk modulus expressed in c_{ij}. As an illustration we check that the $c_{\alpha\beta}$ from (3.84) are consistent with the result $K = (c_{11}+2c_{12})/3 = -V(\partial P/\partial V)$ for cubic lattices, eq. (3.30). We get

$$\begin{aligned} c_{11}+2c_{12} &= (1/V)(\partial^2 U/\partial\varepsilon_1^2 + 2\partial^2 U/\partial\varepsilon_1\partial\varepsilon_2) \\ &= (1/V)(\partial/\partial\varepsilon_1)(\partial U/\partial\varepsilon_1 + \partial U/\partial\varepsilon_2 + \partial U/\partial\varepsilon_3) \\ &= (1/V)(\partial/\partial\varepsilon_1)(\partial U/\partial V)(\partial V/\partial\varepsilon_1 + \partial V/\partial\varepsilon_2 + \partial V/\partial\varepsilon_3) \\ &= (\partial/\partial V)(-3pV) = -3V(\partial p/\partial V). \end{aligned} \tag{3.91}$$

Here we have used $c_{12} = c_{13}$, $dV/V = d\varepsilon_1 + d\varepsilon_2 + d\varepsilon_3$, $p = -\partial U/\partial V$ and, in the last step, that $p \to 0$ ($c_{\alpha\beta}$ is calculated for a stress-free sample).

6.2. Higher-order elastic constants

The energy of a crystal can be expanded in powers of the strains ε_i as

$$\begin{aligned} U = U(\varepsilon_i = 0) &+ V_0 \sum c_i\varepsilon_i + (1/2)V_0 \sum c_{ij}\varepsilon_i\varepsilon_j \\ &+ (1/6)V_0 \sum c_{ijk}\varepsilon_i\varepsilon_j\varepsilon_k + \cdots. \end{aligned} \tag{3.92}$$

Indices i, j, k run from 1 to 6 in the summations. The definition of the expansion coefficients is exemplified by

$$c_{ijk} = (1/V_0)(\partial^3 U/\partial\varepsilon_i\partial\varepsilon_j\partial\varepsilon_k)_{S,\varepsilon'}, \tag{3.93}$$

which is called a third-order adiabatic elastic coefficient. (Some authors call it second order and let $c_{\alpha\beta}$ be first order.) $U(\varepsilon_i = 0)$ is the energy of the reference state from which the strains are counted. This energy is uninteresting in our discussion. All terms $c_i\varepsilon_i$ vanish if $\varepsilon_i = 0$, which refers to a stress-free equilibrium. The subscript S, ε' has the same meaning as earlier. In addition to the adiabatic third-order elastic coefficients, (3.93), one can start from the Helmholtz free energy F and define isothermal coefficients. There are also third-order compliances s_{ijk} and higher-order elastic coefficients. See, for example Brugger (1964), Thurston and Brugger (1964), Wallace (1970) or Hearmon (1979, 1984) for details.

As for $c_{\alpha\beta}$, symmetry reduces the number of linearly independent third order elastic coefficients, and many of them vanish. In fcc and bcc lattices one is left with

$$c_{111},\ c_{112},\ c_{123},\ c_{144},\ c_{155},\ c_{456}. \tag{3.94}$$

For isotropic crystals, only three of these are independent. The isotropy conditions are

$$c_{112} = c_{123} + 2c_{144}; \qquad c_{155} = c_{144} + 2c_{456}; \tag{3.95}$$

$$c_{111} = c_{123} + 6c_{144} + 8c_{456}. \tag{3.96}$$

The Cauchy (central force) conditions are

$$c_{112} = c_{155}; \qquad c_{144} = c_{123} = c_{456}. \tag{3.97}$$

Tables of relations between c_{ijk} for various crystal structures are found in the Landolt–Börnstein tables (Hearmon 1979).

7. Pressure dependence

In a material which strictly obeys Hooke's law, the elastic coefficients $c_{\alpha\beta}$ do not vary with the strain. In that case the total energy U, given by (3.92), has no terms of higher order than $c_{\alpha\beta}\varepsilon_\alpha\varepsilon_\beta$. In a real material, however, c_α and $c_{\alpha\beta}$ are strain dependent and therefore also vary with the external pressure.

The elastic constants are often determined from experiments on ultrasonic waves. Then it is natural to speak of effective elastic constants, to be distinguished from the thermodynamic definition, as in (3.84). As an example, we consider the pressure dependence of c_{11} determined from experiments as the ratio

$$(\partial c_{11}/\partial p) = [\rho C_L^2 \,(\text{at } p) - \rho_0 C_L^2 \,(\text{at } p = 0)]/p. \tag{3.98}$$

C_L is the sound velocity of the longitudinal branch, in the [100] direction of a cubic lattice, and ρ and ρ_0 are the mass densities at p and $p = 0$ respectively. For cubic lattice symmetry, one has (Thurston and Brugger 1964, Brugger 1965b)

$$-(\partial c_{11}/\partial p)_{p=0} = 1 + (1/3K)(c_{11} + c_{111} + 2c_{112}), \tag{3.99}$$

$$-(\partial c_{12}/\partial p)_{p=0} = -1 + (1/3K)(c_{12} + c_{123} + 2c_{112}), \tag{3.100}$$

$$-(\partial c_{44}/\partial p)_{p=0} = 1 + (1/3K)(c_{44} + c_{144} + 2c_{166}). \tag{3.101}$$

Then, by $K = (c_{11} + 2c_{12})/3$, one has

$$-(\partial K/\partial p)_{p=0} = (c_{111} + 6c_{112} + 2c_{123})/9K. \tag{3.102}$$

The difference in the pressure dependence of adiabatic and isothermal elastic constants is discussed by Barsch (1967), with numerical applications to 25 materials (Barsch and Chang 1967). Typically, the two pressure derivatives differ by a few percent at ambient temperature.

Example: $\partial c_{ij}/\partial p$ for aluminium and iron. With experimental numbers inserted from the Landolt–Börnstein tables (Hearmon 1979, 1984), eqs. (3.99)–(3.101) take the following form for aluminium:

$$\begin{aligned}
\partial c_{11}/\partial p &= -1 - (108 - 1080 - 2\times 315)/232 = 5.9,\\
\partial c_{12}/\partial p &= +1 - (62 + 36 - 2\times 315)/232 \quad = 3.3,\\
\partial c_{44}/\partial p &= -1 - (28 - 23 - 2\times 340)/232 \quad = 1.9.
\end{aligned} \tag{3.103}$$

For bcc iron, we get $\partial c_{11}/\partial p = 6.7$, $\partial c_{12}/\partial p = 4.5$, $\partial c_{44}/\partial p = 2.7$. Further, $(\partial K/\partial p) = 4.2$ (Al) and 5.3 (Fe).

8. Volume dependence

The volume dependence of elastic properties is, of course, closely related to their pressure dependence. In the theory of anharmonic lattice vibrations, ch. 5, we introduce the Grüneisen parameter $\gamma(\boldsymbol{q},\lambda) = -(\partial \ln \omega(\boldsymbol{q},\lambda)/\partial \ln V)$, where $\omega(\boldsymbol{q},\lambda)$ is the frequency of a phonon with wave vector $\boldsymbol{q}$ and mode index λ. In the long-wavelength limit, $\omega = C_\lambda(\boldsymbol{q})|\boldsymbol{q}|$, where $C_\lambda(\boldsymbol{q})$ is the sound velocity. Consider now a longitudinal sound wave ($\lambda = \mathrm{L}$) in the [100] direction of a cubic lattice. Then, $C_\mathrm{L} = (c_{11}/\rho)^{1/2}$, where ρ is the mass density. Since $|\boldsymbol{q}|/\sqrt{\rho}$ varies with the volume V as $V^{1/6}$, we get for the corresponding Grüneisen parameter (cf. Brugger and Fritz (1967))

$$\gamma_\mathrm{L}[000] = -\left(\frac{\mathrm{d}\ln \omega_\mathrm{L}}{\mathrm{d}\ln V}\right) = -\frac{1}{2}\left(\frac{\mathrm{d}\ln c_{11}}{\mathrm{d}\ln V}\right) - \frac{1}{6}$$

$$= \frac{K}{2c_{11}}\left(\frac{\partial c_{11}}{\partial p}\right) - \frac{1}{6}. \qquad (3.104)$$

From the sound velocities expressed in c_{ij} (table 3.6), we can define Grüneisen parameters $\gamma_\mathrm{L}[hkl]$, $\gamma_\mathrm{T1}[hkl]$ and $\gamma_\mathrm{T2}[hkl]$, referring to the longitudinal and the two transverse modes, respectively. If the system is isotropic, we are left with only two Grüneisen parameters:

$$\gamma_\mathrm{L} = (K/2c_{11})(\partial c_{11}/\partial p) - 1/6$$

$$= -K/2c_{11} - (c_{111} + 2c_{112})/6c_{11} - 1/3, \qquad (3.105)$$

$$\gamma_\mathrm{T} = (K/2c_{44})(\partial c_{44}/\partial p) - 1/6$$

$$= -K/2c_{44} - (c_{144} + 2c_{166})/6c_{44} - 1/3. \qquad (3.106)$$

Further, we can write

$$\left(\frac{\partial \ln K}{\partial \ln V}\right)_T = \frac{V}{K_T}\left(\frac{\partial K}{\partial V}\right)_T = \frac{V}{K_T}\left(\frac{\partial K}{\partial p}\right)_T\left(\frac{\partial p}{\partial V}\right)_T = -\left(\frac{\partial K}{\partial p}\right)_T. \qquad (3.107)$$

Note that for an elastically isotropic system, a knowledge of the two sound velocities suffices to determine the two linearly independent elastic constants (e.g., c_{11} and c_{44}), but there are three linearly independent third-order elastic constants and only two Grüneisen parameters. To obtain all c_{ijk} one has to measure the sound velocities not only for varying crystal volume but also for, say, uniaxial tension (cf. Thurston and Brugger 1964). Elastic Grüneisen parameters in non-cubic systems have been calculated by Gerlich (1969).

The elastic isotropy in a single crystal depends on the accidental equality $c_{11}-c_{12}=2c_{44}$. As the following example shows, such an isotropy does not imply that the Grüneisen parameter is isotropic. On the other hand, there are truly isotropic systems, e.g. glasses, for which (3.105) and (3.106) hold.

Example: Sound-velocity Grüneisen parameters in aluminium. Using the relation (3.104), and the analogous results for other directions [*hkl*] and for the two transverse modes, and with the data for c_{ij} and $\partial c_{ij}/\partial p$ of aluminium (the example on p. 52), we get the sound velocities C_λ and the Grüneisen parameters γ_λ of table 3.8. Note that although C_T is almost isotropic, γ_{T1} and γ_{T2} vary considerably with [*hkl*].

Table 3.8
Sound velocity and sound wave Grüneisen parameters in aluminium

[*hkl*]	C_L (m/s)	γ_L	C_{T1} (m/s)	γ_{T1}	C_{T2} (m/s)	γ_{T2}
[100]	6320	1.95	3220	2.47	3220	2.47
[110]	6470	2.06	3220	2.47	2920	2.03
[111]	6520	2.10	3020	2.20	3020	2.20

9. Temperature dependence

Wachtman et al. (1961) (see also Durand (1936)) noted empirically that the temperature dependence of Young's modulus E of several oxides could be well fitted to

$$E(T)=[1-bT\exp(-T_0/T)]\,E(0). \tag{3.108}$$

An expansion of the exponential term for $T>T_0$ gives, to leading order,

$$E(T)=[1+bT_0-bT]\,E(0). \tag{3.109}$$

This temperature dependence is observed for many systems other than

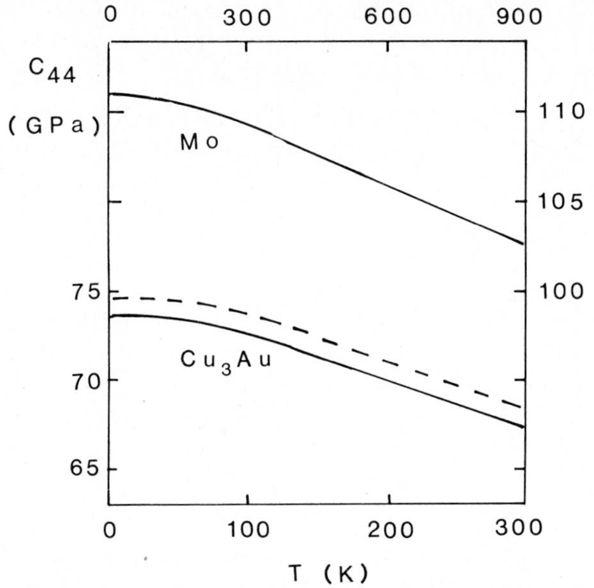

Fig. 3.5. The temperature dependence of c_{44} in Mo (top and right scales; data from Hearmon (1984)) and in Cu_3Au (bottom and left scales; data from Flinn et al. (1960). The Cu_3Au data refer to a disordered (dashed line) and an ordered alloy (solid line).

oxides, and for elastic constants other than E; see, for instance, fig. 3.5. Anderson (1966a) has shown how eq. (3.108) can be understood from the quasi-harmonic model of lattice vibrations. We shall follow the essential points of that work and make contact with the treatment in ch. 5, of anharmonic lattice vibrations. The elastic constants can be related to the long wavelength limit of phonons. Anharmonicity shifts a phonon frequency $\omega(\boldsymbol{q}, \lambda)$ by an amount $\Delta\omega(\boldsymbol{q}, \lambda)$. In a simple model, we may write for the shift $\Delta\omega$ relative to ω at $T = 0$, (5.31)–(5.33),

$$\frac{\Delta\omega(\boldsymbol{q}, \lambda)}{\omega(\boldsymbol{q}, \lambda)} = \frac{k(\boldsymbol{q}, \lambda)}{e^{\Theta_E/T} - 1}. \qquad (3.110)$$

Here, $\boldsymbol{q}$ is the wave vector of the phonon, λ is a mode index (longitudinal and transverse modes etc.), $k(\boldsymbol{q}, \lambda)$ is a dimensionless proportionality constant and Θ_E is an Einstein temperature characteristic of the entire phonon spectrum. Let Y be an elastic constant. A dimensionally correct relation between Y and ω is $\sqrt{(Y/\rho)} = \omega/|\boldsymbol{q}|$. The mass density ρ varies as $1/V$ and $|\boldsymbol{q}|$ as $V^{-1/3}$, where V is the sample volume. Within our simple model, the thermal expansion has the same temperature dependence as $\Delta\omega/\omega$. (In fact, $\Delta\omega$ is to a large extent due to the thermal expansion.) Thus an expression of the type $(\omega/|\boldsymbol{q}|)\sqrt{\rho}$ has a temperature-dependent shift which varies with T as $[\exp(\Theta_E/T)-1]^{-1}$. The same temperature dependence must enter all (small) shifts in the elastic constants. We summarise this by the relation

$$\frac{\Delta Y}{Y}=\frac{a_Y}{\Theta_E}\frac{\Theta_E}{e^{\Theta_E/T}-1}, \tag{3.111}$$

where a_Y is proportionality constant varying with the elastic parameter Y under consideration. Comparing the high-temperature expansions $T\exp(-T_0/T)=T-T_0+T_0^2/2T-\dots$ and $\Theta_E[\exp(\Theta_E/T)-1]^{-1}=T-\Theta_E/2+(1/12)\Theta_E^2/T+\dots$, we note that the two leading terms are identical if $T_0=\Theta_E/2$. This gives a theoretical justification for the empirical rule (3.108) at intermediate and high temperatures, and for the fact that T_0 was observed to be about half the Debye temperature. Since the Debye temperature is less than 500 K for most solids, we also understand why the linear temperature dependence expressed by the series expansion (3.109) is such a good approximation at ambient and higher temperatures.

The energy (enthalpy, free energy) of an insulator varies as T^4 at low T (since C_p varies as T^3). In metals, the temperature dependence of E, H and F is $\sim T^2$. From the fundamental definition of elastic constants, §6, it follows that $\mathrm{d}Y/\mathrm{d}T\sim T^3$ or $\sim T$, i.e. a power law in T. The Einstein model used to get the temperature dependent factor, (3.111), as well as the empirical relation (3.108), gives too rapid (exponential) a temperature dependence of Y at low T. However, the absolute magnitude of ΔY at low temperatures is so small that this discrepancy is of no practical importance. Finally we stress that although a temperature dependence roughly as (3.108) is often observed, there are many exceptions.

10. Dependence on lattice structure and order

There are only a few measurements of the engineering elastic constants in both phases of a polymorphic system. E and G in iron and some Fe–Ni alloys change by less than 10% when the crystal structure changes from bcc to fcc (Ledbetter and Reed 1973). There seems to be no measurements of c_{ij} in pure fcc iron. A weak dependence of the bulk modulus on the crystal structure is expected for non-transition metals, since our simple approach to the cohesive energy (ch. 1 §2.2) showed a very small variation in the structure-dependent Madelung constant α_C.

Some compounds have an order–disorder transformation. Experiments on Cu_3Au (Flinn et al. 1960) and $\beta-CuZn$ (McManus 1963), show that the elastic constants c_{ij} of the ordered and the disordered states typically differ by at most a few percent. An isotopic disorder does not affect the electronic structure and therefore has a negligible influence on c_{ij}. This has been verified in experiments on ^6Li and ^7Li (Felice et al. 1977).

The molecular crystal KCN has a NaCl-type lattice with a dumbbell-like $[CN]^{-1}$ ion which is orientationally disordered at high temperatures, but orders partially at lower T, eventually leading to a transition to a rhombic phase for KCN. The elastic coefficient c_{44} of the cubic lattice decreases strongly as T is lowered (Haussühl 1973). There has been much interest in the elastic properties of this and similar molecular crystals; e.g., Rowe et al. (1978), Sahu and Mahanti (1983) and Strössner et al. (1983). A soft-mode behaviour is also well-known from materials like $BaTiO_3$, which show a displacive transformation.

Metallic glasses are statistically isotropic, and therefore described by two independent engineering elastic constants. We consider K and G. The bulk modulus usually is at most a few percent lower than in the crystalline phase. The shear modulus G, on the other hand, may be typically 30% lower in the glassy state (Logan and Ashby 1974). For instance, in $Pd_{0.775}Si_{0.165}Cu_{0.06}$, Golding et al. (1972) found that K was lower by 6% and G was lower by 35% in the amorphous state. The small change in K is consistent with the small volume difference, i.e. a few percent lower mass density in the amorphous state. If only the volume difference is considered, we expect that $\Delta K/K = -2\gamma_G(\Delta V/V)$, where γ_G is a Grüneisen parameter of the order of 2. The large shift in G may be understood from the fact that the atoms in the glass do not take positions in deep symmetric potential wells but can be displaced a substantial amount under shear forces. This qualitative picture is supported by numerical simulations using a Morse interaction between the atoms (Weaire et al. 1971). The relation $E = 3G[1+G/3K]^{-1}$, and the fact that usually $G < K$, implies that E and G co-vary. Hence also Young's modulus may be typically 30% lower in the amorphous than in the crystalline state. A similar but much larger effect has been observed in amorphous films. The sound velocity in a gallium film was smaller by a factor of 2.8 in the amorphous state, while there was no significant change in the longitudinal sound velocity (Dietsche et al. 1980). The heat capacity Debye temperature in amorphous films of Si and Ge is lower than the bulk value by $\sim 20\%$ (Mertig et al. 1984). This is due to a lowering of transverse acoustic modes, i.e. a lowering of G.

Example: c_{44} in Mo and in ordered and disordered Cu_3Au. Figure 3.5 shows c_{44} as a function of T for Mo (Hearmon 1984) and for Cu_3Au in an ordered and a disordered state (Flinn et al. 1960).

11. Influence of lattice defects

11.1. Point defects

We shall take several different approaches to an estimation of how the elastic properties are affected by point defects; an atomistic nearest-neighbour force-constant description, a conduction electron consideration in metals and an elastic continuum theory.

Atomistic force-constant models. In a simple picture of a solid, atoms are connected by springs. The introduction of a small number of point-like defects means that certain springs are changed. The elastic properties depend on some average over all the springs and therefore should vary smoothly with the defect concentration. Let there be N atoms in a monatomic solid, with p_i springs of type i and strength f_i attached to each atom. A small fraction, c, of all atoms are replaced. Then, cNp_i springs have their strength changed by Δf_i. Averaged over the crystal, the force constant of type i is

$$\bar{f}_i = f_i(1+2c\Delta f_i/f_i). \tag{3.112}$$

For a vacancy, we can take $\Delta f_i = -f_i$ and get

$$\bar{f}_i = (1-2c)f_i. \tag{3.113}$$

Such considerations can be put on a firmer mathematical basis. Consider an fcc lattice with only nearest-neighbour central interactions. The force constants attached to a substitutional defect are changed from f to

$f+\Delta f$. Neglecting relaxation around the defect, one obtains (Dederichs and Zeller 1980, Leibfried and Breuer 1978)

$$\frac{\Delta(c_{11}+2c_{12})}{c_{11}+2c_{12}}=\frac{2c(\Delta f/f)}{1+0.24\Delta f/f}, \tag{3.114}$$

$$\frac{\Delta(c_{11}-c_{12})}{c_{11}-c_{12}}=\frac{2c(\Delta f/f)}{1+0.38\Delta f/f}, \tag{3.115}$$

$$\frac{\Delta c_{44}}{c_{44}}=\frac{2c(\Delta f/f)}{1+0.33\Delta f/f}. \tag{3.116}$$

For a vacancy, $\Delta f/f=-1$. Then all relative shifts of c_{ij} are of the order of $3c$.

If there are only central nearest-neighbour interactions, expressed by a potential $\phi(r)$, equilibrium requires that all $d\phi(r)/dr=0$ at the lattice sites. The change of such bonds leaves no unbalanced forces on the atoms and hence there is no relaxation around the defect in this model. In a real solid, one has to consider relaxations. We can identify two contributions. First, there is an inhomogeneous relaxation in the immediate neighbourhood of the defect. For a statistical distribution of defects, there is also a uniform volume change (V_{def} per defect) of the sample. From a knowledge of V_{def}, the compressibility K and the anharmonic parameters $\partial c_{ij}/\partial p$, and with eq. (3.99) we get a shift $(\Delta c_{11})^*$ per defect, where

$$(\Delta c_{11})^* = V_{\text{def}}\left(\frac{\partial p}{\partial V}\right)\left(\frac{\partial c_{11}}{\partial p}\right) = V_{\text{def}}\frac{K}{V}\left(1+\frac{c_{11}+c_{111}+2c_{112}}{3K}\right), \tag{3.117}$$

with similar expressions for the shifts in c_{12} and c_{44}.

Electron band structure models. In ch. 1 §2.2 we considered a simple model for the bulk modulus K of a free-electron-like metal. There, K was dependent on the number of conduction electrons per volume, expressed through r_s. This suggests a weak dependence of K on the number of point defects and a smooth variation of K in alloys. For transition metals, there is no such simple theoretical description, but empirical (Fisher and Dever 1970) and crude theoretical (Peter et al. 1974) analysis suggest that c_{ij} varies slowly and regularly with the filling of the d-band of conduction electrons.

Elastic continuum model. The elastic sphere model of Eshelby (1975) is another approach to the influence of small defects on the elastic properties. In a homogeneous material, characterised by two of the parameters K, G, E and ν, there are randomly distributed spherical inclusions with elastic properties given by K', G', E' and ν'. In the dilute limit ($c \ll 1$ where c is the volume fraction of inclusions) one obtains

$$\frac{1}{K}\frac{\mathrm{d}K}{\mathrm{d}c} = \left[\frac{1+\nu}{3(1-\nu)} + \frac{K}{K'-K}\right]^{-1}, \tag{3.118}$$

$$\frac{1}{G}\frac{\mathrm{d}G}{\mathrm{d}c} = \left[\frac{2}{15}\frac{4-5\nu}{1-\nu} + \frac{G}{G'-G}\right]^{-1}. \tag{3.119}$$

This model is now applied to inclusions as small as a single atom. With the typical value $\nu = 1/3$ and very stiff inclusions ($K' \gg K$, $G' \gg G$), $|(\mathrm{d}K/\mathrm{d}c)|/K \lesssim 2$ and $|(\mathrm{d}G/\mathrm{d}c)|G \lesssim 2$. The same conclusion is reached for elastically very soft inclusions ($K' \ll K, G' \ll G$). See also ch. 13 §3.1.

Example: c_{ij} in dilute Mg alloys. Eros and Smith (1961) measured the adiabatic elastic stiffnesses c_{ij} in hcp magnesium alloys with up to a few at.% of Ag, In or Sn. The variations in c_{ij} were less than 1%. Table 3.9 gives typical results. Note that the shift in c_{44} is opposite in sign to that of c_{11} and c_{12}.

Table 3.9
The adiabatic c_{ij} at 298 K, in units of GPa

	c_{11}	c_{12}	c_{13}	c_{33}	c_{44}
pure Mg	59.28	25.90	21.57	61.35	16.32
Mg–1.35 at. % In	59.55	26.26	22.01	61.48	16.25

Example: Elastic properties of bcc Nb – Zr and fcc Cu – Ni. Niobium and zirconium form high-temperature bcc solid solutions, $Nb_{1-c}Zr_c$. By quenching, this structure can be retained at room temperature. Figure 3.6 shows c_{11}, c_{12} and c_{44} at $T = 300$ K, as a function of c; from experiments by Walker and Peter (1977).

Nickel and copper form fcc solid solutions, $Ni_{1-c}Cu_c$. Epstein and

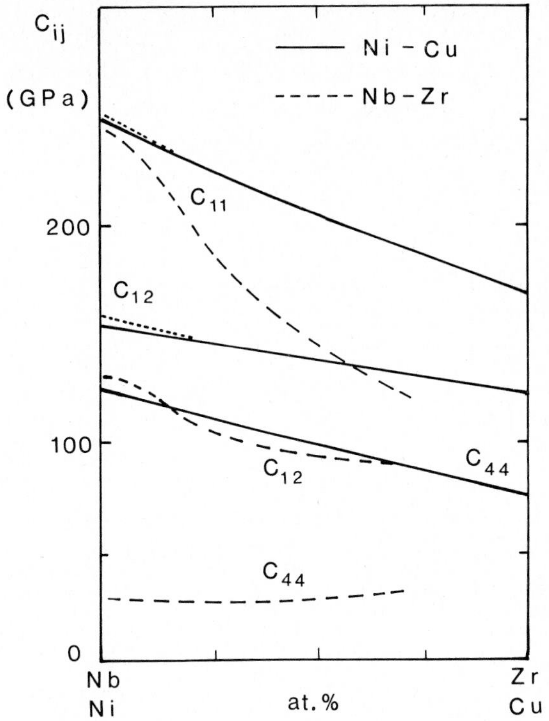

Fig. 3.6. The elastic stiffnesses c_{11}, c_{12} and c_{44}, as a function of alloy composition, for bcc Nb–Zr (Walker and Peter 1977) and fcc Ni–Cu (Epstein and Carlson 1965) solid solutions. The short-dashed curve for Ni–Cu refers to a magnetically saturated state.

Carlson (1965) measured c_{11}, c_{12} and c_{44} of $Ni_{1-c}Cu_c$ at room temperature. For $c < 0.3$ the alloy is ferromagnetic. Then the elastic properties were determined both in a state of magnetic saturation and in an unmagnetised state (i.e. with randomly oriented magnetic domains). The results are given in fig. 3.6.

11.2. Dislocations and grain boundaries

The constants G and E in a cold-worked material may be typically 5% to 20% lower than in the annealed state. The bulk modulus, on the other hand, is not much changed on cold working. Even with a dislocation density Λ as high as 10^{15} m^{-2}, less than 0.1% of all atoms are adjacent to the dislocation cores. It is therefore obvious that the decrease in G and E is not related to "bond cutting" or volume changes (as is the case for the effect of vacancies on elastic properties). Instead, it is caused by a reversible motion of the dislocations (Read 1940, Eshelby 1949, Mott 1952, Friedel 1953, Koehler and DeWit 1959). The shift in Young's modulus can be written

$$\frac{\Delta E}{E} = -k\Lambda L^2. \tag{3.120}$$

Here k is a dimensionless constant ~ 0.1, Λ is the dislocation density and L the dislocation line length. Since L depends on the pinning of dislocations, impurities may have an appreciable indirect effect on E and G. There may also be a reversible grain boundary sliding, which affects the apparent E and G.

In this context we should remark on the measurement of elastic parameters. Young's modulus is defined as the ratio of uniaxial stress to strain. In many experiments, the stress varies sinusoidally with a frequency ω. If the material is strictly elastic, the strain will follow the stress without any phase lag. In anelastic (i.e. dissipative) materials, the strain still varies with the frequency ω, but there is a phase difference between stress and strain. Then E is not uniquely defined. The measured E (or G) has one truly elastic part E_0 and one anelastic part (Zener 1948). We can write

$$E(\omega) = E_0 - \frac{E_0 - E_R}{1+(\omega\tau)^2}. \tag{3.121}$$

E_R includes relaxations and τ is a characteristic time for stress and strain relaxation. The reader is also reminded of the difference between isothermal and adiabatic elastic constants, see §6.1.

When a polycrystalline material is deformed, it may develop a texture, i.e. a statistically anisotropic distribution of the crystallographic orientation of the individual crystallites. If the single-crystal elastic properties are anisotropic, this leads to an apparent change in the engineering elastic constants of the polycrystal which may completely mask the changes due to dislocations (Weiner et al. 1975). It may also yield an apparent Poisson ratio $\nu > 1/2$, in violation of the condition (3.77) on ν (Ledbetter and Reed 1973).

12. Dependence on magnetic fields

An external magnetic field affects Young's modulus of ferromagnetic materials. This is often referred to as the ΔE effect, a name given by its discoverers, Honda et al. (1902). Since E and G co-vary, G is also affected, while K depends only weakly on magnetic fields. The physical basis of the effect is the same as for magnetostriction. A (non-saturated) magnetic material has more or less randomly oriented magnetic domains. The magnetic energy of these domains couples to the elastic strain energy. Typically, E and G of an annealed specimen (i.e. with very low residual elastic strains) may be 10% to 20% less than in the unannealed state (Ledbetter and Reed 1973). This is also the order of magnitude of the shifts in E and G when a strong magnetic field is applied. In some metallic glasses, Young's modulus may change by more than a factor of two. The ΔE effect decreases as the temperature is increased, and is zero above the Curie temperature. It may happen that the temperature dependence of the ΔE effect cancels the ordinary decrease in E caused by anharmonicity. The result is a Young's modulus which is almost temperature independent over a certain temperature interval. This is called the elinvar effect. The ΔE effect is unusually large in some Fe–B amorphous alloys (Kikuchi et al. 1978, Mitchell et al. 1979). Reviews of magnetoelastic phenomena have been given by Steinemann (1978, 1979).

Some metals are strongly paramagnetic, i.e. they come close to the Stoner criterion for an ordered magnetic state (ch. 15 §4). The Debye temperature $\Theta(0)$, and hence the elastic parameters, was found to be insensitive to an external magnetic field of strength $\approx 10\,\mathrm{T}$, for $LuCo_2$ (Ikeda and Gschneidner 1980) and Pd (Hisang et al. 1981).

13. Theoretical calculations of elastic parameters

There are two different approaches to a theoretical calculation of elastic parameters: (i) one considers the total energy in different static atomic configurations or, (ii) one obtains the elastic properties from the long-wavelength limit of lattice vibrations. The static approach to the bulk modulus was discussed in ch. 1 and lattice vibrations are dealt with in ch. 4.

It was stated in ch. 1 that the total energy of metals depends crucially on the atomic volume, but not very much on the crystal structure. That might seem to imply that the shear modulus G should be very low, since shear does not affect the volume. However, the ratio $G/K \sim 0.5$ for most metals. The reason is that although the Madelung constants are almost identical for several different crystal structures (at constant volume), the changes in the Madelung constants on shear are not small (Ling and Gelatt 1980).

A brief account of theoretical calculations of third-order elastic constants is found in the Landolt–Börnstein tables (Hearmon 1979, 1984). Nielsen and Martin (1983) have performed an ab initio pseudopotential calculation of the elastic properties of Si, including several third- and fourth-order elastic constants. The results are in excellent agreement with experiments.

CHAPTER 4

HARMONIC LATTICE VIBRATIONS

1. Introduction

The simplest description of lattice vibrations is the Einstein model (Einstein 1907). All atoms are assumed to vibrate as independent harmonic oscillators with the frequency ω_E. In such a model, the heat capacity C_E per atom (index E for "Einstein") is

$$C_E = 3k_B\left(\frac{\hbar\omega_E}{k_BT}\right)^2 \frac{\exp(\hbar\omega_E/k_BT)}{[\exp(\hbar\omega_E/k_BT)-1]^2}. \tag{4.1}$$

Figure 4.1 shows that this extremely simple one-parameter model accounts qualitatively for the heat capacity of solids. Figure 4.2 gives the Einstein-model result and experimental data for the entropy $S(T)$ and the enthalpy $H(T)$. The discrepancies in figs. 4.1 and 4.2 between the model and the measurements at high temperatures are mainly due to anharmonic effects. They are treated in ch. 5. Non-vibrational

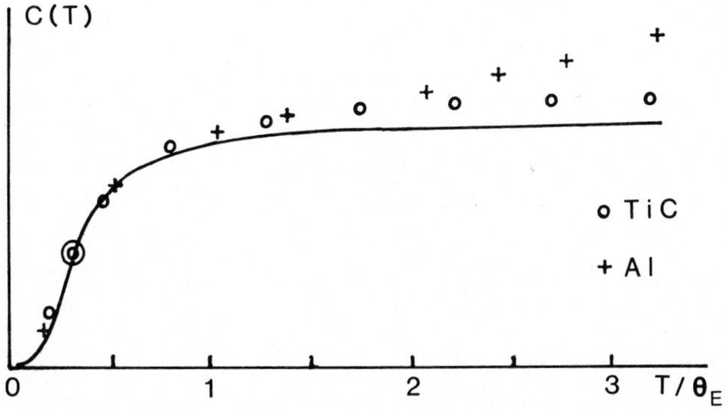

Fig. 4.1. The measured heat capacity $C(T)$ may be qualitatively represented by an Einstein model for the lattice vibrations. The Einstein temperatures for TiC and Al are determined by fits of the measured $C(T)$ to the Einstein model at the point marked with a double circle.

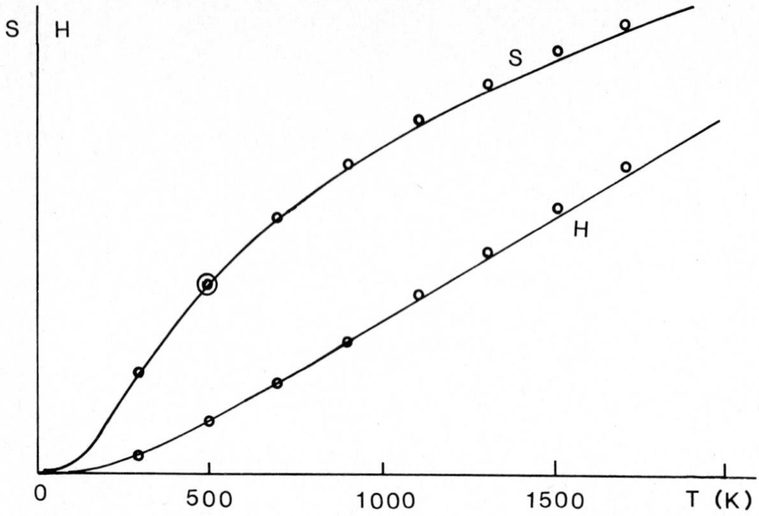

Fig. 4.2. The measured entropy $S(T)$ and enthalpy $H(T)$ of TiC (circles, arbitrary units) may be qualitatively represented by an Einstein model for the lattice vibrations (solid lines). The Einstein temperature is fixed by a fit of the measured $S(500\,\mathrm{K})$ to the Einstein model.

contributions to the heat capacity are discussed in chs. 7 and 8. In this chapter we are primarily concerned with a description of the thermal properties of solids which goes beyond the simple Einstein model.

2. Phonon dispersion curves

It is assumed that the reader is familiar with elementary aspects of phonons in solids. We first recapitulate some important results and introduce the notation. The reader who wants a more thorough presentation of general lattice dynamics is referred to reviews by, e.g., Maradudin et al. (1963), Maradudin (1974), Reissland (1973) and Brüesch (1982). Plots of dispersion curves and the density-of-states function for insulators are found in Bilz and Kress (1979) while the Landolt–Börnstein tables (Schober and Dederichs 1981) have similar data for a large number of solids.

A phonon state is labelled by $(\boldsymbol{q}, \lambda)$, where $\boldsymbol{q}$ is a wave vector in the first Brillouin zone and λ is an index which refers to longitudinal and transverse branches, as well as to optical and acoustic branches. The phonon frequencies $\omega(\boldsymbol{q}, \lambda)$, and the corresponding eigenvectors $\boldsymbol{\varepsilon}(\boldsymbol{q}, \lambda)$, are obtained from the dynamical matrix D (appendix C):

$$\omega^2(\boldsymbol{q},\lambda)\varepsilon_\alpha(\boldsymbol{q},\lambda) = \sum_\beta D_{\alpha\beta}(\boldsymbol{q})\varepsilon_\beta(\boldsymbol{q},\lambda). \tag{4.2}$$

Indices α and β denote the Cartesian components (x, y, z). The eigenvectors are orthonormal:

$$\sum_\alpha \varepsilon_\alpha(\boldsymbol{q},\lambda)\varepsilon_\alpha(\boldsymbol{q},\lambda') = \delta_{\lambda\lambda'}. \tag{4.3}$$

Theoretical accounts of phonon dispersion curves may have two different aims; to reproduce experimental data or to give an ab initio calculation with a minimum of input information. In the first case, one has some experimental knowledge about $\omega(\boldsymbol{q},\lambda)$, perhaps only through the elastic constants, and the aim is to map $\omega(\boldsymbol{q},\lambda)$ for all $\boldsymbol{q}$ and λ as accurately as possible so that the phonon density of states, for example, can be calculated. The standard procedure is to fit a set of force constants which represent the interatomic forces. Such models may be elementary, e.g. the Born–von Kármán model, or more sophisticated such as shell models describing ionic compounds. Ab initio calculations of phonon frequencies are much more difficult, and lie outside the scope of this book. One has been fairly successful for non-transition metals, alkali halides and semiconductors (see Devreese et al. (1983) and Horton and Maradudin (1974) for reviews and Kunc and Martin (1982) for an ab initio calculation in GaAs) while calculations for transition metals are more uncertain (Finnis et al. 1984).

3. The density of phonon states

Several thermophysical properties may be calculated without any detailed knowledge of the phonon dispersion curves and the phonon eigenvectors. The only information needed about the phonons is the density of states $F(\omega)$. The quantity $F(\omega)\Delta\omega$ measures the number of phonon states with frequencies in the interval $[\omega, \omega+\Delta\omega]$. We must specify how $F(\omega)$ is normalised, and take

$$\int_0^{\omega_{\max}} F(\omega)\mathrm{d}\omega = 3. \tag{4.4}$$

This choice is motivated by the fact that each atom has three vibrational degrees of freedom. In a solid with N atoms, there are in total $3N$ phonon states. $NF(\omega)\Delta\omega$ is the number of phonon states $(\boldsymbol{q}, \lambda)$ with frequencies in the interval $[\omega, \omega+\Delta\omega]$. From a knowledge of the dispersion relations $\omega(\boldsymbol{q}, \lambda)$, one obtains $F(\omega)$ by

$$F(\omega) = \frac{V}{(2\pi)^3}\frac{1}{N}\sum_{\lambda}\int \mathrm{d}^3\boldsymbol{q}\,\delta(\omega(\boldsymbol{q},\lambda)-\omega)$$

$$= \frac{V}{(2\pi)^3}\frac{1}{N}\sum_{\lambda}\int_{S_{\boldsymbol{q}}} \frac{\mathrm{d}S}{|\nabla_{\boldsymbol{q}}\omega(\boldsymbol{q},\lambda)|}, \tag{4.5}$$

where V is the crystal volume and $\delta(x)$ is the Dirac delta-function. The last integral is over that surface $S_{\boldsymbol{q}}$ on which $\omega(\boldsymbol{q}, \lambda) = \omega$.

Most theoretical calculations of phonon dispersion relations $\omega(\boldsymbol{q}, \lambda)$ are limited to $\boldsymbol{q}$-vectors in the directions of high lattice symmetry, e.g., the [100], [110] and [111] directions of fcc or bcc lattices. That information usually is insufficient for an accurate evaluation of the integral in (4.5). If $\omega(\boldsymbol{q}, \lambda)$ is known in a mesh of $\boldsymbol{q}$-points, interpolation procedures yield $F(\omega)$ with a high numerical accuracy.

Example: $F(\omega)$ of an fcc lattice with nearest-neighbour interactions. Consider an fcc lattice, with only central nearest-neighbour interactions. The dispersion curves, expressed by the single force constant F (cf. appendix C), take a very simple form. For instance, we get for the longitudinal (L) and the two degenerate transverse (T) branches in the [100] direction in reciprocal space

$$\omega(q_x; \mathrm{L}) = 4\sqrt{(F/M)}\,\sin(q_x a/4), \tag{4.6}$$

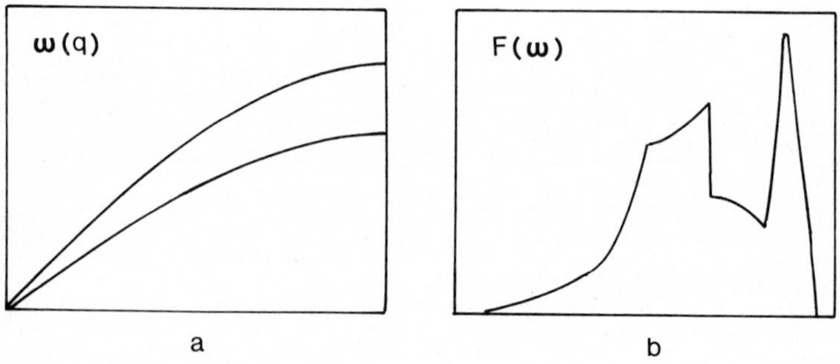

Fig. 4.3. (a) Phonon dispersion curves in the [100] direction of a fcc lattice with central nearest-neighbour interactions, and (b) the phonon density of states $F(\omega)$ obtained by integration over all phonon modes.

$$\omega(q_x; \mathrm{T}) = 4\sqrt{(F/2M)}\,\sin(q_x a/4). \tag{4.7}$$

Similar relations are obtained in other directions. Figure 4.3a shows the dispersion curves in the [100] direction. In spite of this simple form of $\omega(\boldsymbol{q}, \lambda)$, it is not easy to evaluate $F(\omega)$. In fact, $F(\omega)$ must be calculated by numerical integration (Leighton 1948, Maradudin et al. 1958). Figure 4.3b shows the density of states thus obtained.

4. The Debye spectrum

In the long-wavelength limit, i.e. for small $\boldsymbol{q}$, we can write

$$\omega(\boldsymbol{q}, \lambda) = C(\boldsymbol{q}, \lambda)|\boldsymbol{q}|. \tag{4.8}$$

$C(\boldsymbol{q}, \lambda)$ is a directional-dependent velocity, defined for the three λ-values corresponding to the acoustic branches. The simplest Debye model (Debye 1912) assumes a constant value $C(\boldsymbol{q}, \lambda) = C_{\mathrm{D}}$ for all $(\boldsymbol{q}, \lambda)$, with $\omega(\boldsymbol{q}, \lambda)$ linear in $|\boldsymbol{q}|$ for all wave numbers $|\boldsymbol{q}| < q_{\mathrm{D}}$. The maximum frequency (the Debye frequency) is $\omega_{\mathrm{D}} = Cq_{\mathrm{D}}$. From the general expression (4.5) we get the Debye density of states

$$F_{\mathrm{D}}(\omega) = \frac{3V}{(2\pi)^3 N}\,\frac{4\pi q^2}{C_{\mathrm{D}}}\bigg|_{C_{\mathrm{D}}q=\omega} = \frac{3V\omega^2}{2\pi^2 N C_{\mathrm{D}}^3}. \tag{4.9}$$

The value of q_{D} is fixed by the normalisation condition

$$\int_0^{Cq_{\mathrm{D}}} F_{\mathrm{D}}(\omega)\,\mathrm{d}\omega = 3. \tag{4.10}$$

This yields

$$q_{\mathrm{D}} = (6\pi^2 N/V)^{1/3} = (6\pi^2/\Omega_{\mathrm{a}})^{1/3}. \tag{4.11}$$

In applications to thermophysical properties, it is convenient to introduce the Debye temperature Θ_{D} as a measure of the maximum frequency ω_{D}. The two parameters are related by

$$\hbar\omega_{\mathrm{D}} = k_{\mathrm{B}}\Theta_{\mathrm{D}}. \tag{4.12}$$

In a real solid $C(\boldsymbol{q}, \lambda)$ is anisotropic, and different for the longitudinal and the transverse acoustic branches. We write $C(\boldsymbol{q}, \lambda) = C(\theta, \phi, \lambda)$ and $dS = q^2\,d\Omega = [\omega^2/C^2(\theta, \phi)]\,d\Omega$ where (θ, ϕ) are angular coordinates for q and $d\Omega = \sin\theta\,d\theta\,d\phi$. Equation (4.5) becomes

$$F(\omega) = \frac{V}{(2\pi)^3 N} \sum_{\lambda=1}^{3} \int \frac{\omega^2\,d\Omega}{C_\lambda^3(\theta, \phi)} = \frac{\Omega_a \omega^2}{2\pi^2} \sum_{\lambda=1}^{3} \int \frac{1}{C_\lambda^3(\theta, \phi)} \frac{d\Omega}{4\pi}. \tag{4.13}$$

This agrees with (4.9) if we define the Debye sound velocity C_D by

$$\frac{3}{C_D^3} = \sum_{\lambda=1}^{3} \int \frac{1}{C_\lambda^3(\theta, \phi)} \frac{d\Omega}{4\pi}. \tag{4.14}$$

If C_λ is isotropic, but different, for the longitudinal (L) and the two degenerate transverse (T) branches, one has

$$\frac{3}{C_D^3} = \frac{1}{C_L^3} + \frac{2}{C_T^3}. \tag{4.15}$$

The Debye temperature Θ_D can now be expressed as

$$\Theta_D = \frac{\hbar}{k_B}\left(\frac{6\pi^2 N}{V}\right)^{1/3} C_D = \frac{\hbar}{k_B}\left(\frac{6\pi^2 r L \rho}{M}\right)^{1/3} C_D. \tag{4.16}$$

Here r is the number of atoms in a molecule ($r = 1$ for an element, 2 for NaCl, 5 for Al_2O_3), M is the weight of a mole of the material, L is Avogadro's number and ρ is the mass density of the material. Since $\omega_D = C_D q_D = C_D(6\pi^2 N/V)^{1/3}$ we can also write the Debye density of states (4.9) as

$$F_D(\omega) = \frac{3Vq_D^3\omega^2}{2\pi^2 N \omega_D^3} = \frac{9\omega^2}{\omega_D^3}. \tag{4.17}$$

In its usual form, the Debye model has a single cut-off frequency $\omega_D = C_D q_D$. If this cut-off is applied to the longitudinal and transverse branches separately, we get different wavenumber cut-offs $(q_D)_L = \omega_D/C_L$ and $(q_D)_T = \omega_D/C_T$. This is unphysical since $\boldsymbol{q}$ is bounded by the Brillouin zone and thus takes the same values for all branches. As an alternative Debye model we therefore take a maximum $(|\boldsymbol{q}|)_L = (|\boldsymbol{q}|)_T = q_D$, which gives cut-off frequencies $(\omega_D)_L = C_L q_D$ and $(\omega_D)_T = C_T q_D$. We shall now see that the two approaches give essentially equivalent thermophysical results. Let the density of states be $F(\omega) = F_L(\omega) + 2F_T(\omega) = a_L\omega^2 + 2a_T\omega^2$ and determine a_L and a_T by the normalisation conditions on $F(\omega)$, applied to each branch separately; then

$$\int_0^{(\omega_D)_L} a_L\omega^2\,d\omega = \int_0^{(\omega_D)_T} a_T\omega^2\,d\omega = 1, \tag{4.18}$$

and, for $\omega < (\omega_D)_T$,

$$F(\omega) = (3/q_D^3)(1/C_L^3 + 2/C_T^3)\omega^2. \tag{4.19}$$

This is identical to the form (4.9) when C_D is defined by (4.15). Thus, it makes no difference for the low-frequency part of $F(\omega)$ if we take a common Debye-frequency cut-off ω_D for all branches or a common wavenumber cut-off q_D. This conclusion remains valid if C_L and C_T are anisotropic. Figure 4.4 shows $F_L(\omega)$, $2F_T(\omega)$ and $F(\omega)$. For $\omega > (\omega_D)_L$, eq. (4.19) differs from $F_L(\omega) + 2F_T(\omega)$ but then none of the Debye models give a true description of a real solid. In the following we shall use a common frequency cut-off ω_D when we refer to the Debye model.

In a crystal with several atoms per primitive cell there are optical branches, with $\boldsymbol{q}$ in the first Brillouin zone. The corresponding Debye model has only acoustic branches, and q_D lies beyond the first Brillouin

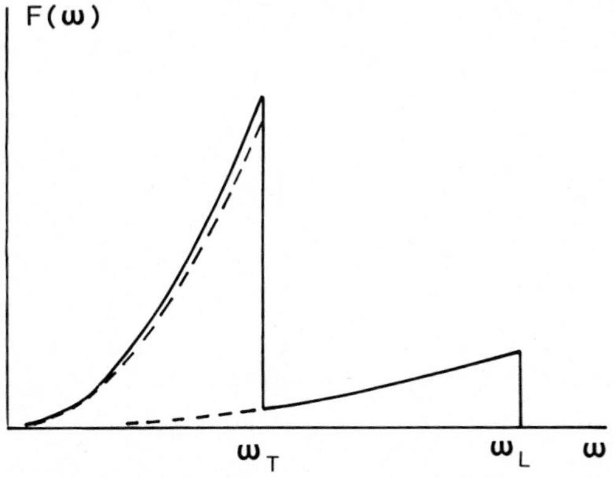

Fig. 4.4. A modified Debye model with a fixed wave-vector cut-off q_D gives different cut-offs ω_L and ω_T for the longitudinal and transverse modes. The solid line is the sum, $F(\omega)$, of the doubly degenerate transverse and the longitudinal branch. The dashed lines give their separate contributions to $F(\omega)$.

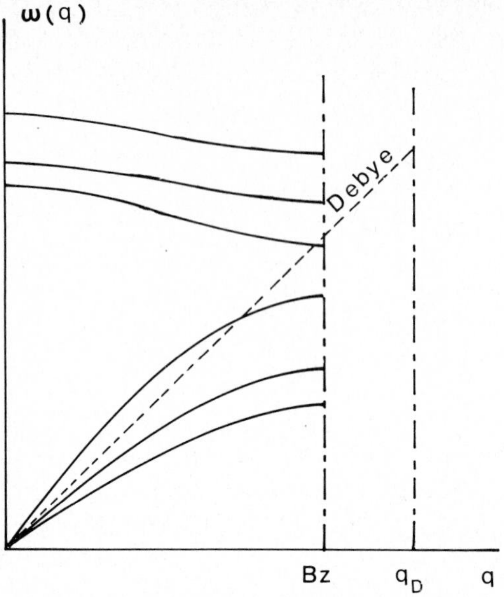

Fig. 4.5. Schematic phonon dispersion curves for a crystal with two atoms per primitive cell (solid lines) and a Debye approximation (dashed line). Bz and q_D mark the limits to the wave vector set by the Brillouin zone and the Debye wave number.

zone. Figure 4.5 shows typical phonon dispersion curves. The main reason for the applicability of the Debye model to thermophysical properties is not that $\omega(\boldsymbol{q})$ is well described but that one often needs to know only a certain average of $F(\omega)$, and not its detailed shape.

A complication may seem to arise in molecular crystals, such as NH_4Cl. In a first approximation, we may regard this as a crystal MCl, where M = NH_4 vibrates as a single mass, and the "internal" vibrations and rotations of NH_4 are neglected. In a complete theory, on the other hand, we should consider the vibrations of all atoms on the same footing. In these two cases, eq. (4.16) for Θ_D contains the same molecular weight, but the number of atoms (rather, mass points), Lr, is different. Since Θ_D is proportional to $(Lr)^{1/3}$, the Debye temperature will also be different. Thermal properties calculated from Θ_D, such as the low-temperature heat capacity and entropy, are proportional to Lr/Θ_D^3. Hence, r cancels and both approaches to Θ_D give correct thermal properties at low temperatures. At high temperatures, there is of course a difference because one of the models excludes rotations and internal vibrations of NH_4.

The Debye model assumes that $\omega(\boldsymbol{q}, \lambda) = C(\theta, \phi)|\boldsymbol{q}|$ is linear in $|\boldsymbol{q}|$ for all wave numbers. In a real solid, this is not the case. With the inclusion of only the first correction term, we can write

$$\omega(\boldsymbol{q}, \lambda) = c_1(\theta, \phi; \lambda)|\boldsymbol{q}| + c_2(\theta, \phi; \lambda)|\boldsymbol{q}|^2. \tag{4.20}$$

It is then not difficult to prove that $F(\omega)$ has the form

$$F(\omega) = a_1\omega^2 + a_2\omega^4 + \ldots, \tag{4.21}$$

i.e. $F(\omega)$ only contains even powers of ω.

Example: Debye temperature from approximate sound velocities. From a knowledge of the elastic coefficients c_{ij}, one obtains the velocities $C(\theta, \phi, \lambda)$ as eigenvalues of a secular equation (eq. 3.54), and then C_D and Θ_D after a numerical integration over θ and ϕ (eq. 4.13). A much simpler, but approximate, method is to estimate the bulk modulus K and the shear modulus G of a polycrystalline material by the Voigt–Reuss–Hill approximation, ch. 14 §4.3, then to obtain the longitudinal sound velocity from $\rho C_L^2 = K + (4/3)G$ and the transverse sound velocity from $\rho C_T^2 = G$ and apply eqs. (4.15) and (4.16) to give Θ_D. Anderson (1963, 1965) investigated how well a calculation of C_D and Θ_D by the latter method approximates the exact Θ_D. For Al_2O_3 (trigonal lattice) he obtained $C_D = 7190$ m/s ("exact" numerical evaluation) and $C_D = 7093$ m/s (Voigt–Reuss–Hill approximation) and for $CaCO_3$ (orthorhombic lattice) $C_D = 3942$ m/s ("exact") while $C_D = 3991$ m/s (VRH). Calculations on a large number of other systems showed that the typical error in Θ_D is less than 2% when the approximate method is used. This also means that an accurate value of Θ_D may be obtained from the measured longitudinal and transverse sound velocities of (statistically isotropic) polycrystalline materials, without recourse to the elastic constants c_{ij} of a single crystal. When the single crystal is isotropic, Θ_D is given exactly by the Voigt–Reuss–Hill expression, since then $K_R = K_H$ and $G_R = G_H$. Using as a measure of elastic anisotropy in a single crystal the quantity $A_H = (G_V - G_R)/(G_V + G_R)$, Anderson (1963) found that when A_H is less than 0.2, the error in Θ_D is less than about 2%. In the example on p. 41 it is argued that Θ_D can be well estimated even if only G is known.

5. Frequency moment representation of $F(\omega)$

5.1. Definitions

In the limit of low or high temperatures, it may happen that only a certain average of $F(\omega)$, in the form of a frequency moment, is needed to calculate a certain thermophysical property. We define such moments $\langle\omega^n\rangle$ by

$$\langle\omega^n\rangle = \mu_n = \int_0^{\omega_{max}} \omega^n F(\omega)\mathrm{d}\omega \Big/ \int_0^{\omega} F(\omega)\mathrm{d}\omega. \tag{4.22}$$

Usually, it is more convenient to work with a frequency $\omega(n)$, related to the nth moment by

$$\omega(n) = [\mu_n]^{1/n}. \tag{4.23}$$

There is no restriction on n to be an integer, but since $F(\omega) \sim \omega^2$ for small ω, (4.22) converges only if $n > -3$. The limiting value $\omega(0)$, i.e. for $n \to 0$, is defined by

$$\ln[\omega(0)] = \int_0^{\omega_{max}} \ln \omega F(\omega)\mathrm{d}\omega \Big/ \int_0^{\omega_{max}} F(\omega)\mathrm{d}\omega. \tag{4.24}$$

It is easy to prove that $\omega(n)$ increases monotonically with n and asymptotically approaches the maximum phonon frequency ω_{max} when $n \to \infty$. Figure 4.6 shows $\omega(n)$ for Ge.

For a Debye spectrum, with $F(\omega)$ given by (4.9), we get

$$\mu_n = \frac{(3V/2\pi^2NC_D^3)\int_0^{\omega_D}\omega^{2+n}\mathrm{d}\omega}{(3V/2\pi^2NC_D^3)\int_0^{\omega_D}\omega^2\,\mathrm{d}\omega} = \frac{3\omega_D^n}{n+3}, \tag{4.25}$$

which yields

$$\omega(n) = \left(\frac{3}{n+3}\right)^{1/n} \omega_D. \tag{4.26}$$

When $n = 0$, eq. (4.24) leads to

$$\ln[\omega(0)] = \int_0^{\omega_D} \omega^2 \ln \omega\,\mathrm{d}\omega \Big/ \int_0^{\omega_D} \omega^2\,\mathrm{d}\omega = \ln(\omega_D) - 1/3, \tag{4.27}$$

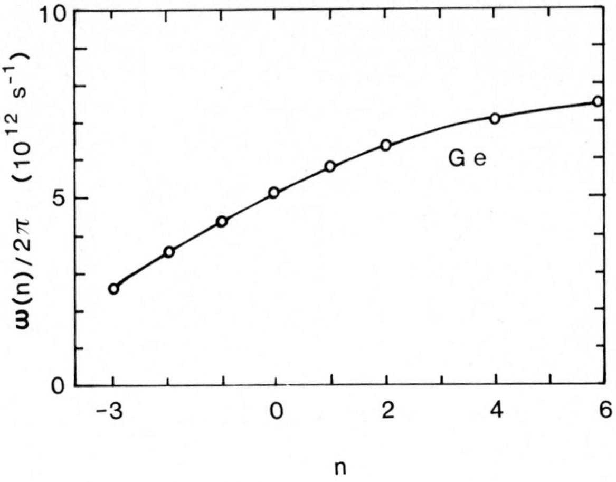

Fig. 4.6. The frequency moment $\omega(n)$, as a function of n, for Ge. Data from Flubacher et al. (1959).

or

$$\omega(0) = \omega_D \exp(-1/3). \tag{4.28}$$

Using a well-known result for $(1+1/N)^N$ when $N \to \infty$, one easily confirms that $\omega(n)$ in (4.26) has the limit $\omega(0)$ of (4.28) when $n \to 0$.

5.2. Relations between $\omega(n)$ and the dynamical matrix

For any $\boldsymbol{q}$, the frequencies $\omega(\boldsymbol{q}, \lambda)$ are related to the trace (Tr) of the dynamical matrix D by (appendix C)

$$\sum_{\lambda} \omega^2(\boldsymbol{q}, \lambda) = \sum_{i=1}^{3r} D_{ii}(\boldsymbol{q}) = \mathrm{Tr}\, D(\boldsymbol{q}). \tag{4.29}$$

Normally, the atomic masses and the interatomic force constants are not separable in $\omega(\boldsymbol{q}, \lambda)$ if there is more than one kind of atomic mass. However, the mass does separate from the forces in the quantity $\omega(0)$ (Grimvall and Rosén 1983). A general theorem in mathematics gives the roots $\omega(\boldsymbol{q}, \lambda)$ to the secular equation as a product,

$$\prod_{\lambda=1}^{3r} \omega^2(\boldsymbol{q}, \lambda) = \det D(\boldsymbol{q}). \tag{4.30}$$

Here r is the number of atoms in a primitive cell, and $\det D$ is the determinant of the dynamical matrix. Consider now a compound A_xB_y, with atomic masses M_A and M_B. Using (C.5), and the fact that the logarithm of a product is a sum of logarithms, one may show that

$$\ln[\omega(0)] = (1/6N)\sum_{\boldsymbol{q}} \ln[\det D_0(\boldsymbol{q})] - (1/2)\ln M_{\text{eff}}. \tag{4.31}$$

Here N is the number of atoms and M_{eff} is an effective mass, $(x+y)\times \ln M_{\text{eff}} = x\ln M_A + y\ln M_B$, and D_0 is the force-constant part of the dynamical matrix. Except for the trivial case of elements, and the elastic-limit quantity $\Theta_D(-3)$, $\omega(0)$ is the only frequency among $\omega(n)$ that allows such a general separation of the mass effect.

5.3. Debye temperatures related to frequency moments

For a Debye spectrum, $\omega(n) = [3/(n+3)]^{1/n}\omega_D$ [eq. (4.26)]. When $F(\omega)$ is not of the Debye form, we can define Debye frequencies $\omega_D(n)$, and corresponding Debye temperatures

$$\Theta_D(n) = \hbar\omega_D(n)/k_B, \tag{4.32}$$

by

$$\int_0^{\omega_D(n)} \omega^n\omega^2\,d\omega \bigg/ \int_0^{\omega_D(n)} \omega^2\,d\omega = \int_0^{\omega_{\max}} \omega^n F(\omega)\,d\omega \bigg/ \int_0^{\omega_{\max}} F(\omega)\,d\omega. \tag{4.33}$$

Thus we let $\omega_D(n)$ be the cut-off frequency in a Debye model which reproduces correctly the nth moment of ω for a given density of states $F(\omega)$. The right-hand side of (4.33) is, by definition, equal to $[\omega(n)]^n$. The left-hand side is $3[\omega_D(n)]^3/(n+3)$. Then, if $n > -3$ and $n \neq 0$,

$$\omega_D(n) = \left(\frac{n+3}{3}\right)^{1/n} \omega(n). \tag{4.34}$$

When $n = 0$, ω^n is replaced by $\ln\omega$. The right-hand side of (4.33) is then given by (4.24) and the left-hand side by (4.27). Hence,

$$\omega_D(0) = \exp(1/3)\,\omega(0). \tag{4.35}$$

If $n = -3$, the integral (4.22) for $\omega(n)$ diverges at $\omega = 0$. We then use the fact that $F(\omega) = a_1\omega^2 + a_2\omega^4 + \ldots$, (4.21), and define $\omega_D(-3)$ so that it has the same divergent behaviour as $\omega(-3)$ in (4.33). If we replace the lower integration limit 0 in (4.33) by an infinitesimal quantity ε, take $n = -3$ and recall the normalisation $\int F(\omega)\mathrm{d}\omega = 3$, the right-hand side of (4.33) diverges as $(a_1/3)\ln(1/\varepsilon)$ and the left-hand side diverges as $\{3/[\omega_D(-3)]^3\}\ln(1/\varepsilon)$. Hence,

$$\omega_D(-3) = (9/a_1)^{1/3}. \tag{4.36}$$

In a strict Debye model, a_1 follows from (4.9), giving, as expected,

$$\omega_D(-3) = C_D q_D, \tag{4.37}$$

where C_D is the average (4.14) over the acoustic branches and $q_D = (6\pi^2/\Omega_a)^{1/3}$.

It is now natural to ask what is the use of the Debye parameters $\omega_D(n)$ and $\Theta_D(n)$, since they are just other ways of expressing the magnitude of the frequency moments and do not contain any information that is not already in μ_n or $\omega(n)$. We note that in a strict Debye model, all $\omega_D(n)$ are constant, equal to the cut-off frequency ω_D, and all $\Theta_D(n)$ are equal to a single Debye temperature Θ_D. In a real solid, $\omega(n)$ varies considerably, while $\omega_D(n)$ and $\Theta_D(n)$ are fairly constant. Figure 4.6 shows $\omega(n)$ for Ge and figs. 4.7 and 4.8 show $\Theta_D(n)$ for Al, Mo, Mg, Nb, Ge and CsCl. Often $\Theta_D(n)$ has a minimum near $n = 0$.

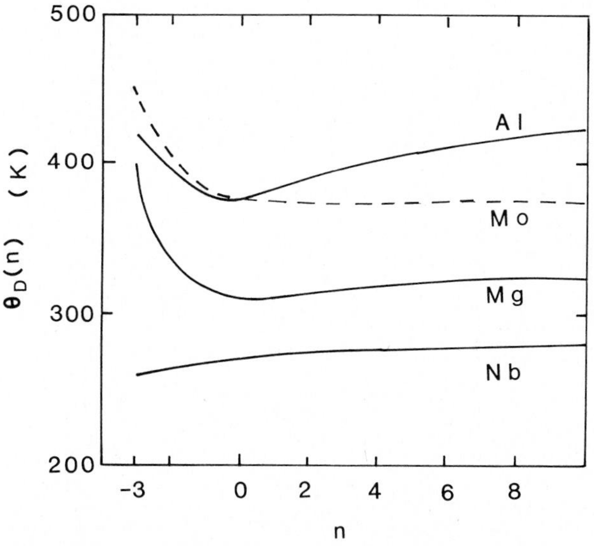

Fig. 4.7. The experimentally determined Debye temperature $\Theta_D(n)$, as a function of n, for Al, Mo, Mg and Nb. Data from Schober and Dederichs (1981).

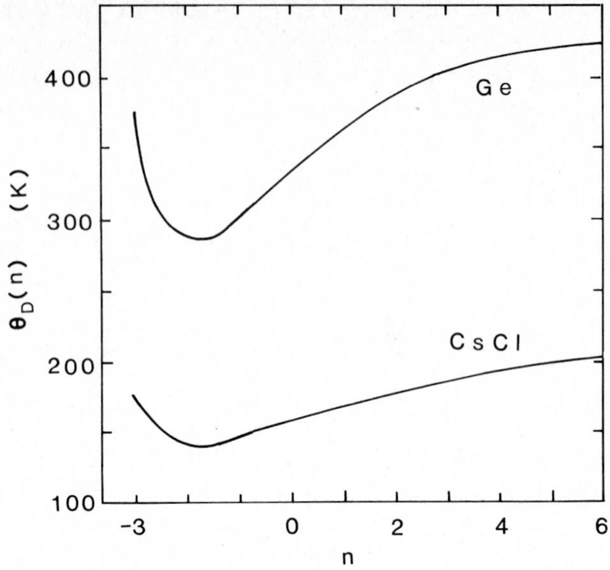

Fig. 4.8. The experimentally determined Debye temperature $\Theta_D(n)$, as a function of n, for Ge (Flubacher et al. 1959) and CsCl (Bailey and Yates 1967).

In conclusion, the density of states of a real solid may be far from that of a Debye spectrum, but still some averages over $F(\omega)$, expressed by $\omega_D(n)$ or $\Theta_D(n)$, are fairly well represented by a single parameter, ω_D or Θ_D. This is the reason for the wide applicability of the Debye model. However, there are examples when $F(\omega)$ is so different from a Debye spectrum that the concept of a Debye temperature is less applicable. For instance, in graphite and boron nitride, $\Theta_D(n)$ increases from about 500 K when $n = -3$ to about 2200 K for large n (Yates et al. 1975).

6. The energy, free energy and entropy of phonons

6.1. General relations

In thermal equilibrium, the energy associated with a phonon mode $(\boldsymbol{q}, \lambda)$ is

$$E(\boldsymbol{q}, \lambda) = \hbar\omega(\boldsymbol{q}, \lambda)[n + 1/2], \tag{4.38}$$

where n is the Bose–Einstein statistical factor

$$n(\boldsymbol{q}, \lambda) = \frac{1}{\exp(\hbar\omega/k_{\mathrm{B}}T) - 1}. \tag{4.39}$$

The Helmholtz free energy $F(\boldsymbol{q}, \lambda)$ may be obtained from (4.38) through

$$F = E - TS = E - T\int_0^T \left(\frac{\partial E}{\partial T'}\right)\frac{\mathrm{d}T'}{T'}, \tag{4.40}$$

or from the partition function Z,

$$Z(\boldsymbol{q}, \lambda) = \sum_{n=0}^{\infty} \exp[-\hbar\omega(\boldsymbol{q}, \lambda)(n+1/2)/k_{\mathrm{B}}T], \tag{4.41}$$

giving

$$F(\boldsymbol{q}, \lambda) = -k_{\mathrm{B}}T \ln Z = k_{\mathrm{B}}T \ln[2\sinh(\hbar\omega(\boldsymbol{q}, \lambda)/2k_{\mathrm{B}}T)]. \tag{4.42}$$

Convenient expressions for the entropy are

$$S(\boldsymbol{q}, \lambda) = k_{\mathrm{B}}[(1+n)\ln(1+n) - n\ln(n)], \tag{4.43}$$

with $n = n(\boldsymbol{q}, \lambda)$, and

$$S(\boldsymbol{q}, \lambda) = k_{\mathrm{B}}\left\{\frac{\hbar\omega(\boldsymbol{q}, \lambda)}{2k_{\mathrm{B}}T}\coth\left(\frac{\hbar\omega(\boldsymbol{q}, \lambda)}{2k_{\mathrm{B}}T}\right) - \ln\left[2\sinh\left(\frac{\hbar\omega(\boldsymbol{q}, \lambda)}{2k_{\mathrm{B}}T}\right)\right]\right\}. \tag{4.44}$$

The total energy of a crystal with N atoms is

$$E(T) = \sum_{\boldsymbol{q}} E(\boldsymbol{q}, \lambda) = N\int_0^{\omega_{\max}} E(\omega; T)F(\omega)\,\mathrm{d}\omega, \tag{4.45}$$

where

$$\begin{aligned} E(\omega; T) &= \hbar\omega\left[\frac{1}{\exp(\hbar\omega/k_{\mathrm{B}}T) - 1} + 1/2\right] \\ &= (\hbar\omega/2)\coth(\hbar\omega/2k_{\mathrm{B}}T), \end{aligned} \tag{4.46}$$

with analogous integral expressions for the total free energy and the entropy.

If $F(\omega)$ is represented by the Debye spectrum (4.17), one obtains

$$E(T) = E_{\rm D}(T) = (9/8)N\hbar\omega_{\rm D} + \frac{9N\hbar}{\omega_{\rm D}^3}\int_0^{\omega_{\rm D}} \frac{\omega^3\,{\rm d}\omega}{\exp(\hbar\omega/k_{\rm B}T)-1} \tag{4.47}$$

and

$$S(T) = S_{\rm D}(T) = \frac{12Nk_{\rm B}}{\omega_{\rm D}^3}\int_0^{\omega_{\rm D}} \frac{(\hbar\omega^3/k_{\rm B}T)\,{\rm d}\omega}{\exp(\hbar\omega/k_{\rm B}T)-1}$$

$$-3Nk_{\rm B}\ln[1-\exp(-\hbar\omega_{\rm D}/k_{\rm B}T)]. \tag{4.48}$$

The Debye-model functions $E_{\rm D}(T)$ and $S_{\rm D}(T)$ are tabulated, e.g., in the American Institute of Physics Handbook (1972).

Example: Zero-point vibrations and the cohesive energy. The energy of the zero-point vibrations per atom is

$$E_{\rm vib}(T=0) = \int_0^{\omega_{\max}} (\hbar\omega/2)F(\omega)\,{\rm d}\omega = (3/2)\hbar\omega(1) = (9/8)k_{\rm B}\Theta_{\rm D}(1). \tag{4.49}$$

$E_{\rm vib}(T=0)$ is less than 1% of the cohesive energy (cf. table 1.1). Usually it does not affect the relative stability of different crystal structures at $T = 0$ K, because $\Theta_{\rm D}(1)$ of different structures, with the same type of chemical bonding, typically differ by less than 5% (cf. §11). However, graphite seems to be stabilised relative to diamond at $T = 0$ K, due to the zero-point vibrational energy (Yin and Cohen 1984).

6.2. High-temperature expansions

The thermal energy has the high-temperature expansion (Thirring 1913)

$$E(\boldsymbol{q},\lambda) = k_{\rm B}T\left[1-\sum_{n=1}^{\infty}(-1)^n\frac{B_{2n}}{(2n)!}\left(\frac{\hbar\omega_{\boldsymbol{q}\lambda}}{k_{\rm B}T}\right)^{2n}\right], \tag{4.50}$$

where B_n are the Bernoulli numbers: $B_2 = 1/6$, $B_4 = 1/30$, $B_6 = 1/42$, $B_8 = 1/30$. (There are different conventions for B_n.) The expansion (4.50) converges when $(\hbar\omega/k_{\rm B}T) < 2\pi$, which is approximately equivalent to $T > 0.2\Theta_{\rm D}$. For a crystal with N atoms we get

$$E(T) = N\int_0^{\omega_{\max}} E(\omega;T)F(\omega)\mathrm{d}\omega$$
$$= 3Nk_BT\{1+(1/12)[\hbar\omega(2)/k_BT]^2$$
$$-(1/720)[\hbar\omega(4)/k_BT]^4+\ldots\}. \quad (4.51)$$

Expressed in the Debye temperatures $\Theta_D(n)$, (4.51) becomes

$$E(T) = 3Nk_BT\{1+(1/20)[\Theta_D(2)/T]^2$$
$$-(1/1680)[\Theta_D(4)/T]^4-\ldots\}. \quad (4.52)$$

The Thirring expansion for the entropy is

$$S(\boldsymbol{q},\lambda) = k_B\Bigg[1-\ln(\hbar\omega_{\boldsymbol{q}\lambda}/k_BT)$$
$$-\sum_{n=1}^{\infty}(-1)^n\frac{(2n-1)B_{2n}}{(2n)(2n)!}\left(\frac{\hbar\omega_{\boldsymbol{q}\lambda}}{k_BT}\right)^{2n}\Bigg]. \quad (4.53)$$

For a crystal with N atoms,

$$S(T) = Nk_B\int_0^{\omega_{\max}}[1+\ln(k_BT/\hbar\omega)+(1/24)(\hbar\omega/k_BT)^2$$
$$-(1/960)(\hbar\omega/k_BT)^4+\ldots]F(\omega)\mathrm{d}\omega, \quad (4.54)$$

which can also be written

$$S(T) = 3Nk_B\{1+\ln[k_BT/\hbar\omega(0)]+(1/24)[\hbar\omega(2)/k_BT]^2$$
$$-(1/960)[\hbar\omega(4)/k_BT]^4+\ldots\}. \quad (4.55)$$

Expressed in Debye temperatures $\Theta_D(n)$, the entropy has the form

$$S(T) = 3Nk_B\{4/3+\ln[T/\Theta_D(0)]+(1/40)[\Theta_D(2)/T]^2$$
$$-(1/2240)[\Theta_D(4)/T]^4+\ldots\}. \quad (4.56)$$

Example: Latimer's rule for standard entropies of the elements. The standard entropy $S^\circ_{298} = S(T = 298.15\text{ K})$ of the non-magnetic elements is almost entirely due to the lattice vibrations. (The electronic part typically is $1-2\%$ of S° for simple metals and 10% for transition metals.) Latimer (1921, 1951) noted that S°_{298}, defined to be per mole, is well described by the simple expression

$$S_L = 3Nk_B[k_L + (1/2)\ln(M/M_0)]. \tag{4.57}$$

Here k_L is an empirical dimensionless constant of the order of unity, N is the number of atoms in the solid and M/M_0 is the atomic weight (i.e. $12M_0$ is the mass of the atom ^{12}C). When all the masses M in the lattice are equal, we can write $\Theta_D^2(0) = \alpha/M$, where $\alpha k_B^2/\hbar^2$ has the dimension of a force constant and is related to the force constant part of the dynamical matrix. Then, the leading terms of (4.56) give

$$S^\circ_{298} \approx 3Nk_B\{4/3 + \ln[298.15\sqrt{(M_0/\alpha)}] + (1/2)\ln(M/M_0)\}. \tag{4.58}$$

This is of the form suggested by Latimer, provided that the interatomic forces, expressed by α, do not vary as much as M does, through the periodic table. Figure 4.9 shows that this may be a reasonable assumption. But elements with low atomic masses have a high $\Theta_D(0) = \sqrt{(\alpha/M)}$ and then the high-temperature limit (4.58) has not yet been reached at $T = 298$ K. (For instance, $\Theta_D(0) = 974$ K for Be.) Therefore Latimer's rule tends to overestimate S°_{298} for the elements of low atomic number. The dashed line in fig. 4.9 has slope 1/2, as required by (4.58). The result (4.31) can be pursued to argue for Latimer's rule in compounds, a case which is of practical interest; see Grimvall (1983).

Example: Excess entropy of mixing in liquid alloys. Let $S^{liq}(A;T)$ and $S^{liq}(B;T)$ be the entropies of the pure elements A and B in their liquid states and $S^{liq}(c;T)$ be the entropy of the liquid alloy $A_{1-c}B_c$ $(0 < c < 1)$. The excess entropy $\Delta S^{liq}(c;T)$ is defined

$$\Delta S^{liq}(c;T) = S^{liq}(c;T) - (1-c)S^{liq}(A;T) - cS^{liq}(B;T). \tag{4.59}$$

It is known from experiments (fig. 4.10 and Hultgren et al. (1973b)) that $\Delta S^{liq}(c;T)$ is well approximated by

$$S^{dis}(c) = -Nk_B\{c\ln c + (1-c)\ln(1-c)\}, \tag{4.60}$$

where N is the total number of atoms. The configurational, or "disorder", entropy S^{dis} is the same as for a random distribution of

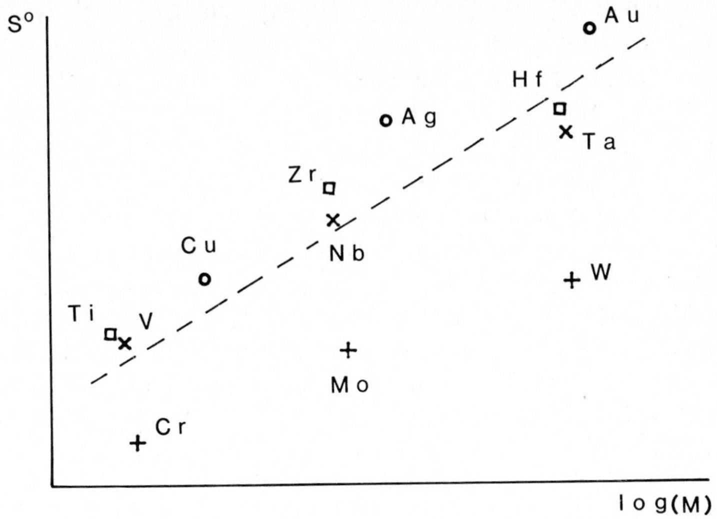

Fig. 4.9. The standard entropy S^0_{298} plotted versus log(M) where M is the atomic mass. The straight dashed line has the slope required by eq. (4.57). Data from Hultgren et al. (1973a).

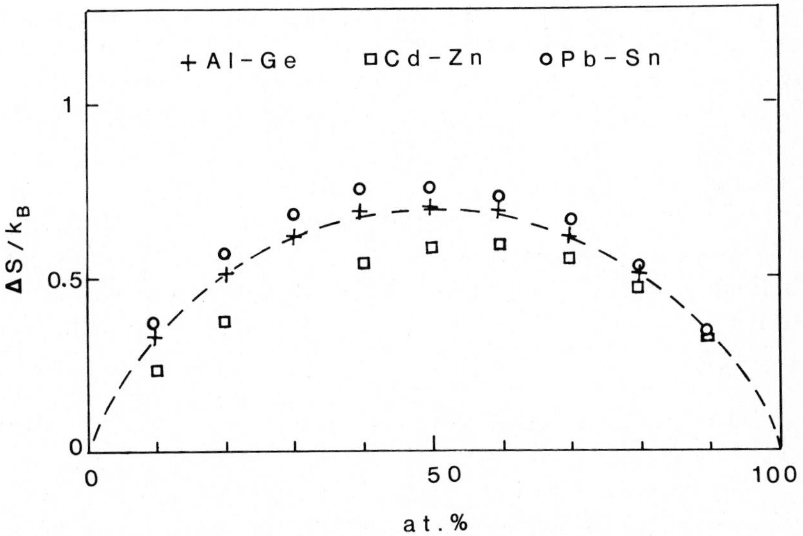

Fig. 4.10. The excess entropy of mixing, $\Delta S(c)/k_B$ (per atom), in liquid alloys $A_{1-c}B_c$. The dashed line is the entropy of ideal mixing; $-k_B[c\ln c+(1-c)\ln(1-c)]$ per atom. Data from Hultgren et al. (1973b).

atoms A and B over lattice sites. Thus there seems to be no significant contribution to ΔS^{liq} of a vibrational nature. This is in line with Latimer's rule. Let the right-hand side of (4.59) refer to the corresponding solid phases, thus defining $\Delta S^{\mathrm{sol}}(c;T)$. By (4.31), the atomic masses in $\Delta S^{\mathrm{sol}}(c;T)$ cancel at high temperatures. If the average force constant parameter α varies as $\ln[\alpha(A_{1-c}B_c)] = (1-c)\ln[\alpha(A)] + c\ln[\alpha(B)]$, there is no vibrational part in ΔS^{sol}. If we now also refer to Richard's rule (Gschneidner 1964), which says that the entropy of fusion of a solid is approximately k_B per atom, we understand why ΔS^{liq} is not much affected by the thermal vibrations. For a discussion of ΔS^{sol}, see Moraitis and Gautier (1977a) and Gachon et al. (1980). When size effects or magnetic or electronic contributions to the entropy are important, one gets deviations from the simple rule (4.60) (Kleppa and Watanabe 1983).

6.3. Polymorphism

Many solids have a polymorphic transformation, i.e. a temperature- or pressure-induced change of the crystal structure. (For elements, the word allotropy is used instead of polymorphism.) Here we shall assume that the temperature-dependent part of the free energy is well described by harmonic lattice vibrations. A transformation from phase 1 to phase 2 takes place at a temperature T_t and a pressure p_t, for which $G_1 = G_2$ (cf. fig. 4.11). Hence,

$$S_{\mathrm{vib},2} - S_{\mathrm{vib},1} = (H_2 - H_1)/T_t = Q/T_t, \tag{4.61}$$

where Q is the latent heat at the transformation.

Consider the high-temperature limit, $T > \Theta(0)$. Then, $E_{\mathrm{vib}} \approx 3Nk_BT$ is independent of the crystal structure, and we can take the two leading terms of $S(T)$ in (4.55). If we also restrict to $p = 0$, and consider the free energy per atom,

$$G_1(T) \approx E_{\mathrm{coh},1} + 3k_BT\ln[\hbar\omega_1(0)/k_BT]. \tag{4.62}$$

$E_{\mathrm{coh},1}$ is the cohesive energy, or any other structure-dependent energy, measured relative to some reference state. $G_1(T) = G_2(T)$ gives

$$3k_B\ln[\omega_1(0)/\omega_2(0)] \approx Q/T_t \approx [E_{\mathrm{coh},2}(T=0) - E_{\mathrm{coh},1}(T=0)]/T_t. \tag{4.63}$$

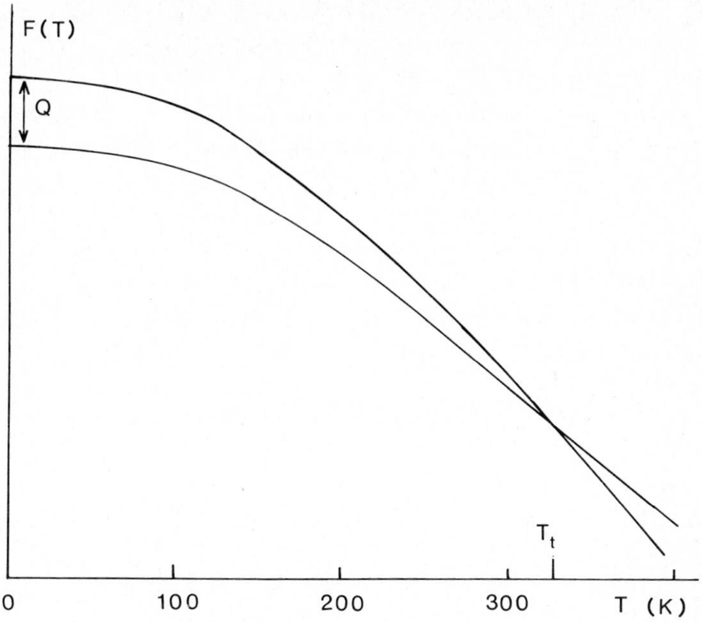

Fig. 4.11. The Helmholtz free energy $F(T)$, in two Debye models with $\Theta_{D1} = 250$ K and $\Theta_{D2} = 200$ K. Q is the latent heat of transformation at T_t.

If the difference $\Delta\omega(0) = \omega_1(0) - \omega_2(0)$ is small, we get

$$\frac{\Delta\omega(0)}{\omega_2(0)} = \frac{\Delta\Theta_D(0)}{\Theta_{2D}(0)} = \frac{Q}{3k_B T_t}. \tag{4.64}$$

Since $Q/k_B T_t$ is positive, the high-temperature phase has the lower Debye temperature $\Theta_D(0)$. (We have then assumed that $T_t > \Theta_D$ and that there are only vibrational structure-dependent contributions to $G(T) - G(0)$. That excludes solids like iron, for which the magnetic part of the free energy is important (Hasegawa and Pettifor 1983).) More than 25 of the metallic elements transform from a low-temperature close-packed (fcc or hcp) to a bcc structure. This appears in most cases to be caused by the somewhat lower $\Theta_D(0)$ for the bcc structure; see §11.

Example: Pressure- and temperature-induced polymorphism in TlI. Thallium iodide has some interesting properties (Samara et al. 1967). At ambient pressure and $T_t = 429$ K, it transforms from a low temperature orthorhombic structure to a more dense cubic CsCl-type structure. The same structural transformation takes place if, at ambient temperature, the pressure is increased to $p_t = 2.9$ kbar. The heat of transformation at 1 bar is $\Delta H = 1230 \pm 160$ J/mol K. The molar volume is lower in the CsCl structure by 3.3%, or 1.5 cm^3/mol. We shall analyse these results within a simple model. Since temperature effects cannot be neglected, eq. (1.20) must be replaced by an equality of the Gibbs free energies: $E_1 - TS_1 + pV_1 = E_2 - TS_2 + pV_2$. For the entropy difference $S_1 - S_2$ we take the high temperature result, (4.63), $3k_B \ln(\Theta_2/\Theta_1)$ where $\Theta_{1(2)}$ are entropy-related Debye temperatures. We neglect the volume dependence of $E_1 - E_2$ and $S_1 - S_2$. (In an accurate calculation this is not satisfactory.) Then we get, with $2N$ atoms per mole.

$$p_t = \frac{E_2 - E_1 - 6Nk_B T \ln(\Theta_1/\Theta_2)}{V_1 - V_2}. \tag{4.65}$$

For the temperature-induced transformation, $p_t \approx 0$ (more precisely, 1 bar). With $E_2 - E_1 = \Delta H = 1230$ J/mol K and $T = T_t = 429$ K, we get $\Theta_2/\Theta_1 = 0.94$. We now let $T = 298$ K and keep the values for Θ_2/Θ_1 and for $E_2 - E_1 = \Delta H$. With $V_1 - V_2 = 1.5$ cm^3/mol, (4.65) yields $p_t = 2.6$ kbar, to be compared with the experimental value 2.9 kbar. The discrepancy is primarily due to our use of the high-temperature limit of the entropy expression.

7. The heat capacity

7.1. General relations

Since we deal with strictly harmonic vibrations, the heat capacities C_p and C_V, taken at constant pressure and constant volume respectively, are equal ($= C_{\text{har}}$). From macroscopic thermodynamics

$$C_{\text{har}} = (\partial E_{\text{har}}/\partial T) = T(\partial S_{\text{har}}/\partial T). \tag{4.66}$$

The heat capacity of a single phonon mode, $C_{\text{har}}(\boldsymbol{q}, \lambda)$, is

$$C_{\text{har}}(\boldsymbol{q}, \lambda) = k_B \left(\frac{\hbar\omega(\boldsymbol{q}, \lambda)}{k_B T}\right)^2 \frac{\exp[\hbar\omega(\boldsymbol{q}, \lambda)/k_B T]}{\{\exp[\hbar\omega(\boldsymbol{q}, \lambda)/k_B T)] - 1\}^2}$$

$$= k_B \left(\frac{\hbar\omega(\boldsymbol{q}, \lambda)}{k_B T}\right)^2 \frac{1}{4\sinh^2[\hbar\omega(\boldsymbol{q}, \lambda)/2k_B T]}. \tag{4.67}$$

The total heat capacity $C_{har}(T)$ of a solid is

$$C_{har}(T) = Nk_B \int_0^{\omega_{max}} \left(\frac{\hbar\omega}{k_B T}\right)^2 \frac{\exp(\hbar\omega/k_B T)}{[\exp(\hbar\omega/k_B T) - 1]^2} F(\omega) d\omega. \tag{4.68}$$

When the density of states $F(\omega)$ is that of a Debye model, we obtain the Debye heat capacity $C_D(T)$

$$C_D(T) = 9Nk_B \left(\frac{T}{\Theta_D}\right)^3 \int_0^{\Theta_D/T} \frac{x^4 e^x}{(e^x - 1)^2} dx. \tag{4.69}$$

7.2. Low temperatures

At low temperatures, the upper integration limit Θ_D/T in the Debye expression (4.69) for the heat capacity can be replaced by ∞. Then the integral becomes a constant, $4\pi^4/15$, and

$$C_D(T) = Nk_B \frac{12\pi^4}{5} \left(\frac{T}{\Theta_D}\right)^3. \tag{4.70}$$

We recall that Θ_D is defined from an average over the anisotropic and mode-dependent sound velocities, eq. (4.14), so eq. (4.70) is valid for any solid in the low-temperature limit (but see ch. 7, §2.1, for a comment on glasses). The lowest-order correction term to (4.70) is easily obtained if in (4.69), we add and subtract the same integral taken from 0 to ∞. We are left with (4.69) plus an integral from Θ_D/T to ∞, where the leading term is obtained from an integration by parts. The result is

$$C_D(T) = Nk_B[(12\pi^4/5)(T/\Theta_D)^3 - 9(\Theta_D/T)\exp(-\Theta_D/T)]. \tag{4.71}$$

We noted in §4 that the density of states $F(\omega)$, for small ω, can be expanded in a series $a_1\omega^2 + a_2\omega^4 + \ldots$, which only contains even powers of ω. If this form of $F(\omega)$ is inserted in the expression (4.68) for $C_{har}(T)$, and we go to the low-temperature limit where terms of the order $\exp(-\hbar\omega_{max}/k_B T)$ can be neglected, the result is

$$C_{har}(T) = Nk_B[(4\pi^4/15)a_1(k_B T/\hbar)^3 + (16\pi^6/21)a_2(k_B T/\hbar)^5 + \ldots]. \tag{4.72}$$

Thus the low-temperature expansion of $C_{har}(T)$ only contains odd powers of T.

7.3. *High temperatures*

The high-temperature expansion of the heat capacity, per mode $(\boldsymbol{q}, \lambda)$, is (Thirring 1913),

$$C_{\mathrm{har}} = k_{\mathrm{B}}\left[1 + \sum_{n=1}^{\infty} (-1)^n \frac{(2n-1)B_{2n}}{(2n)!}\left(\frac{\hbar\omega_{\boldsymbol{q}\lambda}}{k_{\mathrm{B}}T}\right)^{2n}\right]. \tag{4.73}$$

In terms of frequency moments $\omega(n)$, or the related Debye temperatures $\Theta_{\mathrm{D}}(n)$, we get the high temperature limits, for a solid with N atoms,

$$C_{\mathrm{har}}(T) = 3Nk_{\mathrm{B}}\left[1 - \frac{1}{12}\left(\frac{\hbar\omega(2)}{k_{\mathrm{B}}T}\right)^2 + \frac{1}{240}\left(\frac{\hbar\omega(4)}{k_{\mathrm{B}}T}\right)^4 - \dots\right], \tag{4.74}$$

$$C_{\mathrm{har}}(T) = 3Nk_{\mathrm{B}}\left[1 - \frac{1}{20}\left(\frac{\Theta_{\mathrm{D}}(2)}{T}\right)^2 + \frac{1}{560}\left(\frac{\Theta_{\mathrm{D}}(4)}{T}\right)^4 - \dots\right]. \tag{4.75}$$

Example: The Dulong–Petit law. Dulong and Petit (1819) noted empirically that the specific heat (i.e. heat capacity per mass) at room temperature, multiplied by the atomic weight, was approximately a constant for 13 solid elements. Their numbers correspond to a heat capacity of about 6.5 cal/mole. This is consistent with the high temperature limit $3Nk_{\mathrm{B}}$ in eq. (4.75), which yields 5.96 cal/mol. The slightly higher value of Dulong and Petit is due to anharmonic effects. (They used C_p while $3Nk_{\mathrm{B}}$ refers to a harmonic system.) Most solids, and not only the elements, have a heat capacity which is approximately k_{B} per atom at room temperature. Exceptions are materials with a high $\Theta_{\mathrm{D}}(2)$, say $\Theta_{\mathrm{D}}(2) > 400$ K. Such high Debye temperatures are found for solids with strong interatomic forces and light masses, e.g., boron, diamond, silicon and beryllium. Table 4.1 gives $C_p/3Nk_{\mathrm{B}}$ for some elements at 298 K (data from Hultgren et al. 1973a).

Table 4.1
$C_p/3Nk_{\mathrm{B}}$ at 298 K, illustrating the Dulong–Petit rule

Be	B	Al	Cu	Zr	Sn	W	Pb
0.66	0.44	0.98	0.98	1.02	1.08	0.97	1.07

Example: The equipartition theorem and the zero-point vibration energy. The thermal energy of a phonon mode $(\boldsymbol{q}, \lambda)$ is $\hbar\omega(\boldsymbol{q}, \lambda)(n + 1/2)$. We note

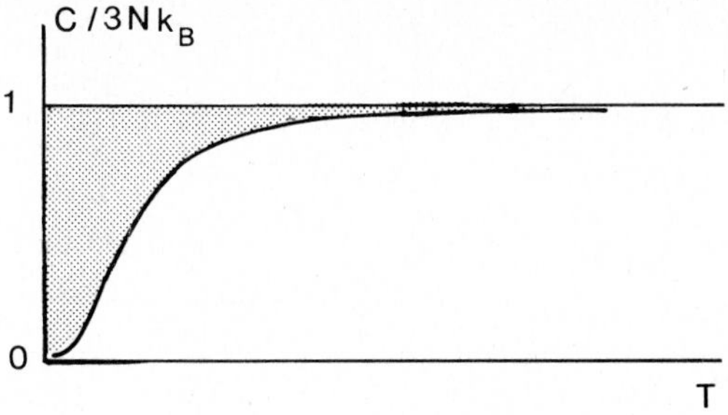

Fig. 4.12. The vibrational heat capacity $C/3Nk_B$. The shaded area corresponds to the zero-point vibrational energy.

from (4.50) that the high-temperature expansion of the Bose–Einstein factor n contains a term $-1/2$ which cancels the explicit term $1/2$ in $\hbar\omega(n+1/2)$. Thus the thermal energy approaches k_BT, which is the well-known result of the equipartition theorem in classical mechanics. This fact can be given a graphical interpretation. The shaded area in fig. 4.12, between the Dulong–Petit value $3Nk_B$ and $C_{har}(T)$, equals the zero-point energy of the vibrations.

8. Thermal atomic displacements

8.1. General relations

Much of the material in the following discussion of thermal atomic displacements has been developed in relation to thermal effects in X-ray crystallography. The monograph by Willis and Pryor (1975) covers these aspects in depth. Butt et al. (1985) have tabulated displacements.

The instantaneous position $\boldsymbol{R}(\kappa, l; t)$, at time t, of the κth atom in the lth unit cell is written

$$\boldsymbol{R}(\kappa, l; t) = \boldsymbol{R}^0(\kappa, l) + \boldsymbol{u}(\kappa, l; t), \tag{4.76}$$

where $\boldsymbol{u}$ is the displacement vector from the equilibrium position $\boldsymbol{R}^0$. We wish to calculate the thermal averages $\langle u^2 \rangle$ and $\langle u_i^2 \rangle$, of $\boldsymbol{u}^2(\kappa, l)$ and its Cartesian components $u_i(\kappa, l)$, respectively. There are N_c primitive cells in the lattice, with r atoms in each cell. The theory of lattice dynamics gives (e.g. Maradudin et al. 1963, Willis and Pryor 1975)

$$u(\kappa, l; t) = (N_c M_\kappa)^{-1/2} \times \sum_{q\lambda} \frac{[E(q, \lambda)]^{1/2}}{\omega(q, \lambda)} \varepsilon(q, \lambda; \kappa) \exp\{i[q \cdot R^0(\kappa, l) - \omega(q, \lambda)t]\}. \tag{4.77}$$

The sum goes over all the $3N = 3rN_c$ degrees of freedom (q, λ), $E(q, \lambda)$ is the average thermal energy of the state (q, λ) and $\varepsilon(\kappa)$ is the κ-part of the total eigenvector ε, i.e. the three components of $\varepsilon(q, \lambda; \kappa)$ which refer to the displacements of the κth atom. We now form $u \cdot u^*$, use that the phases of different modes (q, λ) are uncorrelated, and the fact that

$$\langle \varepsilon(q_1, \lambda_1) \cdot \varepsilon(q_2, \lambda_2) \rangle = \delta_{q_1 q_2}. \tag{4.78}$$

One obtains

$$\langle u_i^2(\kappa, l) \rangle = \frac{1}{N_c M_\kappa} \sum_{q\lambda} \frac{E(q, \lambda)}{\omega^2(q, \lambda)} \varepsilon_i(q, \lambda; \kappa) \varepsilon_i^*(q, \lambda; \kappa). \tag{4.79}$$

Since $E(q, \lambda) = \hbar\omega(q, \lambda)[n(q, \lambda) + 1/2]$, where n is the Bose–Einstein factor, we can find the thermal displacement at any temperature from (4.79). In particular, the high temperature limit is

$$\langle u_i^2(\kappa, l) \rangle = \frac{k_B T}{N_c M_\kappa} \sum_{q\lambda} \frac{1}{\omega^2(q, \lambda)} \varepsilon_i(q, \lambda; \kappa) \varepsilon_i^*(q, \lambda; \kappa), \tag{4.80}$$

for any lattice structure. Here ε^* denotes the complex conjugate of ε. The total mean-square displacement is

$$\langle u^2(\kappa, l) \rangle = \langle u_x^2(\kappa, l) \rangle + \langle u_y^2(\kappa, l) \rangle + \langle u_z^2(\kappa, l) \rangle. \tag{4.81}$$

If the site (κ, l) has cubic lattice symmetry, the average displacement is isotropic, with

$$\langle u^2(\kappa, l) \rangle = 3\langle u_x^2(\kappa, l) \rangle. \tag{4.82}$$

The distribution function $P(u_\alpha, \kappa l)$ for the displacement of the (κl)-atom in the α-direction is Gaussian, i.e.

$$P(u_\alpha) = [2\pi\langle u_\alpha^2(\kappa, l) \rangle]^{-1/2} \exp[-u_\alpha^2 / 2\langle u_\alpha^2(\kappa, l) \rangle]. \tag{4.83}$$

This relation was proved in the classical limit by Debye (1914) and

Waller (1925) and shown by Ott (1935) to be rigorously true also in a quantum mechanical treatment. For a site of cubic lattice symmetry, the distribution function for the root mean-square displacement $u = (u_x^2 + u_y^2 + u_z^2)^{1/2}$ is

$$P(u) = [2\pi\langle u_x^2\rangle]^{-3/2} \exp[-u^2/2\langle u_x^2\rangle]. \tag{4.84}$$

From the distribution functions $P(u_\alpha)$, $P(u)$ we can calculate the probability that an atom is to be found in a certain region near its equilibrium position. For instance,

$$\int_{-s}^{s} P(u_\alpha)\,du_\alpha = 1/2 \tag{4.85}$$

when $s = 0.68\langle u_\alpha^2\rangle^{1/2}$. Thus, there is a 50% probability that the α-component of the displacement vector $\boldsymbol{u}$ is larger than $1.54\langle u_\alpha^2\rangle^{1/2}$. This can be generalised to the following result: Consider an ellipsoidal surface S_{50}, centred at the equilibrium position of a certain atom (κl) and with ellipsoid semi axes $1.54\langle u_\alpha^2\rangle^{1/2}$ $(\alpha = x, y, z)$. This is referred to as the 50% probability ellipsoid of thermal displacements. There is equal probability for the atom to be inside or outside S_{50}, at any arbitrary instant. (See Nelmes (1969) for details regarding these kinds of ellipsoidal surfaces.)

Anharmonic effects have been neglected. They are pronounced for large displacements, and thus make $P(u)$ less reliable when $u^2 \gg \langle u^2\rangle$. In molecular crystals the situation is complicated by the presence of both internal and external modes (Willis and Pryor 1975).

In a monatomic solid with cubic structure, the index κ is irrelevant, and

$$\sum_i \varepsilon_i(\boldsymbol{q}, \lambda)\varepsilon_i^*(\boldsymbol{q}, \lambda) = 1, \tag{4.86}$$

giving

$$\langle u^2\rangle = \frac{1}{M}\int_0^{\omega_{\max}} \frac{E(\omega; T)}{\omega^2} F(\omega)\,d\omega. \tag{4.87}$$

Thus, we have the important result that the mean-square thermal displacement in a monatomic solid with cubic lattice symmetry can be calculated without any knowledge about the eigenvectors $\boldsymbol{\varepsilon}$. We also observe that (4.87) converges at the lower integration limit only if $F(\omega) \sim \omega^p$ for small ω, with $p > 1$. In a one-dimensional chain, which is often used to illustrate important concepts in lattice dynamics, $F(\omega)$ tends to a constant for small ω. Then $\langle u^2 \rangle$ diverges at all temperatures.

The high-temperature asymptote of (4.87) is

$$\langle u^2 \rangle = \frac{3k_B T}{M\omega^2(-2)}. \tag{4.88}$$

The Thirring expansion (4.50) for $E(\omega; T)$ at high temperatures easily gives the correction to (4.88) in terms of moments $\omega(n)$. When $T \to 0$, only the zero point vibrations remain. Then $n + 1/2 \to 1/2$ and

$$\langle u^2 \rangle = \frac{3\hbar}{2M\omega(-1)}. \tag{4.89}$$

Note that the high and low temperature limits of the displacement do not depend on the detailed shape of the density of states $F(\omega)$ but only on a single parameter, $\omega(-2)$ and $\omega(-1)$ respectively.

8.2. Thermal displacements in a Debye model

With a Debye model for $F(\omega)$, (4.87) becomes

$$\langle u^2 \rangle = \frac{9\hbar^2 T}{Mk_B\Theta_D^2}[\Phi(\Theta_D/T) + (1/4)(\Theta_D/T)], \tag{4.90}$$

where Φ is the Debye integral function

$$\Phi(x) = \frac{1}{x}\int_0^x \frac{z}{e^z - 1}\,dz. \tag{4.91}$$

For small x,

$$\Phi(x) = 1 - (x/4) + (x^2/36) - \ldots. \tag{4.92}$$

The high temperature expansion of $\langle u^2 \rangle$ is

$$\langle u^2 \rangle = \frac{9\hbar^2 T}{Mk_B\Theta_D^2}[1 + (1/36)(\Theta_D/T)^2 + \ldots]. \tag{4.93}$$

In the low temperature limit, $\Phi \to 0$ and hence

$$\langle u^2 \rangle = \frac{9\hbar^2}{4Mk_B\Theta_D}. \tag{4.94}$$

Example: The Lindemann melting criterion. Lindemann (1910) suggested that a material melts at a temperature T_m for which the root mean-square displacement $\langle u^2 \rangle^{1/2}$ exceeds a certain fraction of the nearest-neighbour distance. There have been numerous discussions of Lindemann's rule, trying to give it some theoretical justification (e.g., Born 1939, Leibfried 1955, Enderby and March 1966, Jacobs 1983) or merely examining its validity empirically (e.g., Shapiro 1970, Hansen 1970, Hoover et al. 1971, Young and Alder 1974, Stillinger and Weber 1980, Cho 1982). Since $T_m > \Theta_D$, we take the leading term in (4.93) and write Lindemann's rule in the form

$$\xi^2 = \frac{\langle u^2 \rangle}{4a^2} = \frac{9\hbar^2 T_m}{4a^2 Mk_B\Theta_D^2}, \tag{4.95}$$

where $2a$ is some characteristic distance between neighbouring atoms, here defined by $(4\pi/3)a^3 = \Omega_a$. In table 4.2 are given $\xi = [\langle u^2 \rangle]^{1/2}/2a$, calculated from (4.95) and with Θ_D from Rosén and Grimvall (1983). Strictly, (4.95) is derived for monatomic cubic solids, but it is often a reasonable approximation in more complicated lattices. Sometimes one assumes ξ to be an empirically known quantity and uses (4.95) to estimate Θ_D from the measured T_m. It is obvious that this is not a very reliable method to obtain the Debye temperature.

Table 4.2
The ratio $[\langle u^2 \rangle]^{1/2}/2a$ at T_m, illustrating Lindemann's rule

Na	Mg	Al	Cu	Tl	Pb
0.13	0.11	0.10	0.11	0.12	0.10

In this context we remark that ordinary melting is not caused by an instability of the lattice. For instance, the shear modulus of the solid does not vanish as one approaches the melting point, although the liquid has a low shear resistance. A theoretical account of T_m must consider the free energies of both the solid and the liquid phases, and find that temperature at which they are equal (Stroud and Ashcroft 1972, Moriarty et al. 1984). Still it is obvious that atoms cannot remain in a solid lattice when their thermal displacements become comparable with the interatomic distance, so in that respect Lindemann's idea is well founded. (For completeness, it should be remarked that there *is* a lattice instability at high temperatures. In metals, it occurs well above the actual melting temperature (Stroud and Ashcroft 1972).)

8.3. The Debye–Waller factor

The mean-square displacements can be found from the temperature dependence of the intensity in X-ray or neutron scattering experiments (Debye 1914, Waller 1923). Let I be the actual intensity, I_0 the intensity when the lattice is rigid, $\boldsymbol{k}_1$ and $\boldsymbol{k}_2$ the wave vectors and $\lambda = 2\pi/|\boldsymbol{k}|$ the wavelength of the photon (or neutron) before and after the (elastic) scattering, $\boldsymbol{q} = \boldsymbol{k}_1 - \boldsymbol{k}_2$ and 2θ the angle between $\boldsymbol{k}_1$ and $\boldsymbol{k}_2$. One has

$$I = I_0 \exp(-2M). \tag{4.96}$$

In the Deby–Waller factor $\exp(-2M)$ of a cubic lattice,

$$\begin{aligned} M &= \langle(\boldsymbol{u}\cdot\boldsymbol{q})^2\rangle/2 \\ &= (1/6)\langle u^2\rangle q^2 = (1/3)\langle u^2\rangle(8\pi^2 \sin^2\theta/\lambda^2) = B \sin^2\theta/\lambda^2, \end{aligned} \tag{4.97}$$

where B is known as the B-factor of the atom. According to (4.93), a plot of $\ln(I/I_0)$ versus $\sin^2\theta/\lambda^2$ (a Wilson plot) should give a straight line with a slope which is linear in T at high temperatures. At high temperatures, anharmonic effects give rise to a non-linearity in T (Hahn and Ludwig 1961, Maradudin and Flinn 1963, Cowley and Cowley 1966, Mair 1980).

8.4. The interatomic distances

The instantaneous positions of two atoms, labelled 0 and j, are $\boldsymbol{R}_0 = \boldsymbol{R}_0^0 + \boldsymbol{u}_0$ and $\boldsymbol{R}_j = \boldsymbol{R}_j^0 + \boldsymbol{u}_j$. The distance $\boldsymbol{d}$ between the atoms differs from the distance $\boldsymbol{d}_0$ in the static lattice, such that

$$\boldsymbol{d} = (\boldsymbol{R}_j^0 + \boldsymbol{u}_j) - (\boldsymbol{R}_0^0 + \boldsymbol{u}_0) = \boldsymbol{d}_0 + (\boldsymbol{u}_j - \boldsymbol{u}_0), \tag{4.98}$$

Consider the mean-square relative displacement,

$$\sigma^2 = \langle(\boldsymbol{d} - \boldsymbol{d}_0)^2\rangle = \langle \boldsymbol{d}^2\rangle - \boldsymbol{d}_0^2 = \langle \boldsymbol{u}_j^2\rangle + \langle \boldsymbol{u}_0^2\rangle - 2\langle \boldsymbol{u}_j\cdot\boldsymbol{u}_0\rangle. \tag{4.99}$$

Often one is interested in displacements along $\boldsymbol{R}_j - \boldsymbol{R}_0$. We consider

$$\sigma_R^2 = \langle [\hat{\boldsymbol{R}}_j \cdot (\boldsymbol{u}_j - \boldsymbol{u}_0)]^2 \rangle, \tag{4.100}$$

where $\hat{\boldsymbol{R}}_j$ is a unit vector along $\boldsymbol{R}_j - \boldsymbol{R}_0$. If the atoms vibrate independently, as in an Einstein model, $\langle \boldsymbol{u}_j \cdot \boldsymbol{u}_0 \rangle = 0$. In another extreme limit, that of long wavelength vibrations, all atoms move in phase and $\boldsymbol{d} = \boldsymbol{d}_0$. In a real solid we expect the correlation between the atomic motions to be significant when j and 0 are neighbours, but to be small for atoms far apart. A general expression for σ_R is (Grüneisen and Goens 1924, Warren 1969)

$$\sigma_R^2 = \frac{\hbar}{NM} \sum_{q\lambda} \frac{[\boldsymbol{e}(\boldsymbol{q}, \lambda) \cdot \hat{\boldsymbol{R}}_j]^2}{\omega(\boldsymbol{q}, \lambda)} \coth\left[\frac{\hbar\omega(\boldsymbol{q}, \lambda)}{2k_B T}\right] \times [1 - \cos[\boldsymbol{q} \cdot (\boldsymbol{R}_j^0 - \boldsymbol{R}_0^0)]]. \tag{4.101}$$

Without the cosine term in (4.101) $\sigma_R^2 = 2\langle u^2 \rangle/3$; cf. (4.99).

The essential points are illustrated in a calculation by Barsch (quoted by Leibfried (1955)) for a simple cubic lattice treated in a Debye model with $C_L = 2C_T$. Let atom j be the hth neighbour to atom 0 along the [100] axis. To leading order in the parameter $\eta = h(6\pi^2)^{1/3}$,

$$\langle \boldsymbol{d}_x^2 \rangle - \boldsymbol{d}_0^2 = 2\langle u_x^2 \rangle [1 - (4/3\eta)\mathrm{Si}(\eta) + \ldots], \qquad T > \Theta_D; \tag{4.102}$$

$$\langle \boldsymbol{d}_x^2 \rangle - \boldsymbol{d}_0^2 = 2\langle u_x^2 \rangle [1 - (18/5)\eta^{-2} + \ldots], \qquad T = 0. \tag{4.103}$$

Here $\langle u_x^2 \rangle = \langle u_j^2 \rangle/3 = \langle u_0^2 \rangle/3$ and

$$\mathrm{Si}(\eta) = \int_0^\eta [(\sin t)/t] \,\mathrm{d}t. \tag{4.104}$$

The factors [...] above rapidly tend to 1 with increasing h. When $h = 1$, [...] = 0.4 ($T > \Theta_D$) and 0.8 ($T = 0$), respectively.

One may recast (4.101) in a convenient form which makes direct contact with our results for $\langle u^2 \rangle$ (Sevillano et al. 1979, Allen et al. 1970, 1971). Define a density of states $F_R(\omega)$ by

$$F_R(\omega) = \sum_{\boldsymbol{q}, \lambda} [\boldsymbol{e}(\boldsymbol{q}, \lambda) \cdot \hat{\boldsymbol{R}}_j]^2 [1 - \cos(\boldsymbol{q} \cdot \boldsymbol{R}_j)] \delta(\omega - \omega_{\boldsymbol{q}\lambda})/N. \tag{4.105}$$

Then, σ_R^2 is obtained by the expression (4.87) for $\langle u^2 \rangle$, if only $F(\omega)$ is replaced by $2F_R(\omega)$. A spherical average of (4.105) yields the Debye model result

$$F_R(\omega) = \frac{3\omega^2}{\omega_D^3}\left[1 - \frac{\sin(\omega R q_D/\omega_D)}{\omega R q_D/\omega_D}\right]. \tag{4.106}$$

One may even model $F_R(\omega)$ by an Einstein spectrum, $F_R = \delta(\omega - \omega_R)$, with an appropriate ω_R. This is only a way of approximating the temperature dependence of σ_R^2, and it does not mean that $\sigma_R^2 = 2\langle u^2\rangle/3$, as would be the case in a strict Einstein model. Beni and Platzman (1976) evaluated σ_R^2 for a monatomic solid in a Debye model, while Sevillano et al. (1979) used a realistic $F(\omega)$ for Cu, Fe and Pt. The quantity σ_R^2 is accessible in experiments on the extended X-ray absorption fine structures, EXAFS (Greegor and Lytle 1979, Böhmer and Rabe 1979, Marcus and Tsai 1984).

8.5. *A general expression for the thermal displacement*

We now seek a general but tractable expression for $\langle u^2(\kappa, l)\rangle$ in a lattice of arbitrary structure. It is convenient to introduce a matrix B with elements (do not confuse B with the B-factor in (4.97))

$$B_{ij} = \langle u_i u_j \rangle. \tag{4.107}$$

Let the r atoms in the unit cell be numbered by $\kappa = 1,\ldots,r$. Then u_i is a component of the vector $(u_{1x}, u_{1y}, u_{1z}, u_{2x},\ldots,u_{rz})$. For brevity, we restrict to high temperatures. From a result in matrix theory (Born 1942) we can write

$$B(\kappa\kappa) = \frac{k_B T}{N_c M_\kappa} \sum_{\boldsymbol{q}} \Big[D^{-1}(\boldsymbol{q}\,;\kappa\kappa) + \frac{1}{12}(\hbar/k_B T)^2 I - \frac{1}{720}(\hbar/k_B T)^4 D(\boldsymbol{q}\,;\kappa\kappa) + \ldots \Big]. \tag{4.108}$$

Here D is the $3r \times 3r$ dynamical matrix and I is a unit matrix. $B(\kappa\kappa)$ refers to the displacement of atoms of kind κ. This expression is convenient since it allows the evaluation of $\langle u^2\rangle$ directly from the inverted dynamical matrix D. This is a simpler mathematical task than solving for all the frequencies $\omega(\boldsymbol{q}, \lambda)$ from D and then performing the sum over $(\boldsymbol{q}, \lambda)$. However, even the evaluation of (4.108) is far from trivial. An Einstein model version of (4.108) is discussed in §10.

8.6. Two atoms per primitive cell

An important special case of (4.108) is a lattice with a primitive cell containing two atoms, denoted by 1 and 2. Let their masses be M(1) and M(2). From (4.77) we cannot obtain the displacements of each kind of atom without knowing the 6-component eigenvectors

$$\boldsymbol{\varepsilon}(\boldsymbol{q}, \lambda) = [\boldsymbol{\varepsilon}(1; \boldsymbol{q}, \lambda), \boldsymbol{\varepsilon}(2; \boldsymbol{q}, \lambda)]. \tag{4.109}$$

However, we can obtain the weighted sum

$$M(1)\langle u_1^2\rangle + M(2)\langle u_2^2\rangle = 2\int_0^{\omega_{\max}} [E(\omega; T)/\omega^2]F(\omega)\,\mathrm{d}\omega. \tag{4.110}$$

Here $E(\omega, T)$ is the Einstein thermal energy. To obtain the displacement of each atom we turn to (4.108). $D(\boldsymbol{q})$ is now a 6×6 matrix. It can be blocked into four parts, $D(\kappa\kappa')$, with $D(\kappa\kappa') = D_0(\kappa\kappa')/[M(\kappa)M(\kappa')]^{1/2}$, cf. eq. (C.5). Then (4.108) yields, at high T,

$$M(1)\langle u_1^2\rangle_\alpha = (k_\mathrm{B}T/N_\mathrm{c})\sum_{\boldsymbol{q}} D_\alpha^{-1}(11; \boldsymbol{q}), \tag{4.111}$$

$$M(2)\langle u_2^2\rangle_\alpha = (k_\mathrm{B}T/N_\mathrm{c})\sum_{\boldsymbol{q}} D_\alpha^{-1}(22; \boldsymbol{q}). \tag{4.112}$$

The index α refers to a Cartesian component in the displacements and in the block $D(\kappa\kappa)$.

If $D_0(11)$ and $D_0(22)$ are equal, one has the important result that the mean-square thermal displacements in a diatomic solid at high temperatures is the same for both kinds of atoms, irrespective of their mass ratio. The condition $D_0(11) = D_0(22)$ is fulfilled if there are only nearest-neighbour interactions but also for the direct Coulomb forces in an ionic compound A^+B^-.

In the low-temperature limit, only the zero-point vibrations remain. They weight all modes, which usually gives unequal amplitudes for the two kinds of atoms. Note that for the low-temperature excitations (the elastic limit), all atoms move in phase and thus have equal amplitudes. However, these excitations are not important for the displacements.

The monatomic hexagonal lattice is a special case of diatomic lattices.

The B-matrix has the form

$$B = \begin{bmatrix} \langle u_a^2 \rangle & 0 & 0 \\ 0 & \langle u_a^2 \rangle & 0 \\ 0 & 0 & \langle u_c^2 \rangle \end{bmatrix}, \tag{4.113}$$

where u_a refers to vibrations along an a-axis and u_c to vibrations perpendicular to that axis. $D_0(11) = D_0(22)$ and $M(1) = M(2) = M$. From (4.110) we obtain the average $\langle u^2 \rangle$ over all directions,

$$\langle u^2 \rangle = (2\langle u_a^2 \rangle + \langle u_c^2 \rangle) = \frac{1}{M} \int \frac{E(\omega; T)}{\omega^2} F(\omega) \mathrm{d}\omega. \tag{4.114}$$

In an arbitrary direction $\boldsymbol{q}$, making an angle θ with the c-axis, one has

$$\langle (\boldsymbol{u} \cdot \hat{\boldsymbol{q}})^2 \rangle = \cos^2 \theta \langle u_c^2 \rangle + \sin^2 \theta \langle u_a^2 \rangle. \tag{4.115}$$

Example: Mean-square displacements in NaCl lattices. Since the essential interatomic forces in alkali halides, A^+B^-, are the Coulomb forces plus interaction from overlap of atomic wavefunctions on neighbouring ions, we expect that $\langle u^2(\mathrm{A}) \rangle \approx \langle u^2(\mathrm{B}) \rangle$ at high temperatures. Calculations confirm this result (Huiszoon and Groenewegen 1972). For work on thermal displacements in $NbC_{0.95}$ (which has NaCl structure), see Kaufmann and Meyer (1984), citing Müllner, Reichardt and Christensen.

Example: Anisotropic vibrations in zinc. Zinc has unusually anisotropic thermal displacements. According to eq. (4.115) it suffices to find $\langle u_c^2 \rangle$ and $\langle u_a^2 \rangle$. A strict treatment requires not only the frequencies $\omega(\boldsymbol{q}, \lambda)$ but also the corresponding eigenvectors. However, it is a useful approximation (Grüneisen and Goens 1924), to introduce two effective density-of-states functions of the Debye type, $F_a(\omega)$ and $F_c(\omega)$, and choose Debye temperatures Θ_a and Θ_c such that the temperature dependence of the displacements is well described. Figure 4.13 shows that this gives a good account of $\langle u_a^2 \rangle$ and $\langle u_c^2 \rangle$ in zinc. The normalisation of $F_a(\omega)$ and $F_c(\omega)$ is defined by fitting theory and experiment at $T = 0\,K$. The inset shows the actual $F(\omega)$ and the two Debye spectra used here.

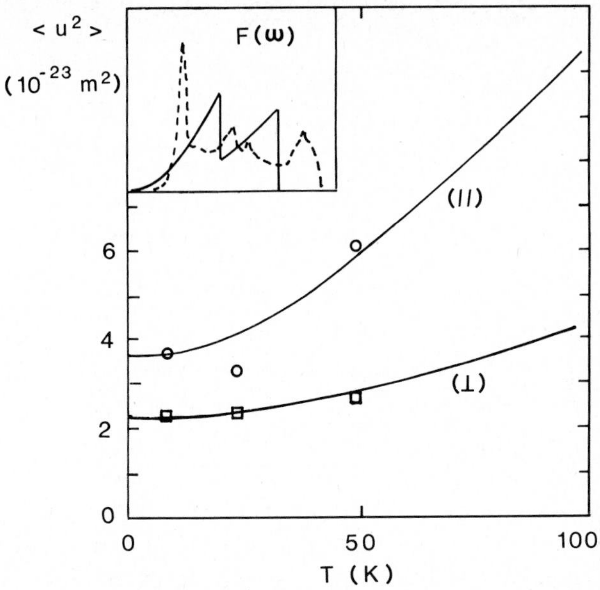

Fig. 4.13. Anisotropic thermal displacements in zinc, calculated with separate Debye models (shown in the inset together with the full $F(\omega)$) for directions parallel (||) and perpendicular (⊥) to the hcp c-axes. Circles and squares are measured values. (Reproduced with permission from Potzel et al. (1984). Phys. Rev. **B30** (1984) 4980.)

8.7. *Vibrational velocity*

The instantaneous velocity $v_\alpha(\kappa, l; t)$, of an atom (κl) in the direction α, is obtained from (4.77) as $\partial u_\alpha(\kappa, l; t)/\partial t$, i.e.

$$v_\alpha(\kappa, l; t) = -\mathrm{i}(N_c M_\kappa)^{-1/2} \sum_{\boldsymbol{q}\lambda} [E(\boldsymbol{q}, \lambda)]^{1/2} \varepsilon_\alpha(\boldsymbol{q}, \lambda; \kappa) \times \exp\{\mathrm{i}[\boldsymbol{q} \cdot \boldsymbol{R}^\circ(\kappa, l) - \omega(\boldsymbol{q}, \lambda)t]\}. \tag{4.116}$$

Proceeding as for the amplitude $u_\alpha(\kappa, l)$, we get the thermal average

$$\langle v_\alpha^2(\kappa, l)\rangle = \frac{1}{N_c M_\kappa} \sum_{\boldsymbol{q}\lambda} \hbar\omega(\boldsymbol{q}, \lambda) \left[\frac{1}{\exp(\hbar\omega_{\boldsymbol{q}\lambda}/k_B T) - 1} + \frac{1}{2}\right] \varepsilon_\alpha \varepsilon_\alpha^*. \tag{4.117}$$

In a monatomic solid, $\langle v^2\rangle = \langle v_x^2\rangle + \langle v_y^2\rangle + \langle v_z^2\rangle$ is isotropic:

$$\langle v^2\rangle = \frac{1}{M} \int_0^{\omega_{\max}} E(\omega; T) F(\omega) \mathrm{d}\omega, \tag{4.118}$$

with the low and high temperature limits

$$\langle v^2 \rangle = 3\hbar\omega(1)/2M = (9k_B/8M)\Theta_D(1), \qquad T = 0; \tag{4.119}$$

$$\langle v^2 \rangle = 3k_BT/M, \qquad k_BT > \hbar\omega_{max}. \tag{4.120}$$

9. Einstein and Debye models – beyond the simplest ideas

9.1. Introduction

In fig. 4.1 we noted how remarkably well the heat capacity $C(T)$ of real materials was represented by an Einstein model at intermediate and high temperatures, in spite of the fact that the Einstein model only contains one free parameter, Θ_E. Similarly, the one-parameter Debye model would give an excellent representation of $C(T)$. Disregarding the fact that $C(T)$ does not tend to a constant at high T, because of anharmonicity, we can understand the success of the simple models, if we consider the high temperature expansion of the heat capacity,

$$C_{har} = 3Nk_B\{1 - (1/12)[\hbar\omega(2)/k_BT]^2 - \cdots\}. \tag{4.121}$$

To leading order, C_{har} only depends on the single parameter $\omega(2)$. The Debye and Einstein models can be made to agree with the first two terms in (4.121) if $k_B\Theta_D = \sqrt{(5/3)}\hbar\omega(2) = 1.29\hbar\omega(2)$ and $k_B\Theta_E = \hbar\omega(2) = \sqrt{(3/5)}\ k_B\Theta_D = 0.77k_B\Theta_D$, respectively. More generally, we can always choose Θ_D (or Θ_E) such that the heat capacities of the Debye (or Einstein) models agree with C_{har} at a particular temperature T_1;

$$C_{har}(T_1) = C_D(\Theta_D^C; T_1), \tag{4.122}$$

$$C_{har}(T_1) = C_E(\Theta_E^C; T_1). \tag{4.123}$$

The solutions to eqs. (4.122) and 4.123), Θ_D and Θ_E, vary with T_1 unless the true $F(\omega)$ happens to be exactly that of a Debye or Einstein spectrum (fig. 4.14). In this way we can define temperature-dependent Debye and Einstein temperatures, $\Theta^C(T)$. The superscript C denotes that the Θ-value refers to the heat capacity. The Einstein model is inadequate at low temperatures, but from (4.37) we know that (we write Θ^C for Θ_D^C)

$$\lim_{T\to 0} \Theta^C(T) = \Theta_D(-3). \tag{4.124}$$

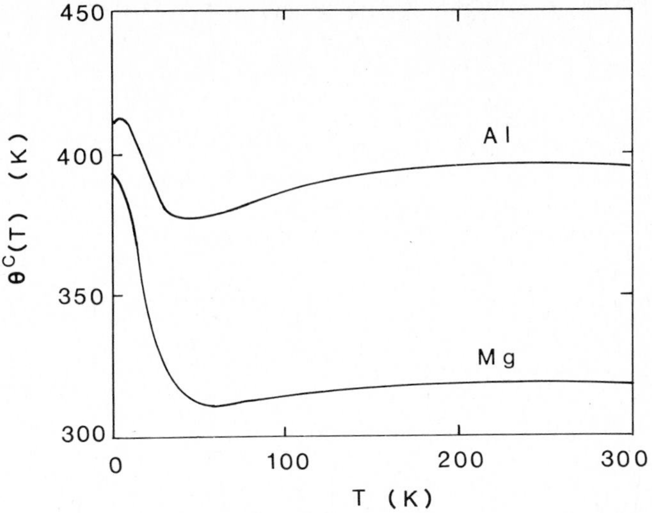

Fig. 4.14. The heat capacity Debye temperature, $\Theta^C(T)$, for Al and Mg. From data in the Landolt–Börnstein tables (Schober and Dederichs 1981).

In analogy to the heat capacity Debye temperature, we can define an "entropy Debye temperature" $\Theta^S(T)$ by

$$S_{\text{har}}(T) = S_D(\Theta^S; T). \tag{4.125}$$

S_{har} is the entropy for some given $F(\omega)$, and S_D is the Debye model entropy. In the same way, one may obtain $\Theta^H(T)$, referring to the zero point energy. A Debye temperature $\Theta^M(T)$ characteristic of the thermal displacements is obtained if we put the quantity M in the Debye–Waller factor of an actual solid equal to the Debye-model result, §8.2. We can write

$$\langle u^2 \rangle_{\text{har}} = \langle u^2(\Theta^M; T) \rangle. \tag{4.126}$$

From Mössbauer experiments, one can derive not only Θ^M but also $\Theta^{MB} = \Theta_D(1)$ related to the zero-point velocity of the atomic motion. In eq. (4.14) we gave an expression for $\Theta_D(-3)$ in terms of the anisotropic sound velocities $C(\theta, \phi)$. Since $C(\theta, \phi)$ is related to the elastic coefficients c_{ij} we can obtain an "elastic" Debye temperature $\Theta^E = \Theta_D(-3)$. For strictly harmonic vibrations, $\Theta^E = \Theta_D(-3)$ at all temperatures. When anharmonicity effects are included, we can still calculate $\Theta^E(T)$ but

$\Theta^E(T) \neq \Theta^C(T=0)$. Anharmonic effects are present also at $T=0$, through the vibrational zero-point motion. It has been discussed whether this affects $\Theta^E(T=0)$ and $\Theta^C(T=0)$ differently. Barron and Klein (1962) and Feldman (1964) showed that the elastic and thermal Debye temperatures agree, when low-order anharmonic effects are included, although Overton (1971) claims that there is a small difference ($<1\%$) from the Δ_3 anharmonic frequency shift.

In ch. 11 we shall introduce the Bloch–Grüneisen formula $\rho_{BG}(\Theta_D; T)$ for the electrical resistivity. If $\rho(T)$ is the true phonon-limited resistivity we can define a Bloch–Grüneisen resistivity Debye temperature $\Theta^{BG}(T)$ by

$$\rho(T) = \rho_{BG}(\Theta^{BG}; T). \tag{4.127}$$

(Actually, one has to fit to ρ_{BG} at two temperatures because ρ is not a normalised quantity.) Sometimes one calculates a Debye temperature Θ^L from the Lindemann melting rule, (4.95). We thus conclude that there are several Debye (or Einstein) temperatures. Whenever Θ is obtained from experimental data, one has to specify which physical quantity and what temperature Θ refers to. Note that Θ^C, Θ^S, Θ^H, Θ^M, Θ^{MB} and Θ^E have exact definitions in terms of the density of states $F(\omega)$ while Θ^{BG} and Θ^L are derived from approximate or empirical relations. Table 4.3 summarises properties described by a single parameter, $\omega(n)$ or Θ_D.

Table 4.3
Vibrational properties described by a single parameter

Physical property	Frequency moment	Debye temperature
heat capacity, low T	—	$\Theta^C(T \to 0) = \Theta_D(-3)$
entropy, low T	—	$\Theta^C(T \to 0) = \Theta_D(-3)$
elastic constants, any T	—	$\Theta^E(T) = \Theta_D(-3)$
thermal displacement, high T	$\omega(-2)$	$\Theta^M(T > \Theta_D) = \Theta_D(-2)$
thermal displacement, $T=0$	$\omega(-1)$	$\Theta^M(T=0) = \Theta_D(-1)$
entropy, high T	$\omega(0)$	$\Theta^S(T > \Theta_D) = \Theta_D(0)$
zero-point energy ($T=0$)	$\omega(1)$	$\Theta^H(T=0) = \Theta_D(1)$
zero-point velocity ($T=0$)	$\omega(1)$	$\Theta^{MB}(T=0) = \Theta_D(1)$
heat capacity, high T	$\omega(2)$	$\Theta^C(T > \Theta_D) = \Theta_D(2)$

Example: Calculation of $\Theta^S(298)$ for NaCl. The "standard entropy" $S° = S(298.15\text{K})$ is given in JANAF Tables (1971) as 17.24 ± 0.05 [cal/mol K]. We divide $S°$ by $3Nk_B$, where N is the number of atoms in the sample. In this case $N = 2L$, where L is Avogadro's number. In the American Institute of Physics Handbook (1972), one finds a table of the Debye model entropy, $S_D(T/\Theta_D)$. Putting that expression equal to $S°/6Lk_B = S°/6R$, we get after interpolation in the table that $\Theta_D/298.15 = 0.911$, which yields $\Theta_D^S(298.15\,\mathrm{K}) = 272\,[\mathrm{K}]$. When $T > \Theta_D/2$, $\Theta^S(T)$ is an excellent approximation to $\Theta_D(0)$ (with temperature dependent anharmonic corrections included). It should be remarked that $S°$ of NaCl only has a vibrational contribution. In metals, one would have to subtract an electronic part of $S°$ before Θ^S is calculated.

9.2. Inequalities

The total heat capacity $C_{\mathrm{har}}(T)$ is a superposition of Einstein-like terms $C_E(T)$ – one for each phonon frequency ω. It is therefore obvious that C_{har} is "broader" than C_E if we think of the width in temperature over which the essential change in C takes place. In fact, one can show rigorously that for any density of states $F(\omega)$, the heat capacity curve C_{har} calculated from $F(\omega)$ crosses the heat capacity curve of a single Einstein model at one and only one temperature, fig. 4.15. There is no similar relation between $C_{\mathrm{har}}(T)$ and the Debye model $C_D(T)$.

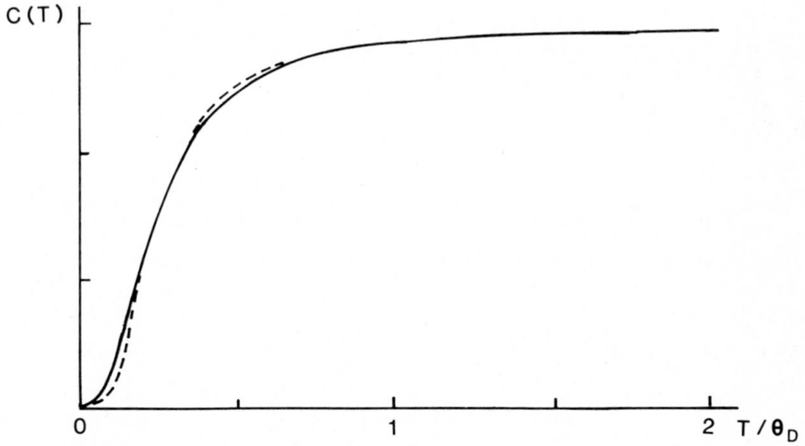

Fig. 4.15. The heat capacities in a Debye model (solid line) and an Einstein model (dashed line) can be made to agree closely over a wide range of temperatures. It can be shown that they cross at one point only.

The frequencies $\omega(n)$ increase monotonically with n,

$$\omega(n) > \omega(n') \text{ when } n > n'. \tag{4.128}$$

For large n, $\omega(n)$ asymptotically approaches $\omega(\infty) = \omega_{\max}$, the maximum frequency of any phonon mode in the specimen. The Debye temperatures $\Theta_D(n)$ are usually not monotonic in n (figs. 4.7 and 4.8), but from (4.128) one may derive inequalities such as

$$\Theta_D(-1) > (\sqrt{3}/2)\Theta_D(-2) \quad \approx 0.87\Theta_D(-2), \tag{4.129}$$

$$\Theta_D(0) > \sqrt{(2/3)e^{1/3}}\,\Theta_D(-1) \quad \approx 0.93\Theta_D(-1), \tag{4.130}$$

$$\Theta_D(1) > (4/3)e^{-1/3}\Theta_D(0) \quad \approx 0.96\Theta_D(0), \tag{4.131}$$

$$\Theta_D(2) > \sqrt{(15/16)}\,\Theta_D(1) \quad \approx 0.97\Theta_D(1). \tag{4.132}$$

Equality signs hold here if $F(\omega)$ represents a single Einstein spectrum. An accidental equality $\omega(n_1) = \omega(n_2)$ is also possible. The low-frequency part of $F(\omega)$, which determines $\Theta_D(-3)$, has no strict mathematical relation to the form of $F(\omega)$ at intermediate and high frequencies. Therefore the ratio $\Theta_D(-2)/\Theta_D(-3)$ may be arbitrarily small or large. The relation $\Theta^M(T) > \Theta^C(T)$, stated by Zener and Bilinski (1936) and based on approximate arguments, is often violated.

The zero-point mean-square thermal displacement $\langle u_x^2 \rangle$ in a monatomic lattice is, (4.89),

$$\langle u_x^2 \rangle_{T=0} = \hbar/2M\omega(-1). \tag{4.133}$$

The asymptotic behaviour at high temperatures is

$$\langle u_x^2 \rangle_T = k_B T/M[\omega(-2)]^2. \tag{4.134}$$

The fact that $\omega(-1) > \omega(-2)$ can be used to put bounds on $\langle u_x^2 \rangle_T$ at an arbitrary temperature T, if this quantity is known at a certain T_1. We first readily prove that

$$\langle u_x^2 \rangle_{T=0} = \frac{\hbar}{2M\omega(-1)} < \frac{\hbar}{2M\omega(-2)} < \frac{\hbar}{2}\left(\frac{\langle u_x^2 \rangle_{T_1}}{Mk_B T_1}\right)^{1/2}. \tag{4.135}$$

From the general curvature of $\langle u^2 \rangle$ as a function of T, and its

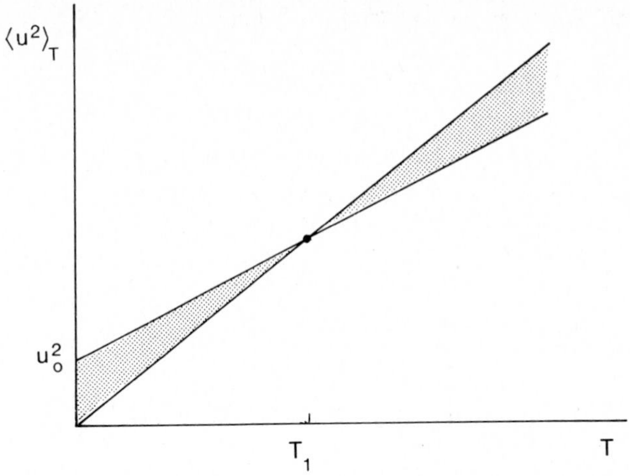

Fig. 4.16. The construction of Housley and Hess (1966), giving bounds (shaded area) to the thermal displacement, $\langle u^2\rangle_T$, when $\langle u^2\rangle_{T_1}$ is known at a certain temperature, T_1. The quantity u_0 is given by $u_0^4 = (\hbar/2)^2\langle u^2\rangle_{T_1}/Mk_BT_1$.

asymptotic behaviour, it is not very difficult to show that $\langle u^2\rangle_T$, for $T \neq T_1$, must lie in the shaded area of fig. 4.16 (Housley and Hess 1966).

In a real crystal there are anharmonic interactions. Their effect on the heat capacity and the mean-square thermal displacement can have any sign (cf. ch. 5.) One should therefore be careful when the inequalities above are applied to data obtained at high temperatures.

Example: Heisenberg's uncertainty relation in solids. In a monatomic solid, the zero-point vibrational displacement and velocity are given by

$$\langle u_x^2\rangle_{T=0} = \hbar/2M\omega(-1), \tag{4.136}$$

$$\langle v_x^2\rangle_{T=0} = \hbar\omega(1)/2M. \tag{4.137}$$

The inequality $\omega(1) > \omega(-1)$ is equivalent with the Heisenberg uncertainty relation $\langle u_x^2\rangle_{T=0}\langle v_x^2\rangle_{T=0} > \hbar^2/4M^2$. A Debye model gives $\langle u_x^2\rangle_{T=0}\langle v_x^2\rangle_{T=0} = (9/8)(\hbar^2/4M^2)$. An analogous inequality holds for a particular atom κ in a lattice with several different atoms, if $\langle u_x^2\rangle$ and $\langle v_x^2\rangle$ refer to the site of that atom and M is replaced by M_κ (Housley and Hess 1966).

9.3. Debye temperatures from the heat capacity

Low temperatures. In §7.2 we obtained an expression for $C_{\mathrm{har}}(T)$ when the phonon density of states has a low-frequency expansion $F(\omega) = a_1\omega^2 + a_2\omega^4 + \cdots$, and we also obtained a result for the Debye heat capacity $C_{\mathrm{D}}(T)$ with the lowest-order correction to the T^3-law. Putting these two expressions equal we get

$$Nk_{\mathrm{B}}\{(12\pi^4/5)(T/\Theta_{\mathrm{D}})^3 - 9(\Theta_{\mathrm{D}}/T)\exp(-\Theta_{\mathrm{D}}/T)\}$$
$$= Nk_{\mathrm{B}}\{(4\pi^4/15)a_1(k_{\mathrm{B}}T/\hbar)^3 + (16\pi^6/21)a_2(k_{\mathrm{B}}T/\hbar)^5\}. \qquad (4.138)$$

This defines a temperature-dependent Debye temperature $\Theta^{\mathrm{C}}(T)$ with $[\Theta^{\mathrm{C}}(0)]^{-3} = 9\hbar^3/a_1k_{\mathrm{B}}^3$. If a_2 is small, we get to the lowest oıder

$$\Theta^{\mathrm{C}}(T) = \Theta^{\mathrm{C}}(0)[1 - (20\pi^2/21)(a_2/a_1)(k_{\mathrm{B}}T/\hbar)^2]. \qquad (4.139)$$

For almost all solids, but with the exception of Au and some of its dilute alloys (Bevk et al. 1977), a_2 is positive. At very low temperatures, $\Theta^{\mathrm{C}}(T)$ therefore decreases quadratically in T^2. A more complicated temperature dependence sets in at slightly higher T. Figure 4.17 shows $\Theta^{\mathrm{C}}(T)$ for Si

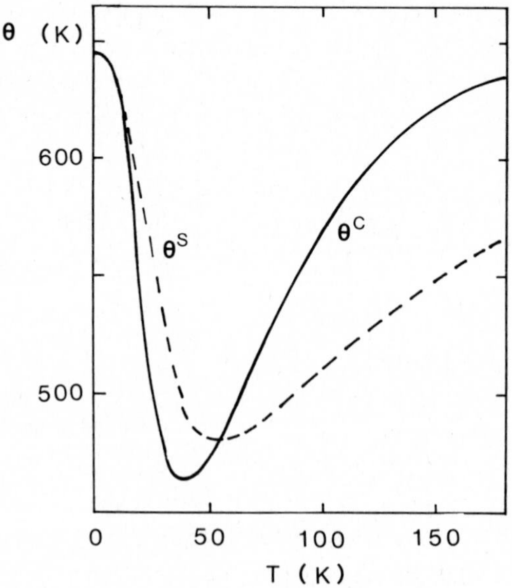

Fig. 4.17. The heat capacity Debye temperature $\Theta^{\mathrm{C}}(T)$ and the entropy Debye temperature $\Theta^{\mathrm{S}}(T)$, calculated from the measured $C_p(T)$ of Si (Flubacher et al. 1959). The curves cross at the minimum of $\Theta^{\mathrm{S}}(T)$.

when $T < \Theta_D/4$. The figure also gives the entropy Debye temperature $\Theta^S(T)$. In the limit $T \to 0$, the two Debye temperatures are equal. Since the entropy $S(T)$ is an integrated quantity over the heat capacity from 0 to T, $\Theta^S(T)$ varies less rapidly than $\Theta^C(T)$ with T. We note in fig. 4.17 that $\Theta^S(T)$ has its minimum where it crosses $\Theta^C(T)$, a feature which is not difficult to prove.

High temperatures. The Debye temperature derived from the heat capacity gets very uncertain with increasing T. For instance, if C_{har} is in error by 1% at $T = \Theta^C/2$, the derived Θ^C is in error by 3%, while a 1% error in C_{har} at $T = \Theta^C$ leads to a 12% error in Θ^C. If Θ^C is derived from the heat capacity at constant pressure, C_p, one usually gets unphysical Θ-values when $T \gg \Theta^C$, since anharmonic effects increase C_p above the asymptotic high-temperature limit $3Nk_B$ of strictly harmonic vibrations. The inset in fig. 4.18 shows $\Theta^C(T)$ obtained in this way for KBr; after Berg and Morrison (1957). Much of the anharmonicity may be suppressed if Θ^C is evaluated from C_V instead of C_p, but still the Debye temperature is very uncertain at high temperatures. However, there are methods to derive $\Theta^C(T \to \infty) = \Theta_D(2)$ from heat capacity data. Barron et al. (1957, 1959) and others have used the following scheme to obtain $\Theta_D(2)$. We can write Θ^C in the form (Tosi and Fumi 1963, Domb and Salter 1952)

$$[\Theta^C(T)]^2 = \Theta_D^2(2)\left(1 + \sum_{n=1}^{\infty} a_{2n} T^{-2n}\right), \tag{4.140}$$

with

$$a_2 = -(1/28)[\Theta_D^4(4) - \Theta_D^4(2)]/\Theta_D^2(2), \tag{4.141}$$

$$a_4 = (5/4536)[\Theta_D^6(6) - \Theta_D^6(2)]/\Theta_D^2(2) + (a_2/14)\Theta_D^2(2). \tag{4.142}$$

$\Theta^C(T)$ is now plotted versus $1/T^2$ at intermediate temperatures (say $\Theta_D/5 < T < 3\Theta_D/4$). The extrapolation in the plot to $1/T^2 = 0$ gives $\Theta_C(2)$. This is illustrated in the main curve of fig. 4.18.

Contrary to the case of $\Theta^C(T)$, the entropy Debye temperature $\Theta^C(T)$ obtained from thermal data is well behaved even in the presence of anharmonicity (ch. 5). Because the entropy is an integrated quantity, $\Theta^S(T)$ varies smoothly with T even when there is some scatter in the heat capacity data used to find the entropy. An error in the experimental

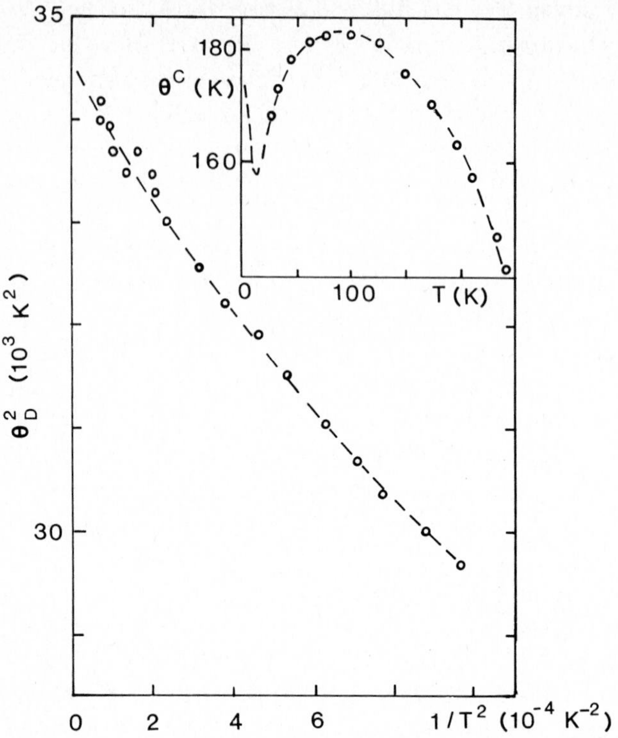

Fig. 4.18. The main dashed curve, extrapolated to $1/T^2 = 0$, gives the Debye temperature $\Theta_D(2)$. The inset shows that the Debye temperature $\Theta^C(T)$, determined directly from the measured C_p, behaves unphysically when anharmonic effects become important. Data (for KBr) and evaluation of $\Theta^C(T)$ from Berg and Morrison (1957). The construction to find $\Theta_D(2)$ is from Barron et al. (1957).

entropy by 1% at $T = \Theta^S$ yields an error of 1.4% in Θ^S. The leading term in the high temperature expansion of the vibrational entropy (4.55) gives the entropy Debye temperature $\Theta^S(T \gg \Theta_D) = \Theta_D(0)$. The expansion analogous to (4.140) is (Tosi and Fumi 1963)

$$\Theta^S(T) = \Theta_D(0)\{1 + \sum_{n=1}^{\infty} a_{2n} T^{-2n}\}, \tag{4.143}$$

where

$$a_2 = (1/40)[\Theta_D^2(0) - \Theta_D^2(2)], \tag{4.144}$$

$$a_4 = (1/2240)[\Theta_D^4(4) - \Theta_D^4(0)] + (a_2/2)[a_2 + (1/10)\Theta_D^2(0)]. \tag{4.145}$$

In a strict Debye model, all $\Theta_D(n) = \Theta_D$. Then all $a_{2n} = 0$, and $\Theta^S(T) = \Theta_D$ is independent of the temperature. In a real solid, a_2 is usually negative (figs. 4.7 and 4.8 or 4.17) and much smaller than $10^{-2}\Theta_D^2(0)$. Therefore $\Theta^S(T)$ often comes close to $\Theta^S(T \to \infty) = \Theta_D(0)$ already when $T \sim \Theta^S/4$. At such low temperatures, anharmonic effects are relatively unimportant, so $\Theta_D(0)$ derived from the measured entropy (i.e. the integrated C_p/T) represents the harmonic spectrum of lattice vibrations at low temperatures. This is to be contrasted to $\Theta_D(-3) = \Theta^E$ taken from room temperature data, or Θ^C evaluated at intermediate temperatures, which may both be significantly influenced by anharmonicity.

Hwang's relation for $\omega(n)$. Hwang (1954) noted that all moments $\omega(n)$, with $-3 < n < 0$, can be obtained from an integral of the heat capacity. One has

$$\int_0^\infty \frac{C_{\text{har}}(T)}{T^{1-n}}\,\mathrm{d}T = 3Nk_B\Gamma(2-n)\zeta(1-n)\left(\frac{\hbar\omega(n)}{k_B}\right)^n, \tag{4.146}$$

$\Gamma(2-n)$ is the gamma function and $\zeta(1-n)$ is Riemann's zeta function ($\Gamma(3) = 2, \zeta(2) = \pi^2/6$ etc.). With a Thirring expansion for C_{har} we can integrate (4.73) to a finite T and get an expansion similar to (4.146) but with a power series in $\omega(n)$ on the right-hand side (Barron et al. 1957).

9.4. Debye temperature from the Debye–Waller factor

The Debye temperature $\Theta^M(T)$ obtained from the thermal displacement $\langle u^2 \rangle$ has a high temperature expansion (Barron et al. 1966)

$$\Theta^M(T) = \Theta_D(-2)\left\{1 + (1/7200)\{[\Theta_D(2)/\Theta_D(-2)]^2 - 1\} \times [\Theta_D^4(-2)/T^4] + \cdots\right\}. \tag{4.147}$$

Thus, unlike $\Theta^C(T)$ and $\Theta^S(T)$ at high T, there is no $1/T^2$-term. Further, $\Theta^M(0) = \Theta_D(-1)$ and $\Theta^M(T) \to \Theta_D(-2)$ when $T \to \infty$. These facts, together with the observation that $\Theta_D(n)$ often has a minimum for n near -1, means that $\Theta^M(T)$ usually varies very little with T. This has been verified in explicit calculations (Barron et al. 1966).

9.5. $F(\omega)$ from the inverted heat capacity

When the density of states $F(\omega)$ has been obtained, theoretically or from experiments (e.g., by neutron scattering), it is common practice to calculate the heat capacity $C_{\mathrm{har}}(T)$ and compare this with its measured value. One may ask if it would not be possible to reverse this scheme, and derive $F(\omega)$ from the measured heat capacity. We shall see that mathematically this is a straightforward task but in practice it is a futile idea. The expression for the heat capacity,

$$C_{\mathrm{har}}(T) = \int_0^{\infty} F(\omega) C_{\mathrm{E}}(\hbar\omega/k_{\mathrm{B}}T)\mathrm{d}\omega, \tag{4.148}$$

can be regarded as an integral equation for the density of states $F(\omega)$, when $C_{\mathrm{har}}(T)$ is known. Mathematically, it can be "inverted", for instance by the method of Fourier transforms (Montroll 1942). Several other inversion formulas have been devised, e.g. by Lifshitz (1954) and Sekimoto (1978). One may also obtain $F(\omega)$ from (4.148) by a numerical iterative process. However, the heat capacity is so insensitive to details in $F(\omega)$ that calculations for a real solid are not very successful. For instance, Malik and Kothari (1980) obtain equally good agreement with the experimental heat capacity of graphite when their $F(\omega)$, obtained by a numerical inversion procedure, had $\hbar\omega_{\mathrm{max}}/k_{\mathrm{B}}$ changed from 1500 K to 2500 K. See also work by Korshunov and Shevchenko (1983).

9.6. $F(\omega)$ from frequency moments

Montroll (1943, 1944) pointed out that $F(\omega)$ is completely specified by all the even positive moments μ_{2n}. If we choose to expand $F(\omega)$ in the even Legendre polynomials P_{2n}, we can write

$$F(\omega) = (3/\omega_{\mathrm{max}}) \sum_{n=0}^{\infty} A_{2n} P_{2n}(\omega/\omega_{\mathrm{max}}), \tag{4.149}$$

If the sum (4.149) is truncated at $n = n'$, $F(\omega)$ is expressed in terms of the moments μ_{2n} $(n < n')$. Usually, even the first 10 even moments are insufficient to reproduce characteristic features in $F(\omega)$ so it is not feasible to obtain the density of states from the moments determined from thermal data. Theoretically, μ_{2n} can be obtained directly from the dynamical matrix, $\mu_{2n} \sim \mathrm{Tr}(D^n)$ but a large n' is required to represent

$F(\omega)$ and the method is still not much used to obtain the bulk $F(\omega)$ (Montroll 1943, Montroll and Peaslee 1944, Garland and Jura 1954, Isenberg 1963, Shapiro 1970). Matsushita and Matsubara (1978) have used the real-space version of Tr D to estimate $F(\omega)$ in small particles. Barron and Morrison (1960) estimated $F(\omega)$ of alkali halides from some moments μ_n derived from C_p-data.

10. The Einstein model and the force constant matrix

In the Einstein model there is no coupling between the motions of the atoms. This means that one neglects the matrix elements $\Phi_{\alpha\beta}(\kappa l, \kappa' l')$, which express the (negative) force on the (κl) atom in the α-direction. Only the diagonal matrix elements $\Phi_{\alpha\alpha}(\kappa l, \kappa l)$ are retained in the dynamical matrix. Equilibrium requires that

$$\Phi_{\alpha\alpha}(\kappa l, \kappa l) = -\sum_{\kappa' l' \neq \kappa l} \Phi_{\alpha\alpha}(\kappa l, \kappa' l'). \tag{4.150}$$

On the right-hand side of (4.150) there is no restriction on the range or the nature of the interatomic forces. It is not even necessary that there is a periodic lattice. In fact, the Einstein model is particularly well suited for a qualitative study of solids with defects such as surfaces or impurities. Then (κl) is replaced by a label n which specifies each individual atom. Note that there is no reciprocal-space description in terms of a dynamical matrix $D(\boldsymbol{q})$, since there are no propagating waves in the Einstein model.

The density of states of the usual Einstein model is a single peak, $F(\omega) = 3\delta(\omega - \omega_E)$. The Einstein model that results from the neglect of the non-diagonal terms in Φ has a density of states with as many separate sharp peaks as there are unequal diagonal elements in Φ. For instance, $F(\omega)$ has only one peak in elements with bcc, fcc or hcp lattices but has two peaks in NaCl and CsCl. Irrespective of how many peaks there are in $F(\omega)$, the Einstein model discussed in this section gives the correct $\omega(2)$. In particular, for elements with bcc, fcc or hcp structures, $\omega_E = \omega(2)$. (We recall that $\omega(2)$ is obtained from Tr Φ.)

Following Salter (1956), the force constant matrix of a real solid is written

$$\Phi = \Phi_d + \Phi_r, \tag{4.151}$$

where Φ_d is the diagonal part (i.e. the part which is kept in the Einstein model) and Φ_r is the remainder. If Φ_r is in some sense small compared to Φ_d, we may express thermodynamic quantities in the diagonal part Φ_d plus correction terms involving Φ_r.

We first consider the thermal displacements. From a real-space version of (4.108) we obtain, for the nth atom in the α-direction and at an arbitrary temperature,

$$\langle u^2\rangle_{n\alpha} = (\hbar/2M_n)[F^{-1/2}\coth(\hbar F^{1/2}/2k_BT)]_{n\alpha,\,n\alpha}, \tag{4.152}$$

where F is the matrix

$$F = M^{1/2}\Phi M^{-1/2} = F_d + F_r, \tag{4.153}$$

and M is the diagonal matrix of atomic masses. The high-temperature limit of (4.152) is

$$\langle u^2\rangle_{n\alpha} = (k_BT/M_n)(F_d^{-1})_{n\alpha,\,n\alpha} + \cdots, \tag{4.154}$$

where the dots denote terms of order $(\Phi_r)^2$ and higher. Masri and Dobrzynski (1971) used this method to estimate $\langle u^2\rangle$ of surface atoms.

An analogous approach to the vibrational entropy of alloys has been worked out by Dobrzynski (1969) (see also Kincaid and Huckaby 1976) and applied by Moraitis and Gautier (1977a). In the high temperature limit one has

$$S = k_B\sum_{n\alpha}\{1 + (1/2)\ln[(k_BT)^2(F_d^{-1})_{n\alpha,\,n\alpha}/\hbar^2]\}. \tag{4.155}$$

The neglect of F_r in (4.153) is equivalent to the approximation of $\omega(0)$ by $\omega(2)$, cf. eq. (C.14). Friedel (1982) applied the method to dislocation cores.

11. Dependence of the Debye temperature on the crystal structure

More than 25 of the metallic elements have a low-temperature close-packed (fcc or hcp) structure and a high-temperature bcc phase. This may be explained if $\Theta_{bcc}(0)$ is lower than $\Theta_{fcc}(0)$ and $\Theta_{hcp}(0)$ (§6.3). Zener (1948) noted that with only nearest-neighbour central forces, the bcc lattice is unstable against shear. This, he argued, would tend to give the bcc lattice a low-lying transverse mode and hence a low Debye

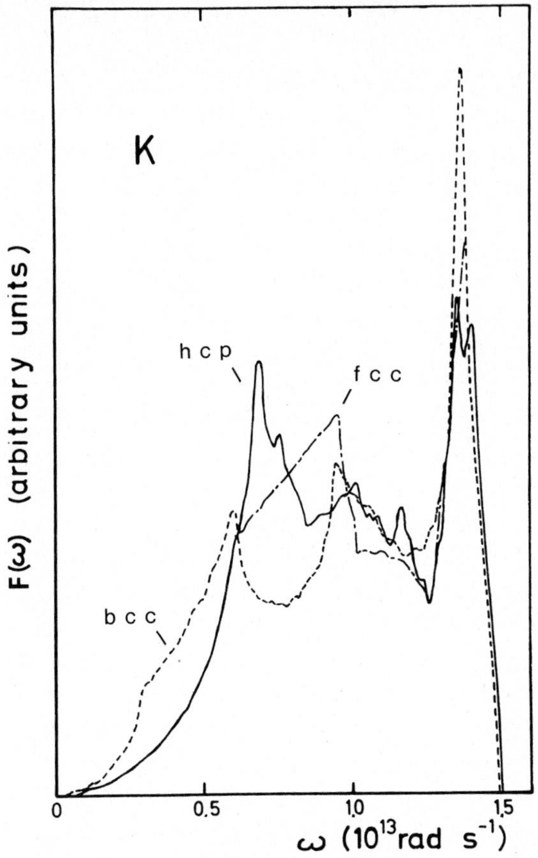

Fig. 4.19. Calculated frequency spectra for potassium in fcc, bcc and hcp structures. From Grimvall and Ebbsjö (1975) (Reproduced with permission from Physica Scripta.)

temperature. Figure 4.19 shows $F(\omega)$ of bcc, fcc and hcp potassium, calculated in a pseudopotential model. The low-lying [110] shear mode does give a strong weight to $F_{bcc}(\omega)$ at low ω, but because $\Theta_D(0)$ is an average over the entire $F(\omega)$, Zener's argument is not sufficient. However, a large number of calculations (Grimvall and Ebbsjö 1975) suggest that for a wide class of interatomic potentials there *is* a tendency for $\Theta_{bcc}(0)$ to be lower than $\Theta_{fcc}(0)$ and $\Theta_{hcp}(0)$ by several percent. A difference of this magnitude is enough to explain the observed allotropy.

Friedel (1974) related $\Theta_D(0)$ of the transition metals to the coordination number z of their crystal structures. With a tight-binding nearest-neighbour interaction, taken to be independent of the crystal structure,

$$\Theta_{\mathrm{bcc}}(0)/\Theta_{\mathrm{fcc}}(0) = \Theta_{\mathrm{bcc}}(0)/\Theta_{\mathrm{hcp}}(0) = \sqrt{(8/12)} \approx 0.82. \qquad (4.156)$$

This is too low a ratio. Moraitis and Gautier (1977b) extended Friedel's model to include also next-nearest neighbours, but $\Theta_{\mathrm{bcc}}(0)/\Theta_{\mathrm{fcc}}(0)$ is still not satisfactorily accounted for.

A relative volume change by $\Delta V/V$ gives rise to a relative change $\gamma_G(0)(\Delta V/V)$ in $\Theta_D(0)$. The Grüneisen parameter typically is $\gamma(0) = 1.5 \pm 1$. If the bcc phase has an atomic volume Ω_a which is significantly higher than that of the fcc or hcp phases, the allotropy of the elemental metals might be due to a lowering of $\Theta_D(0)$ associated with the lattice expansion. However, such an argument fails because $\Delta V/V$ is usually less than 1%.

In amorphous materials, $\Theta_D(-3)$ is typically lower by 15–30% compared with the crystalline state. This is well established, both experimentally (e.g., Golding et al. (1977) for $Pd_{0.775}Si_{0.165}Cu_{0.06}$, Mizutani and Massalski (1980) for $Pd_{0.80}Si_{0.20}$, Suck et al. (1981) for $Mg_{0.70}Zn_{0.30}$) and theoretically (e.g., Hafner 1983). However, it is only the low-frequency part of $F(\omega)$ that is significantly different in the amorphous and the crystalline states. The Debye temperature $\Theta_D(2)$, which measures ω^2 averaged over the entire density of states $F(\omega)$, is essentially the same in both structures (e.g., Suck et al. 1980, 1981) in the case of "dense" amorphous materials, i.e. where the density is only slightly different from that of the crystalline state. The *electronic* states in strongly disordered systems may be localised. In contrast to this, most of the vibrational modes in glasses are not localised (Nagel et al. 1984).

We conclude that the Debye temperatures $\Theta_D(n)$ of a particular substance are only weakly dependent on the crystal structure, provided that the nature of the chemical binding is preserved. (This excludes, e.g., grey and white tin or graphite and diamond.) However, it may happen that $\Theta_D(-3)$, which only samples the elastic limit of the frequency spectrum, is altered by 10–15% or more as the crystal structure is changed. For instance, if a lattice is near an instability under shear, it will have a low $\Theta_D(-3)$. The reader is referred to our discussion in ch. 3 §9 on how the elastic properties vary with the crystal structure.

12. Other factors influencing the Debye temperature

The influence of lattice defects, alloying elements, etc. is considered in ch. 6. The volume dependence of $\Theta_D(n)$ is treated in ch. 10. The Debye temperature of non-magnetic materials is not affected by a magnetic field, not even in the case of strongly paramagnetic metals (Ikeda and Gschneidner 1980, Hsiang et al. 1981), although there are also cases where Θ_D *is* affected by a magnetic field, probably due to induced magnetic moments (e.g., in Sc and YCo_2), see Gschneidner and Ikeda (1983). The Debye temperature of ferromagnets is affected by external fields, in a complicated manner (see, for example, Wohlfarth 1974).

CHAPTER 5

PHONONS IN REAL CRYSTALS: ANHARMONIC EFFECTS

1. Introduction

In the previous chapter we discussed lattice vibrations under very ideal circumstances. The lattice was assumed to be free from defects such as vacancies, impurities, grain boundaries and surfaces. It was also assumed adequate to retain only the harmonic part of the expansion of the potential energy in the atomic displacements. Although these may often be excellent approximations, they completely leave out certain phenomena. For instance, perfectly harmonic vibrations give no thermal expansion. If the vibrations are harmonic and the lattice is also free from defects, the thermal conductivity is infinite.

As a starting point of our theoretical treatment we take the expansion of the total lattice energy in atomic displacements u_i and the corresponding momenta p_i (cf. eq. (C.1); i, j denote atoms and α, β denote Cartesian components)

$$\sum_i \frac{p_i^2}{2M_i} + \Phi_0 + (1/2) \sum_{i,\, j\, \alpha,\, \beta} \Phi_{\alpha\beta}(i, j) u_{i\alpha} u_{j\beta}. \tag{5.1}$$

Corrections to the simple theory of ch. 4 may arise because:

(i) Higher-order powers of u are kept in the expansion of the potential energy Φ.

(ii) The perfect periodicity is destroyed by the presence of impurity atoms, vacancies, grain boundaries, free surfaces, a high concentration of alloying atoms, etc.

Note that if, in the case of (ii), we only retain the quadratic terms in u, the vibrations are still harmonic. However, the solutions to the equations of motion are no longer plane waves characterised by the wave vector $\boldsymbol{q}$ (although that may be a good approximation for most of the vibrational modes). The anharmonic effects mentioned under (i) are present even in a lattice without any defects.

In this chapter we deal with (i), i.e. anharmonic effects. Vibrations in defect lattices, (ii), are considered in ch. 6. Original work covering the main points in this chapter is due, inter alia, to Leibfried and Ludwig (1961), Maradudin and Fein (1962), Cowley (1963) and Barron (1965). There are several reviews, e.g., by Cowley (1968), Wallace (1972) and Barron and Klein (1974).

2. Weakly perturbed harmonic vibrations

Let $\omega_0(\boldsymbol{q}, \lambda)$ be the frequency of the phonon mode $(\boldsymbol{q}, \lambda)$ in the harmonic approximation. In a real system which we assume does not deviate too far from the harmonic conditions the frequency ω_0 of a state $(\boldsymbol{q}, \lambda)$ is shifted to a complex value

$$\omega(\boldsymbol{q}, \lambda) = \omega_0(\boldsymbol{q}, \lambda) + \Delta(\boldsymbol{q}, \lambda) - \mathrm{i}\Gamma(\boldsymbol{q}, \lambda). \tag{5.2}$$

If there had been no imaginary part $-\mathrm{i}\Gamma(\boldsymbol{q}, \lambda)$, the perturbed state $(\boldsymbol{q}, \lambda)$ would have the exact energy eigenvalue $\omega_0(\boldsymbol{q}, \lambda) + \Delta(\boldsymbol{q}, \lambda)$, and a time dependence given by $\exp[-\mathrm{i}(\omega_0 + \Delta)t]$. The imaginary term adds a factor $\exp[-\Gamma t]$ to the time dependence. Hence the state $(\boldsymbol{q}, \lambda)$ decays, but if Γ is small ($\Gamma/\omega_0 \ll 1$), the lifetime of the state is long and it is still meaningful to label it by the $(\boldsymbol{q}, \lambda)$ of the unperturbed state. (The situation is similar to that encountered in atomic physics. The energy levels of the hydrogen atom are well described by the quantum mechanical form of Bohr's simple theory. In a more accurate description, however, there are corrections which shift the energy levels and give them a finite line width.) In this chapter we will assume that Γ is negligible. The most important effect of a finite Γ is to limit the phonon part of the thermal conductivity. This problem is dealt with in ch. 12.

We describe the perturbed state by a spectral function $A(\omega; \boldsymbol{q}, \lambda)$ defined as

$$A(\omega) = \frac{1}{\pi} \frac{4\omega_0^2 \Gamma}{[\omega_0^2 + 2\omega_0 \Delta - \omega^2]^2 + 4\omega_0^2 \Gamma^2}, \tag{5.3}$$

where the label $(\boldsymbol{q}, \lambda)$ on A, ω_0, Δ and Γ has been suppressed. (In a full many-body treatment, there should also be an ω-dependence in Δ and Γ, cf. Barron and Klein (1974).) If Γ is small compared to ω_0, $A(\omega)$ has the form of a sharp peak centered at

$$\begin{aligned} \omega = \omega(\boldsymbol{q}, \lambda) &= [\omega_0^2(\boldsymbol{q}, \lambda) + 2\omega_0(\boldsymbol{q}, \lambda)\Delta(\boldsymbol{q}, \lambda)]^{1/2} \\ &\approx \omega_0(\boldsymbol{q}, \lambda) + \Delta(\boldsymbol{q}, \lambda), \end{aligned} \tag{5.4}$$

where we now consider only $\omega > 0$. Its half-width (the full width at half the peak height) is

$$W = 2\Gamma(\boldsymbol{q}, \lambda). \tag{5.5}$$

When $\Gamma \to 0$, $A(\omega)$ becomes a delta function,

$$A(\omega) = \delta(\omega - \omega_0 - \Delta). \tag{5.6}$$

The quantities $\omega_0 + \Delta$ and Γ can be determined experimentally from the neutron scattering cross section, which is proportional to $A(\omega)[1 - \exp(-\hbar\omega/k_B T)]^{-1}$ (Cowley 1963). The expression for the cross section exemplifies how one should deal with the complex frequencies $\omega_0 + \Delta - i\Gamma$, through $A(\omega)$. Measurable quantities must be real numbers, so it is incorrect to just insert a complex ω in, for example, the Bose–Einstein factors. See, for instance, Pathak and Varshni (1969), who express the free energy as an integral containing $A(\omega)$.

3. Anharmonicity in the quasi-harmonic approximation

3.1. General considerations. Grüneisen parameters

The elements $D_{\kappa\kappa'}(\boldsymbol{q}, \lambda)$ of the dynamical matrix can be expressed in derivatives of an interatomic potential, evaluated at the equilibrium position $\boldsymbol{R}^0(\kappa l)$ of the atoms in the lattice,

$$D_{\kappa\kappa'}(\boldsymbol{q}, \lambda) = (M_\kappa M_{\kappa'})^{-1} \times \sum_{l'} \Phi_{\alpha\beta}(\kappa' l', \kappa 0) \exp[i\boldsymbol{q} \cdot [\boldsymbol{R}^0(\kappa' l') - \boldsymbol{R}^0(\kappa 0)]]. \tag{5.7}$$

If the lattice cell is strained, the positions $\boldsymbol{R}^0(\kappa l)$ change. The derivatives $\Phi_{\alpha\beta} = \partial^2\Phi/\partial x_\alpha \partial x_\beta$ are then to be evaluated at new positions $\boldsymbol{R}^0(\kappa l)$. This alters the elements of the dynamical matrix. Thus $D_{\kappa\kappa'}(\boldsymbol{q}, \lambda)$ depends on the size and the shape of the unit cell in the lattice. However, we still consider the lattice vibrations to be harmonic since only terms quadratic in the displacements of the atoms from their equilibrium positions have been kept in (5.1). This means that we are not quite consistent. If the lattice vibrations are completely harmonic, $\partial^2\Phi/\partial x_\alpha \partial x_\beta$ must be independent of the position $\boldsymbol{R}^0(\kappa l)$ where it is evaluated. We shall therefore refer to our approach as the quasi-harmonic model. It yields a frequency shift Δ, which we denote $\Delta_2(\boldsymbol{q}, \lambda)$, but has $\Gamma_2 = 0$. (Cowley (1963) and some other authors define the term quasi-harmonic to include Δ_3 and Δ_4; cf. § 4. We follow Leibfried and Ludwig (1961), and Barron and Klein (1974).)

The Grüneisen parameter $\gamma(\boldsymbol{q}, \lambda, \varepsilon_i)$ is a measure of how the phonon frequency $\omega(\boldsymbol{q}, \lambda)$ is altered under a small change in ε_i, where ε_i is a strain parameter specifying the geometry of the unit cell; $i = 1, 2, 3$ refer to the lengths a_i of the cell and $i = 4, 5, 6$ refer to the angles between a_i (ch. 3 § 2), and then

$$\gamma(\boldsymbol{q}, \lambda, \varepsilon_i) = -\frac{1}{\omega(\boldsymbol{q}, \lambda)}\left(\frac{\partial \omega(\boldsymbol{q}, \lambda; \varepsilon_i)}{\partial \varepsilon_i}\right)_{\varepsilon_i'}. \tag{5.8}$$

The derivative is taken with all $\varepsilon_i' \neq \varepsilon_i$ kept constant. Usually γ is evaluated at the reference state of zero external tension or shear. If (5.8), for $i = 1, 2, 3$, is expressed as

$$\left(\frac{\Delta \omega(\boldsymbol{q}, \lambda)}{\omega(\boldsymbol{q}, \lambda)}\right) = -\gamma(\boldsymbol{q}, \lambda; \varepsilon_i)\Delta \varepsilon_i = -\gamma(\boldsymbol{q}, \lambda; \varepsilon_i)\left(\frac{\Delta a_i}{a_i}\right), \tag{5.9}$$

it is clear that $\gamma(\boldsymbol{q}, \lambda)$ can be written $-(\partial \ln \omega / \partial \ln a_i)$ and measures the relative change in $\omega(\boldsymbol{q}, \lambda)$ which is associated with a relative change $\Delta a_i / a_i$ in the dimension of the unit cell.

An important case is that of a uniform expansion or contraction of the unit cell, so that the deformation is described by the cell (or crystal) volume alone. Then we can write

$$\gamma(\boldsymbol{q}, \lambda, V) = -\frac{V}{\omega(\boldsymbol{q}, \lambda)}\left(\frac{\partial \omega(\boldsymbol{q}, \lambda; V)}{\partial V}\right) = -\left(\frac{\partial \ln \omega(\boldsymbol{q}, \lambda)}{\partial \ln V}\right). \tag{5.10}$$

This expression will be the corner stone in the simplest theory of thermal expansion, ch. 10 § 7.

In a lattice with cubic symmetry, $a_1 = a_2 = a_3 = a$ and $V \sim a^3$. Further, $\partial \ln a_i / \partial \ln V = 1/3$ if $i = 1, 2, 3$. An infinitesimal shear does not change the volume, so $\mathrm{d}V/\mathrm{d}\varepsilon_i = 0$ if $i = 4, 5, 6$. Hence

$$\begin{aligned}\gamma(\boldsymbol{q}, \lambda; V) &= -\left(\frac{\partial \ln \omega(\boldsymbol{q}, \lambda)}{\partial \ln V}\right) = -\sum_{i=1}^{3}\left(\frac{\partial \ln \omega(\boldsymbol{q}, \lambda)}{\partial \ln a_i}\right)\left(\frac{\partial \ln a_i}{\partial \ln V}\right) \\ &= \gamma(\boldsymbol{q}, \lambda; a).\end{aligned} \tag{5.11}$$

The conventional notation for the lattice parameters in hexagonal lattices is $a_1 = a_2 = a$ and $a_3 = c$. Then

$$\gamma(\boldsymbol{q},\lambda;V) = -2\left(\frac{\partial \ln \omega(\boldsymbol{q},\lambda;a_1)}{\partial \ln a_1}\right)\left(\frac{\mathrm{d}\ln a_1}{\mathrm{d}\ln V}\right) - \left(\frac{\partial \ln \omega(\boldsymbol{q},\lambda;c)}{\partial \ln c}\right)\left(\frac{\mathrm{d}\ln c}{\mathrm{d}\ln V}\right)$$

$$= (1/3)[2\gamma(\boldsymbol{q},\lambda;a_1)+\gamma(\boldsymbol{q},\lambda;c)]. \tag{5.12}$$

Here we have to be careful in the notation. In (5.12), we have written a_1 to denote that we only change the strain in *one* direction, perpendicular to the c-axis. It is now natural to introduce Grüneisen parameters $\gamma_{\|}$ and $\gamma_{\perp}$, such that $\gamma_{\|} = \gamma_{\perp}$ in the special case of cubic symmetry. The obvious choice is

$$\gamma_{\|}(\boldsymbol{q},\lambda) = \gamma(\boldsymbol{q},\lambda;c) = -\left(\frac{\partial \ln \omega}{\partial \ln c}\right)_a, \tag{5.13}$$

$$\gamma_{\perp}(\boldsymbol{q},\lambda) = \gamma(\boldsymbol{q},\lambda;a_1) = -\frac{1}{2}\left(\frac{\partial \ln \omega}{\partial \ln a}\right)_c. \tag{5.14}$$

In the last derivative, it is the cell dimension a that is varied, and this gives rise to the prefactor 1/2.

Usually, the Grüneisen parameters are positive and lie in the range 1.5 ± 1. Negative Grüneisen parameters sometimes occur for low-lying frequencies (long-wavelength transverse modes) in open lattice structures like those of Ge, Si and alkali halides.

Example: Grüneisen parameter for varying c/a in hcp lattices. Some hcp structures, e.g. Cd and Zn, have a c/a ratio which deviates strongly from the "ideal" value 1.63, but the atomic volume is not abnormal (cf. the example on p. 6). It is therefore of interest to consider how the phonon frequencies vary with c/a at fixed volume V. One has

$$\gamma(\boldsymbol{q},\lambda;c/a) = -\left(\frac{\partial \ln \omega(\boldsymbol{q},\lambda)}{\partial \ln (c/a)}\right)_V$$

$$= -\left(\frac{\partial \ln \omega(\boldsymbol{q},\lambda)}{\partial \ln a}\right)_c\left(\frac{\partial \ln a}{\partial \ln (c/a)}\right)_V - \left(\frac{\partial \ln \omega(\boldsymbol{q},\lambda)}{\partial \ln c}\right)_a\left(\frac{\partial \ln c}{\partial \ln (c/a)}\right)_V$$

$$= [2\gamma_{\perp}(\boldsymbol{q},\lambda)-\gamma_{\|}(\boldsymbol{q},\lambda)]/3. \tag{5.15}$$

Here we have used the fact that $ca^2 \sim V = \text{constant}$, which yields $c/a = (\text{constant})/a^3$ and hence $[\partial \ln(c/a)/\partial \ln a]_V = -3$. Similarly, $[\partial \ln(c/a)/\partial \ln c]_V = 3/2$. In zinc, $\gamma_\perp = 2.50$ and $\gamma_\parallel = 1.28$ (fig. 5.1, $n = 1$) when all modes are equally weighted. Then, $\gamma(c/a) = 1.24$.

3.2. Grüneisen parameters for frequency moments and Debye temperatures

Grüneisen parameters can also be defined for the frequency moments $\omega(n)$ or the corresponding Debye temperatures $\Theta_D(n)$. For instance,

$$\gamma(n; V) = -\left(\frac{\partial \ln \Theta_D(n)}{\partial \ln V}\right) = -\left(\frac{\partial \ln \omega(n)}{\partial \ln V}\right). \tag{5.16}$$

In the strict Debye model, $\omega_D = Cq_D = k_B\Theta_D/\hbar$, where C is the sound velocity and $q_D^3 = 6\pi^2 N/V$. Then

$$\gamma(\Theta_D; V) = -\left(\frac{\partial \ln \Theta_D}{\partial \ln V}\right) = -\left(\frac{\partial \ln C}{\partial \ln V}\right) + \frac{1}{3}. \tag{5.17}$$

Only if all $\gamma(\boldsymbol{q}, \lambda; V)$ are equal, is $\gamma(n; V)$ independent of n. Often, $\gamma(\boldsymbol{q}, \lambda; V)$ of different modes differ by a factor of two. The corresponding variation of $\gamma(n; V)$ is shown in fig. 5.1. It is not unusual

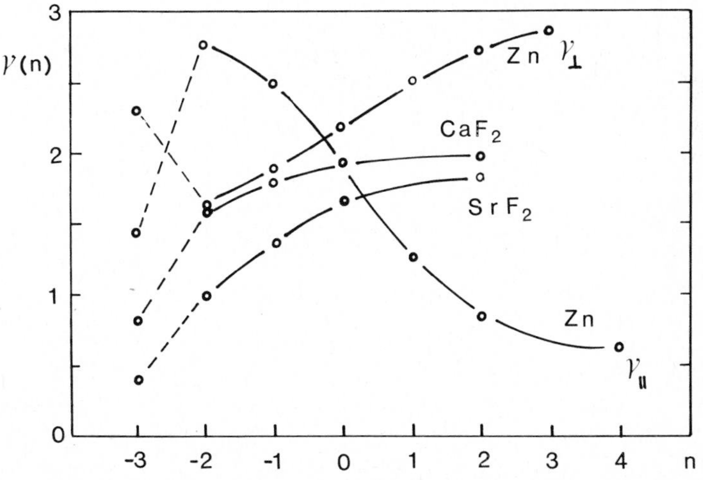

Fig. 5.1. The Grüneisen parameter $\gamma(n)$ as a function of n, for CaF_2 and SrF_2 (Bailey and Yates 1967) and for Zn (Barron and Munn 1967). Since the Zn lattice is hexagonal, there are two Grüneisen parameters, $\gamma_\perp$ and $\gamma_\parallel$.

to approximate $\gamma(n)$ by the thermodynamic Grüneisen parameter γ_G; see ch. 10. This may be too crude an approximation. For instance, in the very anisotropic graphite, $\gamma(n)$ varies very strongly with n (Bailey and Yates 1970).

If $\gamma(n;V)$ depends only weakly on the volume, we can integrate eq. (5.16) and obtain

$$\frac{\Theta_D(n;V)}{\Theta_D(n;V_0)} = \left(\frac{V_0}{V}\right)^{\gamma(n)}. \tag{5.18}$$

In hexagonal lattice structures,

$$\frac{\Theta_D(n;a,c)}{\Theta_D(n;a_0,c_0)} = \left(\frac{a_0}{a}\right)^{2\gamma_\perp}\left(\frac{c_0}{c}\right)^{\gamma_\parallel}. \tag{5.19}$$

Example: The Grüneisen parameter in the Bohm–Staver model. The Bohm–Staver model (Bohm and Staver 1951) gives the (longitudinal) sound velocity in free-electron-like metals as $C_L = v_F(mZ/3M)^{1/2}$, where v_F is the Fermi velocity. With $\omega_D = C_L q_D$ and (5.17) one finds

$$\gamma(\Theta_D;V) = -\left(\frac{\mathrm{d}\ln k_F}{\mathrm{d}\ln V}\right) - \left(\frac{\mathrm{d}\ln q_D}{\mathrm{d}\ln V}\right) = 1/3 + 1/3 = 2/3. \tag{5.20}$$

Within this model, $\gamma(V)$ has the same value, 2/3, for all free-electron-like metals. It is about half of the observed values.

Example: Slater's form of the Grüneisen parameter $\gamma(-3;V)$. Slater (1940) derived an expression for the Grüneisen parameter, essentially as follows. Expand the volume change $V-V_0$, due to an external pressure p, in powers of p and keep only the first two terms;

$$V - V_0 = V_0(a_1 p + a_2 p^2). \tag{5.21}$$

The average sound velocity C to be used in $\Theta_D(-3)$ is given by $3C^{-3} = C_L^{-3} + 2C_T^{-3}$, (4.15). If we neglect the volume dependence of the Poisson ratio and use (3.66) and (3.67) for C_L and C_T expressed in the elastic parameters, then $C \sim (KV)^{1/2}$, where K is the bulk modulus, $K^{-1} = -(1/V)(\partial V/\partial p)_T = -(V_0/V)[a_1 + 2a_2(V-V_0)/a_1 V_0]$. Then, as in (5.17), one obtains Slater's expression

$$\begin{aligned}\gamma_S(-3;V) &= -(1/2)(\mathrm{d}\ln K/\mathrm{d}\ln V) - 1/6 = a_2/(a_1)^2 - 2/3 \\ &= (1/2)(\mathrm{d}K/\mathrm{d}p) - 1/6.\end{aligned} \tag{5.22}$$

The coefficients a_1 and a_2, which yield $\mathrm{d}\ln K/\mathrm{d}\ln V$, can be measured (Gschneidner 1964). At the time of Slater's original derivation, it was not known to what extent $\gamma(\boldsymbol{q}, \lambda)$ varies with the phonon state $(\boldsymbol{q}, \lambda)$ and no distinction was made between $\gamma(n; V)$ for different n. Here we have stated explicitly that Slater's expression is an estimation of $\gamma(-3; V)$. Thus it is not equal to the Grüneisen parameter $\gamma(0; V)$ ($\approx \gamma_G$) which is approximately obtained from the expansion coefficient at moderate and high temperatures (ch. 10 § 7.2).

Example: Internal pressure from zero-point vibrations. Suppose that we have calculated the atomic volume Ω_a from a model which considers only the static lattice. We now estimate how much Ω_a is changed due to the zero-point ($T = 0$) lattice vibrations. The pressure is related to the energy by $p = -(\partial E/\partial V)_S$. If we add to E the zero-point energy $(3/2)\hbar\omega(1)$ per atom, (4.49), the pressure is changed by

$$p_Z = -(V/\Omega_a)(3\hbar/2)[\partial\omega(1)/\partial V]. \tag{5.23}$$

V/Ω_a is the number of atoms in the solid. A pressure p_Z gives rise to a relative change in the atomic volume;

$$\frac{\Delta\Omega_a}{\Omega_a} = -\frac{p_Z}{K_T} = \frac{3\hbar\omega(1)}{2K_T\Omega_a}\gamma(1). \tag{5.24}$$

Consider a lithium crystal of isotopes ^{6}Li or ^{7}Li. The interatomic forces are the same in the two crystals. Due to the mass difference, $\omega(1)$ differs by a factor of $\sqrt{(6/7)}$ while $\gamma(1)$, which measures a relative shift in $\omega(1)$, is not mass dependent. It follows from (5.24) that the lattice parameter a of ^{6}Li is larger than that of ^{7}Li. With $K = 1.1 \times 10^{10}\,\mathrm{N/m^2}$, $\Omega_a = 2.1 \times 10^{-29}\,\mathrm{m^3}$, $\gamma(1) = 1$ and $\hbar\omega(^7\mathrm{Li})/k_B \approx \Theta_D \approx 400\,\mathrm{K}$, we get $\Delta a/a = 0.001$. At high temperatures, we should replace $\hbar\omega(1)\gamma(1)$ in (5.24) by $2k_BT\gamma(0)$. See Johansson and Rosengren (1975) for further aspects of the volume difference between ^{7}Li and ^{6}Li.

As another example (Froyen and Cohen 1984), the bulk modulus K calculated in a static lattice at $T = 0$ differs from K calculated in a vibrating lattice by the relative amount $(p_Z/K)(\partial K/\partial p)$. For NaCl, $\Delta\Omega_a/\Omega_a = 0.02$ and $(\partial K/\partial p) = 5$. Hence, K in the vibrating lattice is lower by 10%.

3.3. *Grüneisen parameter expressed in the dynamical matrix*

Since $\langle\omega^2\rangle$ is directly related to the trace of the dynamical matrix, it is possible to obtain $\gamma(2;V)$ without the explicit evaluation of all the phonon frequencies. From (C.12),

$$\gamma(2;V) = -\frac{1}{2}\frac{\mathrm{d}\ln\omega}{\mathrm{d}\ln V}\left(\frac{1}{3N}\sum_{\boldsymbol{q}}\mathrm{Tr}\,D(\boldsymbol{q})\right). \tag{5.25}$$

When point-ion interactions are present, (5.25) can be further simplified since Laplace's equation in electrostatics implies that the direct Coulomb interaction vanishes when the volume derivative is taken.

4. Explicit anharmonicity

4.1. *Introduction*

The quasi-harmonic frequency shift Δ_2 vanishes if the crystal volume, or rather all ε_i, remain constant. However, there are also what we shall refer to as explicit anharmonic effects. They arise from the higher-than-quadratic terms when the potential energy is expanded in powers of displacements u from the equilibrium positions of the atoms, (5.1). Due to the zero-point motion they are present also at $T = 0$. To see the structure of the associated frequency shifts, it is illuminating first to study an anharmonic one-dimensional harmonic oscillator. Let the mass M move in a potential

$$V(x) = (1/2)M\omega^2x^2 + V_3x^3 + V_4x^4, \tag{5.26}$$

where V_3 and V_4 are in some sense (decreasingly) small. When $V_3 = V_4 = 0$, the eigenvalues are $E_n = \hbar\omega(n+1/2)$. Quantum mechanical perturbation theory, applied to the state n' and $T = 0$, gives energy shifts $\Delta E_{n'}$ which, in conventional notation, can be written

$$\Delta E_{n'} = \langle n'|V_3x^3 + V_4x^4|n'\rangle - \sum_{n\neq n'}\frac{|\langle n|V_3x^3 + V_4x^4|n'\rangle|^2}{E_n - E_{n'}}. \tag{5.27}$$

Since the integrand is an odd function of x, $\langle n'|V_3x^3|n'\rangle = 0$. In the last term of (5.27) we only keep the lowest order part of the denominator, i.e.

$|\langle n|V_3x^3|n'\rangle|^2$. Thus, the term V_4x^4 contributes to the first order, while V_3x^3 contributes to the second order in the perturbation expansion. It is necessary to keep both these terms, even though V_4x^4 was assumed to be much smaller than V_3x^3 in the relevant range of x-values. We shall use the index 2 for the quasi-harmonic shift (because only terms quadratic in the atomic displacements are kept), while frequency shifts originating from u^3 and u^4 terms are denoted Δ_3 and Δ_4. Then, neglecting damping of the phonons,

$$\omega(\boldsymbol{q}, \lambda) = \omega_0(\boldsymbol{q}, \lambda) + \Delta_2(\boldsymbol{q}, \lambda) + \Delta_3(\boldsymbol{q}, \lambda) + \Delta_4(\boldsymbol{q}, \lambda). \tag{5.28}$$

4.2. Temperature dependence of anharmonic frequency shifts

The shifts $\Delta_2(\boldsymbol{q}, \lambda)$, $\Delta_3(\boldsymbol{q}, \lambda)$ and $\Delta_4(\boldsymbol{q}, \lambda)$ can be written in condensed form as (e.g., Maradudin and Fein 1962, Cowley 1963, 1968, 1970, Cowley and Cowley 1965)

$$\Delta_2(\boldsymbol{q}, \lambda) = (2/\hbar) \sum_{\alpha} V_\alpha(\boldsymbol{q}\lambda, -\boldsymbol{q}\lambda)\varepsilon_\alpha, \tag{5.29}$$

$$\Delta_3(\boldsymbol{q}, \lambda) = -(18/\hbar^2) \sum_{\boldsymbol{q}_1\lambda_1, \boldsymbol{q}_2\lambda_2} |V(\boldsymbol{q}\lambda, \boldsymbol{q}_1\lambda_1, \boldsymbol{q}_2\lambda_2)|^2 R(0, 1, 2), \tag{5.30}$$

$$\Delta_4(\boldsymbol{q}, \lambda) = (12/\hbar) \sum_{\boldsymbol{q}_1\lambda_1} V(\boldsymbol{q}\lambda, -\boldsymbol{q}\lambda, \boldsymbol{q}_1\lambda_1, -\boldsymbol{q}_1\lambda_1)[2n(1)+1], \tag{5.31}$$

with

$$\begin{aligned} R(0,1,2) = {} & [n(1)+n(2)+1]\left(\frac{1}{\omega_0+\omega_1+\omega_2} - \frac{1}{\omega_0-\omega_1-\omega_2}\right) \\ & + [n(1)-n(2)]\left(\frac{1}{\omega_0-\omega_1+\omega_2} - \frac{1}{\omega_0+\omega_1-\omega_2}\right). \end{aligned} \tag{5.32}$$

In the Bose–Einstein function $n(i)$, the interaction function $V(i, \ldots)$, the quantities $\Delta(i)$ and ω_i, and in the summations, the index $i = 0, 1, 2$ is short for $(\boldsymbol{q}, \lambda)$, $(\boldsymbol{q}_1, \lambda_1)$ and $(\boldsymbol{q}_2, \lambda_2)$ respectively. In Δ_2, ε_α is a thermal strain, to be calculated as in ch. 10. $V(1)_\alpha$, $V(1, 2)$ and $V(1, 2, 3)$ are short for the Fourier transforms of the interatomic potential. The principal value should be taken in the sum over the singular terms in (5.32).

The temperature dependence of Δ_3 and Δ_4 comes from the Bose–

Einstein factors. At high temperatures, they are linear in T. When $T = 0$, the Bose–Einstein factors n are zero but Δ_3 and Δ_4 are non-vanishing due to the zero-point motion. One can show that Δ_3 is always negative. Δ_4 may have any sign but often cancels much of Δ_3. Typically, $\Delta_3 + \Delta_4$ amounts to several percent of ω_0 near $T = T_m$ (the melting temperature).

Measurements (by neutron scattering) and theoretical calculations of Δ are rather uncertain, sometimes by as much as a factor of two. See Cowley and Cowley (1965) for a calculation in alkali halides. Work by Buyers and Cowley (1969), Copley (1973) and Vaks et al. (1980) exemplify pseudopotential calculations in simple metals. Molecular dynamics (Glyde et al. 1977) offers a way to handle strong anharmonicity.

4.3. Temperature dependence of the frequency shifts in a simple model

The summations in (5.30) and (5.31) cannot be carried out in closed form and one is left with a numerical calculation. However, a simple expression results if we replace ω_2 and ω_3 in the Bose–Einstein factors by the same frequency, ω_E. If we also take an Einstein representation of the strain ε_α (cf. (10.66)), including that due to zero-point vibrations, then

$$\Delta_2(\boldsymbol{q}, \lambda) = k_2(\boldsymbol{q}, \lambda)\omega_0(\boldsymbol{q}, \lambda)\left(\frac{1}{\exp(\hbar\omega_E/k_B T) - 1} + \frac{1}{2}\right), \tag{5.33}$$

$$\Delta_3(\boldsymbol{q}, \lambda) = k_3(\boldsymbol{q}, \lambda)\omega_0(\boldsymbol{q}, \lambda)\left(\frac{1}{\exp(\hbar\omega_E/k_B T) - 1} + \frac{1}{2}\right), \tag{5.34}$$

$$\Delta_4(\boldsymbol{q}, \lambda) = k_4(\boldsymbol{q}, \lambda)\omega_0(\boldsymbol{q}, \lambda)\left(\frac{1}{\exp(\hbar\omega_E/k_B T) - 1} + \frac{1}{2}\right). \tag{5.35}$$

Although the temperature dependence of Δ_2, Δ_3 and Δ_4 is the same as in an Einstein model for the thermal energy, our description is not that of an ordinary Einstein model, since we allow the (dimensionless) constants $k_i(\boldsymbol{q}, \lambda)$ (i = 1, 2, 3) to vary with the mode $(\boldsymbol{q}, \lambda)$.

5. Thermodynamic functions in anharmonic systems

5.1. Introduction

In ch. 4 §§ 6,7 we gave expressions for the energy $E(\boldsymbol{q}, \lambda)$, the entropy $S(\boldsymbol{q}, \lambda)$, the free energy $F(\boldsymbol{q}, \lambda)$ and the heat capacity $C_{\text{har}}(\boldsymbol{q}, \lambda)$ of strictly harmonic systems. In this chapter we have introduced approximate phonon frequencies $\omega_0(\boldsymbol{q}, \lambda)+\Delta_2(\boldsymbol{q}, \lambda)+\Delta_3(\boldsymbol{q}, \lambda)+\Delta_4(\boldsymbol{q}, \lambda)$ of anharmonic systems. A very important question is whether E, S, F and C can be obtained for the anharmonic case just by replacing ω_0 by $\omega_0+\Delta_2+\Delta_3+\Delta_4$ in the usual expressions for E etc. of harmonic vibrations. One can show (Barron 1965) that the entropy is correctly given in this way, but not the energy, the free energy and the heat capacity.

5.2. The quasi-harmonic model

We write

$$\omega(\boldsymbol{q}, \lambda) = \omega_0(\boldsymbol{q}, \lambda)+\Delta_2(\boldsymbol{q}, \lambda; V(T)) \tag{5.36}$$

to indicate that Δ_2 may vary implicitly with the temperature through the thermal expansion (or contraction) of the solid. The entropy is, (4.44),

$$S(\boldsymbol{q}, \lambda) = k_{\text{B}}\{(x/2)\coth(x/2)-\ln[2\sinh(x/2)]\}, \tag{5.37}$$

where $x = \hbar\omega(\boldsymbol{q}, \lambda)/k_{\text{B}}T$. This yields (suppressing the label $(\boldsymbol{q}, \lambda)$ on S)

$$\begin{aligned}\left(\frac{\partial S}{\partial V}\right)_T &= \left(\frac{\partial S}{\partial x}\right)_T\left(\frac{\partial x}{\partial V}\right)_T = \left(\frac{\partial S}{\partial T}\right)_V\left(\frac{\partial T}{\partial x}\right)_V\left(\frac{\partial x}{\partial V}\right)_T = -\left(\frac{C_V}{\omega}\right)\left(\frac{\partial \Delta_2}{\partial V}\right)_T \\ &= -\left(\frac{C_V}{V}\right)\left(\frac{\partial \ln(\omega_0+\Delta_2)}{\partial \ln V}\right)_T = \frac{C_V}{V}\gamma(\boldsymbol{q}, \lambda; V).\end{aligned} \tag{5.38}$$

Thus,

$$\gamma(\boldsymbol{q}, \lambda; V) = \left(\frac{V}{C_V}\right)\left(\frac{\partial S}{\partial V}\right)_T. \tag{5.39}$$

We can generalise (5.39) to arbitrary deformations of the

crystallographic unit cell and write, for strains ε_i ($i = 1$ to 6)

$$\gamma(\boldsymbol{q}, \lambda; \varepsilon_i) = \frac{1}{C_\varepsilon}\left(\frac{\partial S}{\partial \varepsilon_i}\right)_{T, \varepsilon_i'}, \tag{5.40}$$

where C_ε is the heat capacity taken at constant strain.

The heat capacity at constant pressure, C_p, is

$$\begin{aligned} C_p = T\left(\frac{\mathrm{d}S}{\mathrm{d}T}\right)_p &= T\left(\frac{\partial S}{\partial T}\right)_V + T\left(\frac{\partial S}{\partial V}\right)_T\left(\frac{\partial V}{\partial T}\right)_p \\ &= C_V + T\left(\frac{C_V}{V}\right)\gamma(V)\left(\frac{\partial V}{\partial T}\right)_p, \end{aligned} \tag{5.41}$$

i.e.

$$C_p(\boldsymbol{q}, \lambda) = C_V(\boldsymbol{q}, \lambda)[1 + \beta\gamma(\boldsymbol{q}, \lambda; V)T], \tag{5.42}$$

where β is the cubic expansion coefficient.

5.3. Third- and fourth-order anharmonicity

Let ΔF_3 and ΔF_4 be the perturbation correction to the Helmholtz free energy F. With the notation from § 4.2, one has (Ludwig 1958, Maradudin et al. 1961, Cowley 1963, 1968, Barron and Klein 1974)

$$\Delta F_3 = -(6/\hbar) \sum_{\boldsymbol{q}\lambda,\, \boldsymbol{q}_1\lambda_1,\, \boldsymbol{q}_2\lambda_2} |V(\boldsymbol{q}\lambda, \boldsymbol{q}_1\lambda_1, \boldsymbol{q}_2\lambda_2)|^2 R_F, \tag{5.43}$$

$$\Delta F_4 = 3 \sum_{\boldsymbol{q}\lambda,\, \boldsymbol{q}_1\lambda_1} V(\boldsymbol{q}\lambda, -\boldsymbol{q}\lambda, \boldsymbol{q}_1\lambda_1, -\boldsymbol{q}_1\lambda_1)[2n(0)+1][2n(1)+1], \tag{5.44}$$

where

$$\begin{aligned} R_F = {} & \frac{[n(0)+1][n(1)+n(2)+1]+n(1)n(2)}{\omega(0)+\omega(1)+\omega(2)} \\ & + 3\,\frac{n(0)n(1)+n(0)n(2)-n(1)n(2)+n(0)}{\omega(1)+\omega(2)-\omega(0)}. \end{aligned} \tag{5.45}$$

At high temperatures ($T > \Theta_\mathrm{D}$), ΔF_3 and ΔF_4 both vary as T^2. (The term linear in T from the last $n(0)$ in (5.45) cancels against other terms linear in T.) The high-temperature forms are

$$\Delta F_3 = -[24(k_B T/\hbar)^2/\hbar] \sum_{q\lambda,\, q_1\lambda_1,\, q_2\lambda_2} \frac{|V(q\lambda, q_1\lambda_1, q_2\lambda_2)|^2}{\omega(q,\lambda)\omega(q_1,\lambda_1)\omega(q_2,\lambda_2)} \tag{5.46}$$

and

$$\Delta F_4 = 12(k_B T/\hbar)^2 \sum_{q\lambda,\, q_1\lambda_1} \frac{V(q,\lambda, -q\lambda, q_1\lambda_1, -q_1\lambda_1)}{\omega(q,\lambda)\omega(q_1,\lambda_1)}. \tag{5.47}$$

The next-order contributions to F, which are proportional to T^3 at high temperatures, have been written down by Shukla and Cowley (1971) and evaluated numerically for simple models by Shukla and Wilk (1974).

5.4. Thermodynamic functions related to frequency shifts

We may express ΔF and ΔS in the shifts Δ as (Cowley and Cowley 1966)

$$\Delta F = (\hbar/2) \sum_{q\lambda} [n(q,\lambda) + 1/2][2\Delta_2(q,\lambda) + \Delta_3(q,\lambda) + \Delta_4(q,\lambda)], \tag{5.48}$$

$$\Delta S = -\hbar \sum_{q\lambda} \left(\frac{\partial n(q,\lambda)}{\partial T} \right) [\Delta_2(q,\lambda) + \Delta_3(q,\lambda) + \Delta_4(q,\lambda)]. \tag{5.49}$$

It is now obvious that S is correctly obtained when the harmonic frequencies $\omega_0(q,\lambda)$ are replaced by $\omega_0 + \Delta_2 + \Delta_3 + \Delta_4$. The heat capacity is given by $T(\mathrm{d}S/\mathrm{d}T)$. Thus

$$\Delta C_p = -\hbar T \sum_{q\lambda} \{[\partial^2 n(q,\lambda)/\partial T^2][\Delta_2(q,\lambda) + \Delta_3(q,\lambda) + \Delta_4(q,\lambda)]$$
$$+ [\partial n(q,\lambda)/\partial T][\mathrm{d}\Delta_2(q,\lambda)/\mathrm{d}T + \mathrm{d}\Delta_3(q,\lambda)/\mathrm{d}T + \mathrm{d}\Delta_4(q,\lambda)/\mathrm{d}T]\}. \tag{5.50}$$

At constant volume, $\mathrm{d}\Delta_2/\mathrm{d}T = 0$. Thus C_V is given by

$$\Delta C_V = \Delta C_p + \hbar T \sum_{q\lambda} [\partial n(q,\lambda)/\partial T][\mathrm{d}\Delta_2(q,\lambda)/\mathrm{d}T]. \tag{5.51}$$

It remains to show that this is equivalent to the well-known thermodynamic relation (10.30)

$$C_p - C_V = (C_{p,\mathrm{har}} + \Delta C_p) - (C_{V,\mathrm{har}} + \Delta C_V)$$
$$= \Delta C_p - \Delta C_V = \beta \gamma_G C_V T. \tag{5.52}$$

Consider a single mode in (5.52) and neglect the volume dependence of Δ_3 and Δ_4. Let $\gamma_G = -(\partial \ln \omega / \partial \ln V)_p = -(V/\omega)(\partial \Delta_2/\partial V)_T$. Further, $C_V = \partial E/\partial T = \partial[\hbar\omega(n+1/2)/\partial T]_V = \hbar\omega(\partial n/\partial T)$ and $\mathrm{d}\Delta_2/\mathrm{d}T = (\partial \Delta_2/\partial V)_T(\partial V/\partial T)_p = (\partial \Delta_2/\partial V)_T \beta V$. Then $-\hbar T(\partial n/\partial T)(\mathrm{d}\Delta_2/\mathrm{d}T) = (-\hbar T)(C_V/\hbar\omega)(-\omega/V)\gamma_G \beta V = \beta\gamma_G C_V T$, as desired.

5.5. *Self-consistent phonons in strongly anharmonic systems*

As an illustration, consider a particle with mass M, moving in the potential $V_4(x) = a_4 x^4$. In a classical description, the particle would move back and forth with a certain frequency. Also a quantum mechanical description is reminiscent of the motion of a harmonic oscillator. However, $V_4(x)$ is so different from a harmonic potential $V_2(x) = a_2 x^2$ that perturbation theory starting from $V_2(x)$ is inadequate. The self-consistent phonon method allows us to go beyond low-order perturbation theory and make contact between the descriptions of motion in the potentials $V_2(x)$ and $V_4(x)$. Briefly, we seek a harmonic potential which reproduces the thermodynamic properties of the true potential as closely as possible. Let $F_0(\Omega)$ be the free energy of a harmonic oscillator with the characteristic frequency Ω (i.e. $V(x) = M\Omega^2 x^2/2$). This is adjusted at each temperature so that it approximates well the free energy F of the true (i.e. strongly anharmonic) system. This can be done systematically with the help of Bogoliubov's inequality. One can show (e.g. Wallace 1972) that although we vary Ω to minimise the free energy, it is the entropy $S_0(\Omega)$ which is best approximated. The theoretical basis of the self-consistent phonon method is outlined by Werthamer (1970).

Example: The heat capacity in strongly anharmonic systems. Hui and Allen (1975) and Söderkvist and Grimvall (1985) used the self-consistent phonon method to study the thermodynamic properties of a particle in a strongly anharmonic one-dimensional potential. Figure 5.2 exemplifies such a calculation. Note that although $C(T)$ appears to have the same shape as for harmonic vibrations, C/k_B at high temperatures is only about half of the classical Dulong–Petit value.

6. Factors influencing the Grüneisen parameter

Point defects, dislocations, grain boundaries, etc. have a small effect on

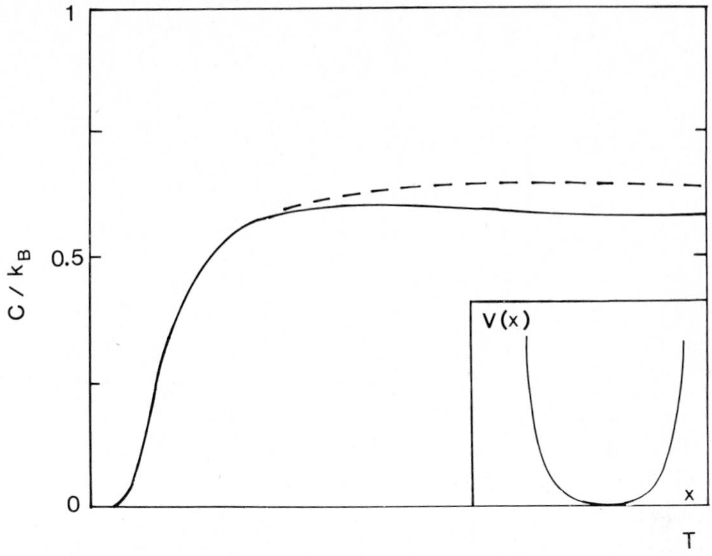

Fig. 5.2. The "exact" heat capacity (solid line) for a particle in an anharmonic potential (potential form shown in the inset), and a calculation using the self-consistent phonon method (dashed line). After Söderkvist and Grimvall (1985).

the Debye temperatures (ch. 4 § 11). In addition, the Grüneisen parameters are also rather insensitive to such defects. In concentrated alloys, one expects that the thermodynamic Grüneisen parameter γ_G, like Θ_D, varies smoothly with the composition (cf. ch. 6 § 8). This has also been found in experiments on solid solutions of Zr, Nb and Mo at 300 K (Smith and Finlayson 1976), where γ_G lies in the range 1.5 ± 0.5. Nagel et al. (1984) noted that although the Grüneisen parameter in glassy materials is affected by phonon localisation, so few modes are localised that the overall Grüneisen parameter is within 10% of its value in the crystalline state.

Moriarty et al. (1984) found, in theoretical calculations for Al, that γ_G decreases smoothly from about 2.0 to about 1.0 when the material is compressed to half its original volume. Analysis of C_p data gives $(\mathrm{d}\ln\gamma_G/\mathrm{d}\ln V) = 3.2 \pm 0.8$ for Ge (Leadbetter and Settatree 1969), 2.1 ± 0.3 for Pb (Leadbetter 1968), 1.4 ± 0.5 for KCl, 1.1 ± 0.4 for NaCl and 0.9 ± 0.6 for KBr (Leadbetter et al. 1969). For NaCl, see also Boehler et al. (1977). In ch. 3 § 8, we showed that the third-order elastic constants were related to the elastic-limit Grüneisen parameter. The fourth-order coefficients, c_{ijkl} are related to the volume dependence of γ.

CHAPTER 6

ATOMIC VIBRATIONS IN DEFECT LATTICES

1. Introduction

We have seen in previous chapters that many thermophysical properties depend on some average of the vibrational density of states $F(\omega)$. Such properties are usually insensitive to the presence of lattice defects like impurities, vacancies, dislocations, grain boundaries, etc. At very low temperatures, however, only modes of low energies are excited. Then, low-frequency defect modes may give a significant contribution to the thermophysical properties of the sample. At room temperature and above, one does not expect any spectacular features caused by a low concentration of imperfections in the lattice. In concentrated alloys and mixed crystals, the vibration modes may be very different from those of the host material, but properties which depend on an average of $F(\omega)$ are expected to vary smoothly with the composition.

2. The vibrational density of states. General aspects

It is convenient to introduce a density of states $\Delta F_{\rm def}(\omega)$ which describes the changes in the atomic vibrations when a defect is created in a lattice. Consider a solid with N lattice sites, of which $N_{\rm d}$ are associated with the defect. For instance, if the defect is a vacancy, $N_{\rm d}/N$ is the vacancy concentration. In the case of a surface, $N_{\rm d}$ is the number of surface atoms. We write

$$F_{\rm tot}(\omega) = F_{\rm bulk}(\omega) + (N_{\rm d}/N)\Delta F_{\rm def}(\omega), \tag{6.1}$$

where $F_{\rm tot}(\omega)$ is the total density of states of the actual specimen and $F_{\rm bulk}(\omega)$ is its density of states in the absence of the defect under consideration. The normalisation relations are $\int F_{\rm tot}(\omega){\rm d}\omega = \int F_{\rm bulk}(\omega){\rm d}\omega = 3$. Then

$$\int_0^{\omega_{\max}} \Delta F_{\text{def}}(\omega)\text{d}\omega = 0, \tag{6.2}$$

i.e. $\Delta F_{\text{def}}(\omega)$ varies in sign. For instance, Rieder and Drexel (1975) found, from neutron scattering experiments, that $\Delta F_{\text{def}}(\omega)$ due to surface effects in TiN had three positive and two negative regions when plotted versus ω.

From $\Delta F_{\text{def}}(\omega)$ we can calculate, for example, the change in the vibrational heat capacity:

$$\Delta C_{\text{har, def}}(T) = \int_0^{\omega_{\max}} C_{\text{har}}(\hbar\omega/k_{\text{B}}T)\Delta F_{\text{def}}(\omega)\text{d}\omega, \tag{6.3}$$

where $\omega_{\max}$ is the maximum frequency in the presence of defects. It may be larger than $\omega_{\max}$ of the defect-free solid. The high-temperature limit ($k_{\text{B}}T \gg \hbar\omega_{\max}$) of (6.3) is

$$\Delta C_{\text{har, def}}(T) = k_{\text{B}} \int_0^{\omega_{\max}} \Delta F_{\text{def}}(\omega)\text{d}\omega = 0. \tag{6.4}$$

The number N_{d} in (6.1) is not always well defined (e.g., for grain boundaries or small dislocation loops). However, we can assume that the ratio N_{d}/N measures the size, or amount, of defect regions so that $\Delta F_{\text{def}}(\omega)$ can be used to calculate thermodynamic properties characteristic of the defect. Note that there may also be a heat capacity associated with the formation of the defect, such as the two-level description of vacancies, eq. (7.8).

Example: Excess heat capacity of a defect in an Einstein solid. Assume that when a certain defect is introduced in a solid with an arbitrary $F_{\text{bulk}}(\omega)$, n_{d} states are pulled out of the bulk density of states at the frequency ω_{b} and reinserted at the frequency ω_{d}. The excess vibrational heat capacity $\Delta C_{\text{def}}(T)$ associated with the defect is

$$\Delta C_{\text{def}}(T) = n_{\text{d}}[C_{\text{E}}(\hbar\omega_{\text{d}}/k_{\text{B}}T) - C_{\text{E}}(\hbar\omega_{\text{b}}/k_{\text{B}}T)], \tag{6.5}$$

where C_{E} is the Einstein heat capacity. Figure 6.1 shows $\Delta C_{\text{def}}/n_{\text{d}}$ for $\omega_{\text{d}}/\omega_{\text{b}} = 1/\sqrt{2}$. A model like this may give a rough representation of ΔC_{def} of vacancies. Let n_{d} be the number of nearest neighbours to the vacant site. The vibrations are described by an Einstein model. The n_{d}

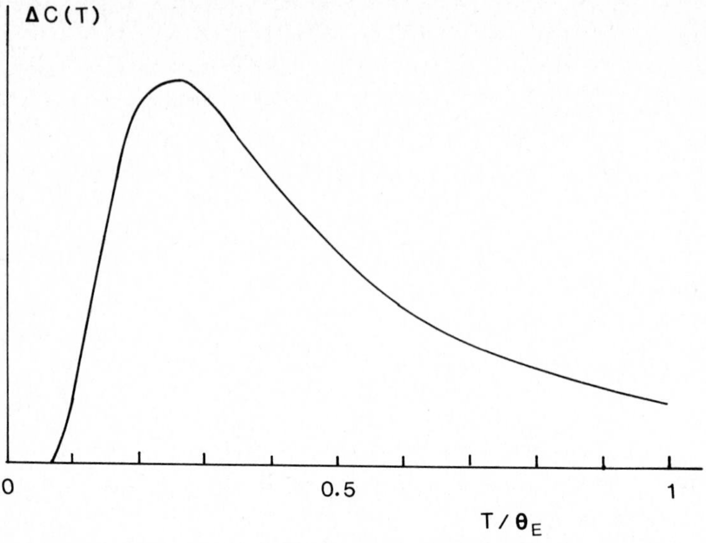

Fig. 6.1. Excess heat capacity $\Delta C(T)$ in an Einstein model (arbitrary units). Modes of energy $k_B\Theta_E$ are withdrawn from $F(\omega)$ and reinserted at $k_B\Theta_E/\sqrt{2}$.

modes per vacancy corresponding to atomic motions towards the empty site have their frequencies lowered from ω_b to ω_d.

3. Surfaces

3.1. Elastic waves in a semi-infinite elastic continuum

In an ordinary Debye model, the lattice vibrations are described by elastic waves propagating in an infinite medium. In a semi-infinite medium, bounded by a free surface, the classical wave equation $\partial^2 u_\beta/\partial t^2 - c_\beta^2 \nabla^2 u_\beta = 0$ has solutions u_β which are the usual bulk waves, but also solutions with an amplitude localised to the surface region. The latter modes, known as Rayleigh waves (Rayleigh 1900), propagate with their wave vector $\boldsymbol{q}$ in the surface plane. Their frequency is

$$\omega = C_T|\boldsymbol{q}|\xi, \tag{6.6}$$

where C_T is the transverse sound velocity in the bulk. The dimensionless parameter ξ depends on the ratio C_T/C_L and lies between 0.874 and

0.955. For a given $\boldsymbol{q}$, the frequency of the Rayleigh wave is thus less than that of the transverse elastic bulk wave. (In a real solid the sound velocity is anisotropic; Gazis et al. (1960).) A brief general discussion of elastic surface waves is found in Landau and Lifshitz (1959). Maradudin (1981) has reviewed the entire field of surface waves.

3.2. *Thermal properties of an elastic-continuum surface*

We shall calculate the heat capacity in the low-temperature limit, i.e. when only elastic waves are excited. The allowed (q_x, q_y) for wave propagation parallel to the surface give a density of states which varies as ω. The number of surface (Rayleigh) states is proportional to the area A of the surface. Their density of states is

$$F_{\mathrm{R}}(\omega) = Ak_1\omega. \tag{6.7}$$

The low-frequency part of the density of states, for the states not localised to the surface, is of the bulk form plus a correction $Ak_2\omega^2$. Hence,

$$F_{\mathrm{tot}}(\omega) = (1 - Ak_3)F_{\mathrm{bulk}}(\omega) + Ak_4\omega, \tag{6.8}$$

This is the "Debye model" in the presence of a surface. For mathematical details, see Wallis (1975), Stratton (1953, 1962), Dupuis et al. (1960), Maradudin et al. (1963) and Maradudin and Wallis (1966). From (6.8) we obtain the low-temperature heat capacity

$$C_{\mathrm{tot}}(T) = (1 - Ak_3)C_{\mathrm{bulk}}(T) + Ak_5T^2, \tag{6.9}$$

where $C_{\mathrm{bulk}} \sim T^3$. The parameters $k_1, \ldots, k_5$ depend on the elastic parameters and the mass density of the material. The T^2-term in (6.9) can be observed only at low temperatures and if the surface-to-volume ratio is large enough, i.e. for very small particles. But then the finite size of the particle is important (§ 3.4). Therefore eq. (6.9) can only be used for very qualitative estimations.

3.3. *Thin slabs*

Consider a thin slab formed by N layers of atoms. This is the three-dimensional generalisation of a finite monatomic linear chain. Such model systems have been extensively studied by Allen et al. (1971). The solutions to the equation of motion are of the form

$$\boldsymbol{u} = \boldsymbol{u}(R_{j(z)})\exp[\mathrm{i}(\boldsymbol{q}\cdot \boldsymbol{R}_{j(xy)} - \omega t)]. \tag{6.10}$$

$R_{j(z)}$ denotes the z-component (i.e. perpendicular to the surfaces) of the position vector $\boldsymbol{R}_j$ of the jth atom. $\boldsymbol{R}_{j(xy)}$ is a position vector along the slab and $\boldsymbol{q} = (q_x, q_y)$ is a two-dimensional wave vector. Almost all of the modes of the form (6.10) have amplitudes $\boldsymbol{u}(R_{j(z)})$ which are appreciable throughout the width of the slab. However, the mode corresponding to Rayleigh waves has displacements $\boldsymbol{u}(R_{j(z)})$ which decrease rapidly as $\boldsymbol{R}_j$ moves inward from either surface. Allen et al. also discovered new surface modes which do not exist in the limit of small $\boldsymbol{q}$ and thus have no elastic-wave counterpart. Similar results have been obtained in a study of the TiN (001) surface (Benedek et al. 1984).

3.4. Small particles

The bulk material has a phonon density of states which is quasi-continuous and varies as ω^2 for small ω. In a very small sample, on the other hand, the eigenfrequencies form a discrete spectrum. In particular, there is a lowest eigenfrequency $\omega_{\min}$ which can be estimated crudely as follows. In the bulk, $\omega = Cq$, where C is a sound velocity. In a small particle, of diameter d, it is meaningless to consider wavelengths larger than d, i.e. $q > 2\pi/d$. With $\hbar\omega_{\mathrm{D}} = Cq_{\mathrm{D}} \sim 2\pi C/d_0$, where d_0 is a diameter of an atom, we get $\omega_{\min} \sim \omega_{\mathrm{D}}(d_0/d)$. The discrete nature of the low-frequency part of the vibrational spectrum means that we must write out explicitly the first terms in the partition function when $T < \Theta_{\mathrm{D}}(d_0/d)$, instead of applying the usual integral approximation. This has been recognised by Burton (1970), Chen et al. (1971) and Baltes and Hilf (1973). Nishiguchi and Sakuma (1981) made an accurate study of the vibrations of a small elastic sphere. The excess heat capacity (above the bulk value) has been measured for ionic solids (NaCl; Barkman et al. (1965)) and metals (Pb, In; Novotny and Meincke (1973)). The theory referred to above is in reasonable agreement with the experiments. Dobrzynski and Leman (1969) developed a frequency-moment representation of the surface phonons and calculated a $\Delta C(T)$ similar to that of fig. 6.1.

4. Point imperfections

4.1. The mass-defect model

The mass defect is the simplest point imperfection in a vibrating lattice. One then assumes that the mass of a particular atom in the perfect lattice is altered from M to M', without any change of the force constants. That is the case if an atom is replaced by one of its isotopes and one therefore also speaks of an isotope defect. The important parameter characterising the impurity is the relative mass difference ε;

$$\varepsilon = \frac{M - M'}{M}. \tag{6.11}$$

In the case of a light impurity ($M' < M$; $\varepsilon > 0$) there may be a localised mode, with a frequency $\omega_i > \omega_{\max}$, i.e. above the highest frequency of the perfect lattice. When the impurity is much heavier than the host atoms ($M' \gg M$; $\varepsilon < 0$) there is a pronounced resonance at a frequency ω_i which is "embedded" in the quasi-continuous spectrum of host lattice vibrations.

Starting from a general expression for ω_i (Kagan and Iosilevskii 1962, 1963, Lifshitz 1956, Brout and Visscher 1962, Dawber and Elliott 1963, Mannheim 1968, Dederichs and Zeller 1980), we restrict the discussion to the case of cubic symmetry and one atom per primitive cell. The impurity mode frequencies ω_i are threefold degenerate (equivalent x, y and z directions) and obtained from

$$\varepsilon\omega_i^2 \int_0^{\omega_{\max}} \frac{F(\omega)\mathrm{d}\omega}{\omega_i^2 - \omega^2} = 3, \tag{6.12}$$

where $F(\omega)$ and $\omega_{\max}$ refer to the unperturbed vibrations of the host.

In the case of a light impurity, $\omega_i > \omega_{\max}$. We then expand the integrand in (6.12) in powers of ω/ω_i and consider $\varepsilon \approx 1$ (i.e. $M' \ll M$). It suffices to keep the first two terms in the expansion, which gives

$$\omega_i^2 \approx \frac{\varepsilon}{3(1-\varepsilon)} \int_0^{\omega_{\max}} F(\omega)\omega^2 \mathrm{d}\omega = \frac{\varepsilon}{1-\varepsilon}[\omega(2)]^2, \tag{6.13}$$

or

$$\omega_i \approx (1-\varepsilon)^{-1/2}\omega(2) = (M/M')^{1/2}\omega(2). \tag{6.14}$$

We define a corresponding characteristic temperature T_i by

$$T_i = \hbar\omega_i/k_B \approx (3/4)^{1/2}(M/M')^{1/2}\Theta_D, \tag{6.15}$$

In the last step, we used a Debye model for the phonon spectrum of the host lattice.

When the impurity is very heavy ($|\varepsilon| \gg 1$; $\varepsilon < 0$) there is a resonance mode $\omega = \omega_i$, where ω_i has a small imaginary part. A Debye model, $F(\omega) = 9\omega^2/\omega_D^3$, gives on the left-hand side of (6.12)

$$\begin{aligned}(9\varepsilon\omega_i^2/2\omega_D^3) & \int_0^{\omega_D} \omega \left\{ \frac{1}{\omega_i - \omega} - \frac{1}{\omega_i + \omega} \right\} d\omega \\ &= (9\varepsilon\omega_i^2/\omega_D^3)\{-\omega_D + (\omega_i/2)\ln[(\omega_D+\omega_i)/(\omega_D-\omega_i)] \\ &\quad - (\omega_i/2)\ln[\omega_i/(-\omega_i)]\}.\end{aligned} \tag{6.16}$$

Using the result that $\omega_D \gg \omega_i$, and since $\ln(-1) = i\pi$, then to lowest order (recall that $\varepsilon < 0$)

$$\omega_i^2 = \frac{\omega_D^2}{3|\varepsilon|}[1 - i(\omega_i/\omega_D)(\pi/2)]. \tag{6.17}$$

We now see explicitly that the heavy impurity gives rise to a damped mode. When $\omega_i/\omega_D \ll 1$ we can neglect $i(\omega_i/\omega_D)(\pi/2)$ compared to 1 in (6.17) and treat the resonance as if it were a true eigenstate with a frequency $\omega_D[3|\varepsilon|]^{-1/2}$. The corresponding characteristic temperature T_i is

$$T_i = |3\varepsilon|^{-1/2}\Theta_D. \tag{6.18}$$

4.2. *Thermal displacements in the mass-defect model*

The theory for the thermal displacement, $\langle u_{\mathrm{def}}^2 \rangle$, of impurity atoms is complicated, but in an approximate theory, one just scales the displacement of the replaced host atom,

$$\frac{\langle u_{\mathrm{def}}^2 \rangle}{\langle u_{\mathrm{host}}^2 \rangle} = \begin{cases} (M'/M)^{1/2} & T \ll \Theta_D, \quad (6.19) \\ 1 & T \gg \Theta_D. \quad (6.20) \end{cases}$$

Calculations by Dawber and Elliott (1963), using a Debye model for the host, show that (6.20) is exact and that (6.19) is correct to within about 5%. These authors also give expressions for the velocity of the defect atom, $\langle v_{\rm def}^2\rangle$. Both $\langle u_{\rm def}^2\rangle$ and $\langle v_{\rm def}^2\rangle$ can be measured in Mössbauer experiments.

4.3. *Debye temperatures in the mass-defect model*

Consider a crystal with a low concentration, c, of impurities. When their vibrations are described by the mass-defect model, the Debye temperature $\Theta_{\rm D}(-3)$ is

$$\Theta_{\rm D}(-3;c) = \Theta_{\rm D}(-3;c=0)[1-c\varepsilon]^{-1/2}. \tag{6.21}$$

This relation is obtained as follows: We know that $\Theta_{\rm D}(-3) \sim$ [force constant/mass density]$^{1/2}$. The mass density is proportional to $(1-c)M+cM' = (1-c)M+c(1-\varepsilon)M = (1-c\varepsilon)M$. Maradudin et al. (1963) give a detailed mathematical discussion of the isotope model leading to (6.21). The Debye temperatures $\Theta_{\rm D}(0)$ and $\Theta_{\rm D}(2)$ are also easily obtained. It was shown in (4.31) that $\Theta_{\rm D}(0) \sim (M_{\rm eff})^{-1/2}$, where the effective mass $M_{\rm eff}$ of a compound A_xB_y is given by $(x+y)\ln M_{\rm eff} = x\ln M_{\rm A} + y\ln M_{\rm B}$. This is readily generalised in the impurity case to give $\Theta_{\rm D}(0;c) \sim (M_{\rm eff})^{-1/2}$ with $M_{\rm eff} = M(1-\varepsilon)^c$, i.e.

$$\Theta_{\rm D}(0;c) = \Theta_{\rm D}(0;c=0)[1-(c/2)\ln(1-\varepsilon)]. \tag{6.22}$$

$\Theta_{\rm D}(2)$ is obtained from the dynamical matrix D; $\sum\omega^2 = {\rm Tr}\,D$, eq. (C.12). It is convenient to consider D in real space (instead of reciprocal space). The diagonal elements of D are of the form [force constant/atomic mass]. If there are no changes in the force constants, it follows that $\sum\omega^2 \sim 1/M_{\rm eff}$ with $1/M_{\rm eff} = (1-c)/M + c/M'$. Hence

$$\Theta_{\rm D}(2;c) = \Theta_{\rm D}(2;c=0)[1+(c\varepsilon/2)/(1-\varepsilon)]. \tag{6.23}$$

Comparing (6.21)–(6.23) in the limit of small ε, we see that all the Debye temperatures vary as $(1+c\varepsilon/2)$.

4.4. *Force constant changes*

The equation (6.12) for the localised or resonance mode frequencies ω_i can be generalised to include force constant changes at the impurity (Kagan and Iosilevskii 1962, Mannheim 1968). Work by Tiwari and Agrawal (1973a, b, c) and Tiwari et al. (1981) exemplify theoretical calculations of resonance states, with allowance for both mass and force constant changes.

The simple approach taken above to estimate changes in the Debye temperatures is now generalised to include, in a very rough way, force constant changes. Neglecting volume changes, we can write $[\omega_D(-3)]^2 \sim Y/M$, where Y is a combination of elastic constants c_{ij}. In ch. 3 § 11, c_{ij} in a simple model was shown to vary approximately as $1+2c(\Delta f/f)$. Combining this with the mass defect result (6.21) we get

$$\Theta_D(-3;c) = \Theta_D(-3;c=0)\{[1+2c(\Delta f/f)]/[1-c\varepsilon]\}^{1/2}. \tag{6.24}$$

The force constant changes discussed above were implicitly assumed to occur without any changes in the lattice parameters. Obviously, the defects distort the host lattice, which then alters the effective forces acting on an atom. Assume that a single impurity changes the volume of the specimen by the amount $\Delta\Omega_a$, and that Ω_a is the atomic volume of the (monatomic) perfect lattice. The total relative volume change of the specimen is $\Delta V/V = c\Delta\Omega_a/\Omega_a$. According to Grüneisen's relation, the host frequencies shift by an average relative amount

$$\langle \Delta\omega/\omega \rangle = -\gamma(\Delta V/V). \tag{6.25}$$

Then,

$$\Theta(\text{dilated}) = \Theta(\text{undilated})[1-c\gamma(\Delta\Omega_a/\Omega_a)], \tag{6.26}$$

where Θ(undilated) includes mass defects and those force constant changes which assume unaltered host lattice positions. Because of the crudeness of the model, we have not discriminated between the various Grüneisen parameters $\gamma(n)$ and Debye temperatures $\Theta_D(n)$ in (6.26). Typically, $\gamma = 2$. With the realistic value $\Delta\Omega_a/\Omega_a = 0.2$ for solid solution we see that relaxation effects may be significant (cf. Tiwari et al. (1981) on Cu–Sn).

4.5. Heat capacity

We now consider the relative change, $\Delta C/C$, in the vibrational heat capacity associated with a low concentration of impurity atoms. Very roughly, one obtains a contribution from the $3Nc$ localised or resonance modes, superimposed on a change in the heat capacity of the $3N(1-c)$ extended modes. The latter term is calculated from $\Theta_D(-3;c)$. In the case of heavy impurities this leads to a $\Delta C/C$ which is peaked at low temperatures (well below T_i of (6.18)). A light impurity has no significant $\Delta C/C$ since the heat capacity of the host has reached its (large) classical value at the temperatures when the localised modes start to be significantly excited. The low temperature peak in $\Delta C/C$ has been observed in several systems, for example for Pb in Mg (Panova and Samoilov 1965, Cape et al. 1966). To achieve a quantitative account of such measurements one must go beyond our simple idea of a sharp resonance mode (e.g., Tiwari and Agrawal 1973a, b, c).

Although it is very difficult to obtain a precise expression for ΔC, there is an integral relation which links ΔC to the excess entropy ΔS. By using (4.55) we can write for $\Delta S(\infty)$:

$$\Delta S(\infty) = \int_0^\infty \frac{\Delta C(T)}{T}\,\mathrm{d}T = 3Nk_B \ln[\Theta_D(0;0)/\Theta_D(0;c)]. \tag{6.27}$$

5. Vacancies

The atoms surrounding a vacant site are more loosely bound than those in the bulk and therefore given an increased vibrational entropy. At high temperatures, we can write (cf. (6.3))

$$S_{vac} = k_B \int_0^{\omega_{max}} \ln(k_B T/\hbar\omega)\Delta F_{vac}(\omega)\mathrm{d}\omega$$

$$= -k_B \int_0^{\omega_{max}} \ln(\omega)\Delta F_{vac}(\omega)\mathrm{d}\omega. \tag{6.28}$$

where we used the result that $\ln(k_B T/\hbar)\int \Delta F_{vac}(\omega)\mathrm{d}\omega = 0$, by (6.2). The physical dimensions of ω in $\ln\omega$ in the last part of (6.28) are reasonable since by (6.2), $\int \ln(\omega/\omega_1)\Delta F_{vac}(\omega)\mathrm{d}\omega$ is independent of the choice of frequency unit ω_1. Adopting the model in the example on page 133, $S_{vac} = n_d k_B \ln(\omega_b/\omega_d)$. The ratio ω_b/ω_d is estimated as follows. Consider a simple cubic lattice with central nearest-neighbour interactions. The 6 atoms adjacent to the vacancy have their force constants reduced by a factor of 2, for vibrations towards the vacant site. Since vibration frequencies vary as [force constant]$^{1/2}$, we reduce the corresponding frequencies by a factor of $1/\sqrt{2}$. Then the vibrational excess entropy in the high-temperature limit becomes

$$S_{\mathrm{vac}} \approx 6k_{\mathrm{B}}\ln(\sqrt{2}) = 3k_{\mathrm{B}}\ln 2 \sim 2k_{\mathrm{B}}. \tag{6.29}$$

This is, of course, much too simplified a picture, but it gives the right order of magnitude of S_{vac}. Similar bond-cutting models have been applied to an fcc lattice (Stripp and Kirkwood 1954), a simple cubic lattice (Mahanty et al. 1960) and fcc Cu (Huntington et al. 1955). They all give $S_{\mathrm{vac}} \approx 1.7\,k_{\mathrm{B}}$ to $2.0\,k_{\mathrm{B}}$ per atom. Experimental values of $S_{\mathrm{vac}}/k_{\mathrm{B}}$ in metals, compiled by Brudnoy (1976) and Wollenberger (1982), usually lie in the range 1 to 3, but with large scatter between different measurements on the same element. The situation is similar in ionic crystals (Harding and Stoneham 1981, Sahni and Jacobs 1982).

An accurate calculation of S_{vac} must include several features in addition to the bond-cutting approach. First, the atoms near the vacancy will relax to new equilibrium positions which changes the effective force constants acting on them. Then there is a dilatation of the lattice even far away from the vacant site. The corresponding shifts in the phonon frequencies can be handled using the Grüneisen model, see, for example, Mott and Gurney (1940), Vineyard and Dienes (1954) and Huntington et al. (1955). The dilatation term may also be obtained from the macroscopic relation (Huntington et al. 1955)

$$(\partial S/\partial V)_T = (\partial p/\partial T)_V = K_T\beta. \tag{6.30}$$

6. Dislocations

The dislocation core has a more open structure than the perfect lattice. One therefore expects a softening of the atomic vibrations near the core. Simple estimates (Friedel 1982) show that, for the core of a dislocation, the vibrational entropy $S_{\mathrm{disl}} \sim 0.5\,k_{\mathrm{B}}$ (or less) per atom in the core. The strain field surrounding a dislocation is of long range and has regions of compression as well as expansion, where a Grüneisen description should be applicable. There seems to be no estimation of the overall effect of dislocations on the vibrational spectrum. In comparison with the large energy E_{disl} associated with a dislocation, the term $-TS_{\mathrm{disl}}$ in the free energy is negligible.

Vibrations of the dislocations as such should also be considered.

When a dislocation line is pinned at its ends, it can vibrate much like a string under tension. Granato (1958) and Ohashi and Ohashi (1980) have considered a theory for the contribution of such vibrations to the heat capacity of a solid, and Bevk (1973) performed experiments on copper. The theory is very similar to that of vibrations in a finite one-dimensional string of mass points connected with springs, although we now have transverse vibrations instead. Let L be the length of the dislocation line and a be a typical lattice spacing. The total number of modes (degrees of freedom) is $\sim L/a$, corresponding to L/a masses in the chain. The thermal energy of a single dislocation loop, and in the classical limit, was estimated by Granato (1958) to be

$$E_{\text{disl}}(T) \approx (\pi/6)L(\rho/G)^{1/2}(1/\hbar)(k_B T)^2. \tag{6.31}$$

G is the shear modulus and ρ is the mass density of the solid. It follows that the heat capacity $C_{\text{disl}} = \partial E_{\text{disl}}(T)/\partial T$ is linear in T. C_{disl} is exceedingly small compared to the lattice part C_{ph} of the total heat capacity C_p, except at very low temperatures. Theory (Granato 1958) and experiments (Bevk 1973) show that $C_{\text{disl}}/C_{\text{ph}} \sim 10^{-3}$ at $T/\Theta_D \sim 10^{-2}$ in heavily cold-worked samples (dislocation density $\sim 10^{14}\,\text{m}^{-2}$).

7. Grain boundaries

The grain boundary energy (per area) is usually determined from the surface tension. The temperature is high and in the surface tension $\gamma_{\text{gr}} = E_{\text{gr}} - TS_{\text{gr}}$, TS_{gr} may amount to $E_{\text{gr}}/4$ or more. There is both a static (configurational) and a dynamic (vibrational) contribution to S_{gr}. They may be of the same order of magnitude (Ewing 1971, Ewing and Chalmers 1972, Hasson et al. 1972).

8. Concentrated alloys and mixed crystals

In this section we go beyond the dilute limit of impurities and consider (one-phase) concentrated alloys and mixed crystals (e.g. $Na_cK_{1-c}Cl$). The lattice vibrations are assumed to be harmonic, but since there is no translational invariance, the eigenstates are no longer plane waves, and the eigenfrequencies cannot be mapped as sharp dispersion curves

$\omega(\boldsymbol{q}, \lambda)$. Even if the dispersion curves are strongly modified, or give a completely inadequate description, there is a well-defined density of states $F(\omega)$. In the long wavelength limit, $\omega(\boldsymbol{q}, \lambda) = C(\boldsymbol{q}, \lambda)|\boldsymbol{q}|$, i.e. $F(\omega) \sim \omega^2$. A strict Debye model, which assumes that the elastic limit is extrapolated to the energy ω_D, may be an equally good approximation in a pure element as in a concentrated alloy or a mixed crystal. In, say, an alloy with one atom per primitive cell, $F(\omega)$ usually retains the general shape with two humps, corresponding to transverse and longitudinal modes. However, the sharp structures in $F(\omega)$ characteristic of a perfect periodic lattice are smoothened out.

Our discussion in § 4 of the Debye temperatures $\Theta_D(-3)$, $\Theta_D(0)$ and $\Theta_D(2)$ can be immediately extended to the present case. Consider a compound, or a solid solution, with the composition $A_{1-c}B_c$. Here $0 < c < 1$, but it is no longer required that c or $1-c$ is small. The masses of the constituents are M_A and M_B. Let $\Theta_D^A(n) = \Theta_D(n; c = 0)$ and $\Theta_D^B(n) = \Theta_D(n; c = 1)$ be Debye temperatures of the pure constituents A and B. Then, in the mass-defect model,

$$\Theta_D(-3; c) = \Theta_D^A(-3)[(1-c)+cM_B/M_A]^{-1/2}, \tag{6.32}$$

$$\Theta_D(0; c) = \Theta_D^A(0)[M_A/M_B]^{c/2}, \tag{6.33}$$

$$\Theta_D(2; c) = \Theta_D^A(2)[(1-c)+cM_A/M_B]^{1/2}. \tag{6.34}$$

Figure 6.2 shows how $\Theta_D(n; c)$ varies with c for $M_B/M_A = 4$ (i.e. $\varepsilon = -3$).The figure has a region with heavy impurities (for small c) as well as a region with light impurities (for $c \approx 1$). At the $c \approx 1$ end, the impurities are characterised by $\varepsilon = 0.75$.

The relations (6.32)–(6.34) can also be written as interpolation formulas:

$$[\Theta_D(-3; c)]^{-2} = (1-c)[\Theta_D^A(-3)]^{-2} + c[\Theta_D^B(-3)]^{-2}, \tag{6.35}$$

$$\Theta_D(0; c) = [\Theta_D^A(0)]^{(1-c)}[\Theta_D^B(0)]^c, \tag{6.36}$$

$$[\Theta_D(2; c)]^2 = (1-c)[\Theta_D^A(2)]^2 + c[\Theta_D^B(2)]^2. \tag{6.37}$$

Within the mass defect model, these expressions are exact. They are also valid if $M_A = M_B$ and the effective force constant f (properly averaged) varies as: (i) $1/f(c) = c/f_B + (1-c)/f_A$; (ii) $f(c) = [f_B]^c[f_A]^{1-c}$; (iii)

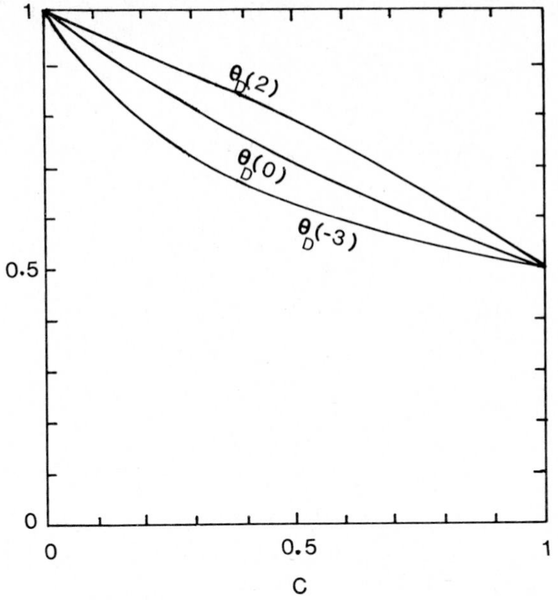

Fig. 6.2. The Debye temperatures $\Theta_D(-3;c)$, $\Theta_D(0;c)$ and $\Theta_D(2;c)$, according to eqs. (6.32–6.34), when $M_B/M_A = 4$. All Θ_D are normalised to 1 for $c = 0$.

$f(c) = cf_B + (1-c)f_A$, where (i)–(iii) refers to (6.35)–(6.37). In the limit of a small $|f_A - f_B|/f_A$ (or better: a small $|f(c) - f_A|/f_A$), the three conditions (i)–(iii) are equivalent. Thus, (6.35)–(6.37) go beyond the mass defect model and are applicable even when there are (moderate) force constant changes. This includes the changes caused by volume effects, cf. (6.26). Therefore, (6.35)–(6.37) may give a good account of the concentration dependence of the Debye temperatures in many real systems.

Theoretical calculations of phonon modes in concentrated alloys are difficult and require drastic approximations (Soma et al. 1984).

Example: $\Theta_D(-3)$ in $KCl_{1-c}Br_c$. Karlsson (1970) measured the low temperature heat capacity of the mixed crystal $KCl_{1-c}Br_c$ for several values of c. Figure 6.3 shows that his data are in excellent agreement with the interpolation formula (6.35). The dashed curve gives the theoretical result in the mass-defect model (eq. 6.32), when the point at $c = 0$, i.e. $\Theta_D(-3; KCl)$, is matched to the experiment. $M_{Br}/M_{Cl} = 2.0$.

Example: $\Theta_D(-3)$ *in Nb–Mo alloys.* Niobium and molybdenum form a

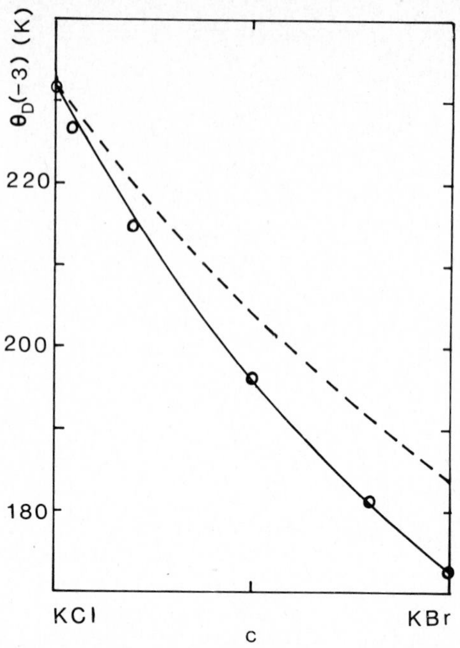

Fig. 6.3. The measured Debye temperature $\Theta_D(-3) = \Theta^C(T = 0)$ in $KCl_{1-c}Br_c$ (Karlsson 1970), the interpolation formula (6.35) (solid line) and an extrapolation from $\Theta_D(-3)$ of pure KCl using the mass defect model only (eq. 6.32), (dashed line).

continuous solid solution. Figure 6.4 shows experimentally determined values of $\Theta_D(-3)$. The data are scattered in the shaded band. Since the atomic masses of Nb and Mo are almost equal, the variation of $\Theta_D(-3)$ with composition must be due to force constant changes. The interpolation formula (6.35), the dashed line in the figure, is not applicable.

9. The Neumann–Kopp rule for C_p

The Neumann–Kopp rule expresses the (molar) heat capacity $C_p(A_xB_y)$ of a compound A_xB_y (or an alloy) as the sum of C_p for the elements forming A_xB_y:

$$C_p(A_xB_y) = xC_p(A) + yC_p(B). \tag{6.38}$$

Table 6.1 suggests that the rule is astonishingly well obeyed but we shall

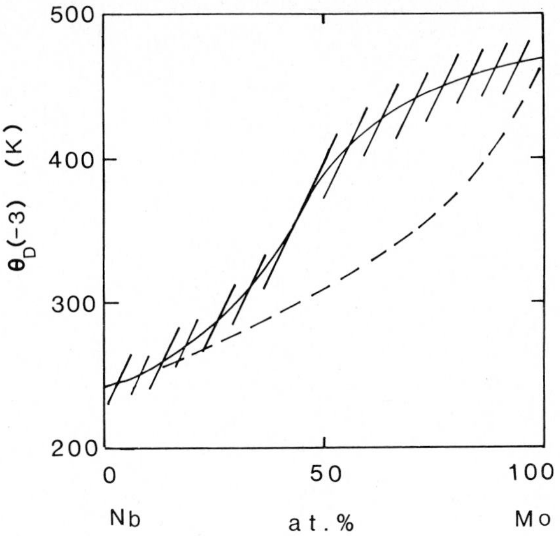

Fig. 6.4. The measured Debye temperature $\Theta_D(-3)$ in the alloy Nb–Mo; data from White et al. (1978).

see that the predictive power is rather limited. First, we note that several of the most successful examples in the table have C_p near the classical high-temperature limit k_B per atom. That value is independent of the atomic masses and the interatomic forces, and only requires harmonic vibrations. We next observe that the first two terms in the high temperature expansion $C_{har} = 3Nk_B\{1-(1/20)[\Theta_D(2)/T]^2+\ldots\}$, (4.75), when combined with the interpolation formula (6.37) for $\Theta_D(2)$, gives exactly the Neumann–Kopp rule. Further, if the Neumann–Kopp rule is valid at all temperatures, we may use the expression $S = \int C_p(\mathrm{d}T'/T')$ for the entropy and obtain $S(A_xB_y) = xS(A)+yS(B)$, which is exactly Latimer's high-temperature rule (p. 81). On the other hand, Latimer's rule does not mathematically imply the validity of the Neumann–Kopp rule at low and intermediate temperatures. At low temperatures, the Neumann–Kopp rule implies that $[\Theta_D(-3;A_xB_y)]^{-3} = x[\Theta_D(-3;A)]^{-3}+y[\Theta_D(-3;B)]^{-3}$. This is not consistent with the interpolation formula (6.35) for $\Theta_D(-3)$, which is of the same structure but has exponents -2 instead of -3.

We conclude that the major reason for the apparent success of the Neumann–Kopp rule is that it expresses the trivial Dulong–Petit results. It also reflects the importance of the atomic mass on the characteristic vibrational frequencies. The rule certainly fails when there are large non-

Table 6.1
Illustration of the Neumann–Kopp rule

Compound	T [K]	Heat capacity C_p [J/mol K]			
A_xB_y		A	B	xA + yB	A_xB_y (exp.)
$Cu_{0.8}Zn_{0.2}$	100	16.0	19.4	16.7	16.7
$Cu_{0.8}Zn_{0.2}$	298	24.5	25.4	24.2	24.7
$MgZn_2$	298	24.9	25.4	75.5	74.6
AlSb	298	24.4	25.2	49.6	47.0
Mg_2Si	298	24.9	20.0	69.8	67.9
Mg_2Si	1000	33.0	26.4	92.3	87.5
B_4Mg	298	11.1	25.0	69.4	70.4
B_4Mg	1000	24.9	33.0	132.6	115.8
TiB_2	100	14.3	1.1	16.4	7.5
TiB_2	298	25.0	11.1	47.3	44.3
TiB_2	1000	32.5	24.9	82.3	76.9
HfC	298	25.7	6.1 [a]	31.9	37.5
HfC	1000	31.1	21.6 [a]	52.7	51.3
Fe_3C	100	12.1	1.7 [b]	37.8	41.9
Fe_3C	298	25.0	8.5 [b]	83.5	106.3
Fe_3C	1000	54.5	21.6 [b]	185.0	125.6

[a] Diamond. [b] Graphite.
The classical heat capacity of an element is $3R = 24.94$ J/mol K.

vibrational contributions to the heat capacity (e.g., the magnetic term in C_p for Fe at 1000 K; table 6.1), when one is well below the Dulong–Petit limit (TiB_2) or when there are large anharmonic contributions in one of the components (e.g., Mg at 1000 K). Data in table 6.1 are from Hultgren et al. (1973a, b) and JANAF Tables (1971).

CHAPTER 7

THERMAL PROPERTIES OF FEW-LEVEL SYSTEMS AND SPIN WAVES

1. Introduction

Several important thermophysical properties may be described by simple models with only two, or a few, energy levels. Of particular interest are vacancy formation, localised f-electrons in rare earths, magnetic excitations in insulators and order–disorder transformations in alloys. A dynamical coupling between the discrete spins of each magnetic atom leads to excitations which are propagating waves, closely analogous to phonons. They are called spin waves, or magnons. Since the scope of this book is mainly to consider thermophysical properties of materials which are of practical importance, we leave out many effects which are significant at a few kelvin and below.

2. Systems with few energy levels

?.1. *Two-level systems*

Consider an ensemble of N two-level systems. The energy levels are E_1 and E_2, with

$$E_2 - E_1 = \Delta E = k_B T'. \tag{7.1}$$

The thermodynamic properties of the two-level system are obtained from the partition function Z_2 (index 2 for "two-level"):

$$Z_2 = e^{-E_1/k_B T} + e^{-E_2/k_B T}. \tag{7.2}$$

The Helmholtz free energy F_2 is $F_2 = -k_B T \ln(Z_2)$, from which we get the heat capacity $C_2 = -T(\partial^2 F_2/\partial T^2)$ as

$$C_2(T) = k_B \frac{x^2 e^x}{[e^x+1]^2}, \tag{7.3}$$

with $x = (E_2 - E_1)/k_B T = T'/T$. We can, alternatively and more easily, obtain (7.3) as follows. The probabilities p_1 and p_2 for a two-level system to be found in its two states, 1 and 2, is given by the normalised Boltzmann factors

$$p_n = \frac{e^{-E_n/k_B T}}{e^{-E_1/k_B T} + e^{-E_2/k_B T}}, \qquad n = 1, 2. \tag{7.4}$$

The thermal properties do not depend on the choice of reference level for the energies, and we can take $E_1 = 0$. The total thermal energy is

$$E = p_1 E_1 + p_2 E_2 = \frac{E_2 e^{-E_2/k_B T}}{1 + e^{-E_2/k_B T}} = \frac{\Delta E}{e^{\Delta E/k_B T} + 1}. \tag{7.5}$$

This expression immediately gives $C_2 = \partial E/\partial T$ as in (7.3). $C_2(T)$ has a characteristic form, known as a Schottky peak (fig. 7.1). The behaviour at low and high temperatures is

$$C_2(T) \approx k_B (T'/T)^2 \exp(-T'/T), \qquad (T \ll T'), \tag{7.6}$$

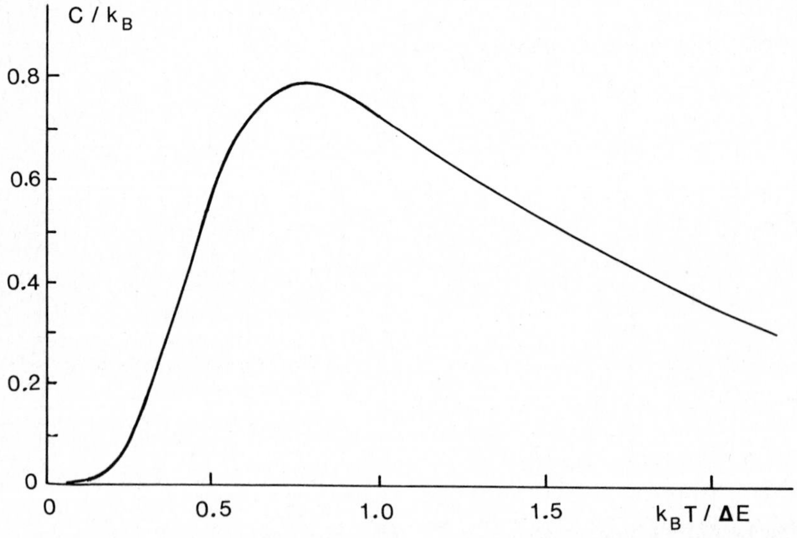

Fig. 7.1. The heat capacity (Schottky peak) for a two-level system with energy spacing ΔE.

$$C_2(T) \approx (1/4)k_B(T'/T)^2, \qquad (T \gg T'). \tag{7.7}$$

Example: Heat capacity from vacancies. Let there be N lattice sites in a solid. A site is either occupied or vacant, so we have N two-level systems. The probability that a site is vacant is $c_{vac}(T)$, eq. (2.10). The energy associated with a thermal-equilibrium concentration of vacancies is $E = NE_{vac}c_{vac}$. The heat capacity, $C_{vac} = \partial E/\partial T$, is

$$C_{vac}(T) = Nk_B \exp(S_{vac}/k_B)(E_{vac}/k_BT)^2 \exp(-E_{vac}/k_BT), \tag{7.8}$$

i.e. of the low-temperature form (7.6). If there is more than one atom per primitive cell, one may have to consider non-equivalent lattice sites separately, with different N, E_{vac} and S_{vac}. Typically, $E_{vac}/k_BT_m = 10$ and $\exp(S_{vac}/k_B) = 3$ for an elemental metal. Then, at T_m, $C_{vac}/3Nk_B \approx 5 \times 10^{-3}$.

Example: Tunnelling states in amorphous materials. Zeller and Pohl (1971) discovered that the low-temperature heat capacity of several non-metallic amorphous materials varied linearly in T. This is surprising, since the Debye T^3-law should hold in the low-temperature limit. The phenomenon, which is present both in insulating glasses and amorphous metals, is still poorly understood (Cibuzar et al. 1984, Graebner and Allen 1983) but it may be qualitatively explained by the two-level tunnelling system model (Anderson et al. 1972, Phillips 1972). The idea is that in a glassy material, there are atoms which can be in any of two neighbouring equilibrium positions, which are almost degenerate on the energy scale k_BT. Let ε be the spacing between two such levels, and $N(\varepsilon)\Delta\varepsilon$ be the number of two-level systems in the specimen with ε lying in $[\varepsilon, \varepsilon + \Delta\varepsilon]$. The total heat capacity is

$$C = \int_0^\infty N(\varepsilon)\, C_2(\varepsilon/k_BT)\, d\varepsilon. \tag{7.9}$$

$C_2(\varepsilon/k_BT)$ is negligible when $k_BT \gg \varepsilon$ or $k_BT \ll \varepsilon$. If $N(\varepsilon)$ is a smoothly varying function of ε we can take $N(\varepsilon^*)$, with $\varepsilon^* \sim k_BT$, outside the integral in (7.9). Then,

$$C = (\pi^2/6)k_B^2 N(\varepsilon^*)T. \tag{7.10}$$

The heat capacity in this model is linear in T, in accordance with the experiments.

2.2. n-level systems

The two-level system is easily generalised to an n-level system for which the energy levels E_i ($1 \leqq i \leqq n$) have degeneracies g_i. The thermodynamics of the n-level system follows from the partition function Z_n:

$$Z_n = \sum_{i=1}^{n} g_i e^{-E_i/k_B T}. \tag{7.11}$$

At high temperatures, i.e. $T \gg (E_n - E_1)/k_B$, all quantum states are equally populated, with a probability $1/G$ where $G = \sum g_i$. Thus the limiting entropy is

$$S = -k_B \sum_{i=1}^{n} g_i[(1/G)\ln(1/G)] = k_B \ln(G). \tag{7.12}$$

Note that this important sum-rule for the entropy holds irrespectively of the detailed nature of the energy levels.

3. Order–disorder transformations

Order–disorder transformations of various kinds form a central part of statistical mechanics. The mathematical complexity is enormous, even for very simplified models. Since the field is well covered in many texts (e.g., Ziman 1979) and since it would not be possible to take a discussion to the same depth as in other chapters in this book, we just mention a few important points.

Two problems of prime importance are the order–disorder transitions of spins in magnetic systems, and of atomic configurations in alloy lattices. In the first case we can (in the simplest Ising model) assign to each lattice site a spin which is either in a “spin-up” or a “spin-down” state. The spins interact with their nearest-neighbours only, and one introduces different interaction energies for the three possible pairs; spin-up–spin-up, spin-down–spin-down and spin-up–spin-down. In the case of atomic ordering in an alloy with atoms A and B, one again assumes interactions with nearest-neighbours and introduces interaction energies E_{AA}, E_{BB} and E_{AB}. One, of several, mathematical solutions to the statistical mechanics of these models makes use of the mean-field approximation. When applied to magnetic systems, it is usually referred to as the Curie–Weiss theory, while in the atomic ordering it is known as the Bragg–Williams model (although priority should perhaps have been given to Borelius; Borelius (1934), Domb (1981)).

The heat capacity C_{dis} of an order–disorder transformation is very difficult to calculate in its full details for a realistic system. Roughly, it has the form of fig. 7.2. Although C_{dis} is poorly known, it always obeys the sum rule related to the entropy $S_{dis}(T)$,

$$S_{dis}(\infty) - S_{dis}(0) = \int_0^\infty [C_{dis}(T)/T]\, dT. \tag{7.13}$$

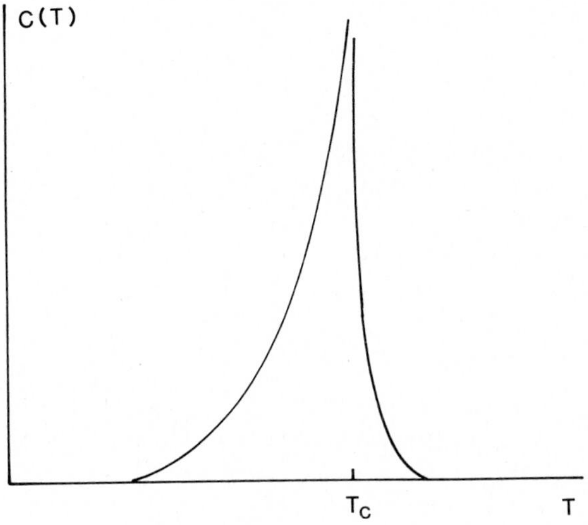

Fig. 7.2. The heat capacity of an order–disorder transformation, with the critical temperature T_c (schematically).

For instance, if an alloy $A_{1-c}B_c$ is completely ordered at low temperatures, the entropy $S_{dis}(0) = 0$. At temperatures so high that the atoms are randomly distributed over the lattice sites, the entropy per site is

$$S_{dis}(\infty) = -k_B\{c \ln c + (1-c)\ln(1-c)\}. \tag{7.14}$$

An analogous relation holds for the spin disorder entropy in magnetic insulators. In metals, the question of persistent spin fluctuations well above the Curie temperature is a matter of controversy. There seem to be such large fluctuations in Fe.

4. Magnons

In the Ising model of magnetism, briefly mentioned above, magnetic excitations correspond to the reversal of the spin at a particular lattice site (for spin 1/2). Such excited states are separated from the ground state by a rather large energy gap, and hence they are frozen out at low temperatures. However, when the spins of adjacent sites are dynamically coupled to each other, there may be wave-like excitations, called magnons or spin waves. This is in close analogy to phonons. In an Einstein model, which is reminiscent of the Ising model, the lowest excited state lies at a finite energy above the ground state. When dynamical coupling between the atoms is introduced, wave-like excitations are possible, in which each atom is displaced only a small amount relative to its neighbours. It lies outside the scope of this book to discuss magnons in detail and the reader is referred to elementary (Kittel 1976) and more advanced (Kittel 1963) accounts. Below, we summarise concepts which are of importance for the thermal properties of solids.

Consider a ferromagnetic insulator, with N regularly ordered atoms which have a spin of magnitude S. (Several properties of metallic magnetic systems are also well described by the model of an insulator.) The lattice may also contain atoms which do not have a moment, but they are of no concern here. In the ground state ($T = 0\,\mathrm{K}$) all spins are aligned. The magnons (spin waves) are excitations characterised by a wave vector $\boldsymbol{q}$, lying in the first Brillouin zone defined by the lattice of magnetic atoms. The energy of a magnon is written $\hbar\omega_{\mathrm{m}}(\boldsymbol{q}, i)$. The index i distinguishes different magnon branches for a given $\boldsymbol{q}$. The terminology is taken over from phonons and one speaks of the acoustic magnon branch and (in the case of several magnetic atoms per primitive cell) optical magnon branches. However, the three polarisation directions for a given $\boldsymbol{q}$ in the case of phonons have no magnetic counterpart and there is only one "polarisation" mode for each $\boldsymbol{q}$

$$\hbar\omega(\boldsymbol{q}, i) = Da^2q^2. \tag{7.15}$$

The index i refers to the acoustic mode and a is the lattice parameter. D

has the dimension of energy, and measures the strength of the magnetic coupling between adjacent spins.

In analogy to the Debye model, we shall assume that (7.15) is valid for all $\boldsymbol{q}$, and evaluate the magnon contribution $C_{\mathrm{magn}}(T)$ to the heat capacity at low temperatures. The density of states in $\boldsymbol{q}$-space is the same as for phonons; $V/(2\pi)^3$. The calculations closely follow those of the Debye model for the heat capacity, but with $\omega(\boldsymbol{q}) \sim q^2$ instead of $\omega(\boldsymbol{q}) \sim |\boldsymbol{q}|$, giving

$$C_{\mathrm{magn}}(T) = \frac{Nk_{\mathrm{B}}}{4\pi^2}\left(\frac{k_{\mathrm{B}}T}{D}\right)^{3/2}\int_0^{\infty}\frac{x^{5/2}\mathrm{e}^x}{(\mathrm{e}^x-1)^2}\,\mathrm{d}x$$

$$= 0.113Nk_{\mathrm{B}}(k_{\mathrm{B}}T/D)^{3/2}, \tag{7.16}$$

where we have written $V = Na^3$. The integral in (7.16) is related to $\zeta\,(2.5)$, where ζ is Riemann's zeta-function. A similar result is obtained for a ferrimagnetic solid, i.e. a ferromagnet where not all the spins are of equal magnitude (Kouvel 1956). In an effective-medium model for a cubic lattice with nearest-neighbour interactions, the Curie temperature T_{c} is related to D by (White 1970)

$$k_{\mathrm{B}}T_{\mathrm{c}} = (S+1)D. \tag{7.17}$$

Within this approximation we can write

$$C_{\mathrm{magn}}(T) = 0.113\; Nk_{\mathrm{B}}(S+1)^{3/2}(T/T_{\mathrm{c}})^{3/2}. \tag{7.18}$$

An entropy argument shows that (7.18) cannot hold all the way up to $T = T_{\mathrm{c}}$. We get, from (7.18),

$$S_{\mathrm{magn}}(T_{\mathrm{c}}) = \int_0^{T_{\mathrm{c}}}[C_{\mathrm{magn}}/T]\,\mathrm{d}T = 0.075Nk_{\mathrm{B}}(S+1)^{3/2}, \tag{7.19}$$

which is incorrect as the theory only applies at temperatures lower than T_{c}. On the other hand, if there is complete spin disorder at $T = T_{\mathrm{c}}$, the entropy is $S_{\mathrm{magn}}(T_{\mathrm{c}}) = Nk_{\mathrm{B}}\ln(2S+1)$. For any reasonable magnitude of the spin S, this is much larger than (7.19) and the discrepancy cannot be explained by the relatively small short-range order that still prevails above T_{c}. Near T_{c}, the heat capacity must be described by the order–disorder models; §3.

Measurements of the heat capacity in rare earth and yttrium garnets (Harris and Meyer 1962, Guillot et al. 1981) exemplify how the interpretation of the data requires a low-temperature collective description (i.e. magnons) and a high-temperature (in fact, at 1 to 10 K) Weiss molecular field approach.

Magnons in antiferromagnets (i.e. with alternating spin directions in the ground state) are similar to the magnons of ferromagnetic insulators, but the dispersion relation corresponding to (7.15) has the form

$$\hbar\omega_{\text{magn}}(\boldsymbol{q}) = [(\hbar\omega_0)^2 + (Daq)^2]^{1/2}, \tag{7.20}$$

where $\hbar\omega_0$ is a characteristic gap energy. In magnetically anisotropic materials, $\omega_0^2 \neq 0$. Then the lowest excitation energy is finite, $\hbar\omega_0$. The heat capacity at temperatures $k_B T \ll \hbar\omega_0$ will be that of two-level systems (§2.1). When $k_B T \gg \hbar\omega_0$, the thermal properties are dominated by the excitations for which $\hbar\omega_0$ can be neglected compared to Daq, i.e. we can make the approximation that $\hbar\omega_0 = 0$ in (7.20). In that case, the heat capacity varies as T^3. This is the same temperature dependence as for phonons, a consequence of the fact that $\omega(\boldsymbol{q}) \sim |\boldsymbol{q}|$ for both phonons and magnons (in antiferromagnetic crystals). At still higher temperatures, it is again necessary to use an order–disorder model.

CHAPTER 8

THERMODYNAMIC PROPERTIES OF CONDUCTION ELECTRONS

1. Introduction

Many thermophysical properties which are related to the electronic structure of metals and alloys depend on the electron density of states $N(E)$ at the Fermi level E_F. For instance, the Sommerfeld formula for the electronic heat capacity, which appears in almost all textbooks on solid state physics, reads (some authors let $N(E)$ refer to both spin directions and then the prefactor of 2 is absent)

$$C_{el} = \gamma T = \frac{2\pi^2}{3} N(E_F) k_B^2 T. \tag{8.1}$$

This expression is qualitatively correct, for simple (i.e. free-electron-like) metals as well as for transition metals and alloys. However, it neglects some important features. There should be an electron–phonon many-body enhancement factor $1+\lambda$ which typically is 1.4 but occasionally (e.g., Pb, Hg) can be as large as 2.5. That correction is temperature dependent and vanishes at high temperatures. Equation (8.1) also assumes that the electron states are probed in such a narrow energy interval around the Fermi energy that the density of states can be regarded as constant and equal to $N(E_F)$. Near the melting temperature, this leads to an error in C_{el} by a factor of two for some transition metals. It is obvious from what has just been said that an accurate account of the thermal properties of conduction electrons has to go beyond the simple text-book formula (8.1). This chapter primarily deals with such aspects.

2. Thermodynamic functions

2.1. The Fermi–Dirac function and the chemical potential

We consider the Sommerfeld model, in which the electrons are assumed to form a gas of fermions, with energies E_k and a density of states $N(E)$. The Fermi–Dirac distribution function is

$$f(E) = \frac{1}{\exp[(E-\mu)/k_B T]+1}, \tag{8.2}$$

where $\mu = \mu(T)$ is the chemical potential. We shall frequently encounter $\partial f/\partial E$ or $\partial f/\partial T$. Some useful expressions are

$$\left(-\frac{\partial f}{\partial E}\right) = \frac{1}{k_B T} f(E)[1-f(E)] = \frac{1}{4k_B T}\frac{1}{\cosh^2[(E-\mu)/2k_B T]}, \tag{8.3}$$

and

$$\frac{\partial f}{\partial T} = -k_B\left(\frac{E-\mu}{k_B T}\right)\left(\frac{\partial f}{\partial E}\right). \tag{8.4}$$

The function $(-\partial f/\partial E)$ is symmetrically peaked around $\mu(T)$, with an approximate width of a few $k_B T$; fig. 8.1.

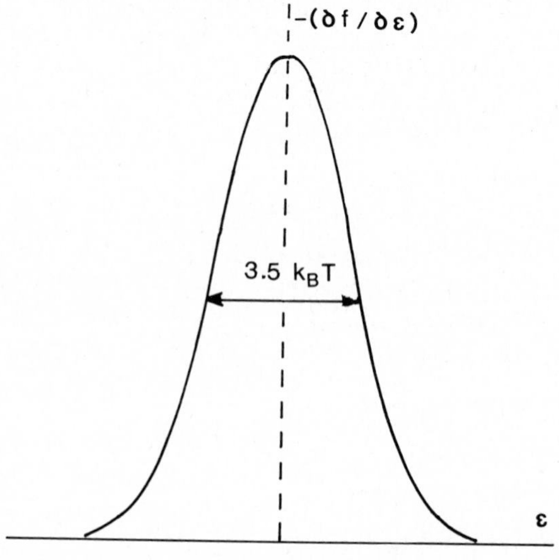

Fig. 8.1. The peaked shape of $-(\partial f/\partial \varepsilon)$ at the Fermi level.

The chemical potential $\mu(T)$ is determined by the condition that the total number of conduction electrons, N_e is conserved, i.e.

$$2\int_{-\infty}^{\infty} N(E)f(E)\mathrm{d}E = N_e. \tag{8.5}$$

The integration limits in (8.5) mean that the integration is over all energies for which the integrand is nonvanishing. Often we let $E = 0$ be the bottom of the electron band, i.e. $N(E) = 0$ for $E < 0$. The lower integration limit then is 0. The Fermi energy (Fermi level) E_F is defined to be the chemical potential at zero temperature, $E_F = \mu(0)$.

The integral in (8.5) is a special case of the general form

$$I = \int_{-\infty}^{\infty} \Phi(E)f(E)\mathrm{d}E. \tag{8.6}$$

It is a standard technique (e.g., Münster 1956, Ashcroft and Mermin 1976) to evaluate (8.6) when Φ is expanded in a Taylor series around the (temperature-dependent) chemical potential μ, giving

$$I = \int_{-\infty}^{\mu} \Phi(E)\mathrm{d}E + \frac{\pi^2}{6}(k_BT)^2 \left.\frac{\mathrm{d}\Phi}{\mathrm{d}E}\right|_{E=\mu(T)} + \frac{7\pi^4}{360}(k_BT)^4 \left.\frac{\mathrm{d}^3\Phi}{\mathrm{d}E^3}\right|_{E=\mu(T)}. \tag{8.7}$$

In practice it is easier to expand Φ around the temperature-independent Fermi level $E_F = \mu(T = 0)$. We write, for the electron density of states,

$$N(E) = N(E_F) + (E - E_F)N' + (1/2)(E - E_F)^2N''. \tag{8.8}$$

$N' = \mathrm{d}N(E)/\mathrm{d}E$ is taken at $E = E_F$ and similarly for N''. The integral (8.5) can be evaluated as in (8.7) if $\Phi\,(= N(E))$ and its derivatives are obtained from (8.8). Then, to lowest order in T,

$$\mu(T) - \mu(0) = \mu(T) - E_F = -(\pi^2/6)(k_BT)^2\, N'/N. \tag{8.9}$$

Since $\partial f/\partial E$ is peaked around $\mu(T)$, with an approximate width $\sim 4k_BT$ (fig. 8.1), it is of interest to know how large is the shift $\mu(T) - \mu(0)$ expressed in the unit k_BT. From (8.9),

$$\frac{\mu(T) - \mu(0)}{k_BT} = -\frac{\pi^2}{6} k_BT \frac{N'}{N} \approx -\frac{T}{T_v'} \operatorname{sgn}(N'). \tag{8.10}$$

T'_v is a characteristic temperature such that $N(E)$ varies significantly (e.g., a variation comparable to $N(E)$ itself) when E is altered by an amount $k_B T'_v$. The function sgn (x) is ± 1, depending on the sign of x. To be more specific, we choose to define

$$k_B T'_v |N'| = N/2. \tag{8.11}$$

For free electrons, (8.11) gives $T'_v = T_F$, the Fermi temperature. Since $T_F \sim 10^4$ to 10^5 K and T is lower than the melting temperature T_m, the shift $\mu(T) - \mu(0)$ is usually negligible in free-electron-like metals. In a transition metal, however, T'_v is often considerably smaller than the temperature defined by E_F/k_B. The temperature dependence of $\mu(T)$ may then be significant.

2.2. *The heat capacity*

The total conduction-electron energy is (in the single-particle description, i.e. with the neglect of certain many-body corrections)

$$E_{el} = 2\int_{-\infty}^{\infty} E\, N(E)\, f(E)\, dE, \tag{8.12}$$

which gives the heat capacity

$$C_{el} = \frac{\partial E_{el}}{\partial T} = 2\int (E - E_F)\, N(E) \left(\frac{\partial f(E)}{\partial T}\right) dE$$

$$= 2k_B^2 T \int \left(\frac{E - E_F}{k_B T}\right)^2 N(E) \left(-\frac{\partial f}{\partial E}\right) dE. \tag{8.13}$$

In the first line of (8.13) a term $\partial(E_F N_e)/\partial T = 0$ was subtracted. If $N(E)$ is slowly varying with E near the Fermi level, $N(E) = N(E_F)$ can be taken outside the integral. Then,

$$C_{el} = \frac{2\pi^2}{3} N(E_F) k_B^2 T. \tag{8.14}$$

It is common practice to write $C_{el} = \gamma T$. This form allows for electron–phonon many-body corrections etc. in the parameter γ. For the Sommerfeld result (8.14) we use the notation

$$C_{el} = \gamma_b T = (m_b/m) C_{fe}. \tag{8.15}$$

The index b means that band structure effects are included. C_{fe} is the heat capacity calculated in a free-electron model. The band mass m_{b} is defined in appendix B.

If the energy dependence of the density of states near the Fermi level is represented by the series expansion (8.8) one has

$$C_{\mathrm{el}} = \frac{2\pi^2}{3} N(E_{\mathrm{F}})k_{\mathrm{B}}^2 T \left\{1-(k_{\mathrm{B}}T)^2 \frac{\pi^2}{2}\left[\left(\frac{N'}{N}\right)^2 - \frac{7}{5}\frac{N''}{N}\right]\right\}. \tag{8.16}$$

This result is easily obtained if one takes $\Phi = (E-E_{\mathrm{F}})N(E)$ in (8.7), considers $[E_{\mathrm{el}}(T+\Delta T)-E_{\mathrm{el}}(T)]/\Delta T$ and excludes some terms of higher order. If the temperature dependence of $\mu(T)$ is neglected, or if $N' = 0$, the resulting C_{el} is

$$C_{\mathrm{el}} = \frac{2\pi^2}{3} N(E_{\mathrm{F}})k_{\mathrm{B}}^2 T \left\{1+\frac{7\pi^2}{10}(k_{\mathrm{B}}T)^2 \frac{N''}{N}\right\}. \tag{8.17}$$

In analogy to the parameter T_{v}' introduced in (8.11), T_{v}'' is defined such that $(k_{\mathrm{B}}T_{\mathrm{v}}'')^2 N''$ is significant, e.g., comparable to N. We take

$$4(k_{\mathrm{B}}T_{\mathrm{v}}'')^2|N''| = N. \tag{8.18}$$

With this choice, $T_{\mathrm{v}}'' = T_{\mathrm{F}}$ in a free-electron model. The parenthesis $\{\ldots\}$ in (8.16) can be written

$$\{\ldots\} = 1+\frac{7\pi^2}{40}\left(\frac{T}{T_{\mathrm{v}}''}\right)^2 \operatorname{sgn}(N'') - \frac{\pi^2}{8}\left(\frac{T}{T_{\mathrm{v}}'}\right)^2. \tag{8.19}$$

For a free-electron density of states, $N(E) \sim \sqrt{E}$, we get $(N'/N)^2 = 1/4E_{\mathrm{F}}^2$ and $N''/N = -1/4E_{\mathrm{F}}^2$. Then (8.16) is the well-known result from textbooks (e.g., Wilson 1954),

$$C_{\mathrm{el}} = \frac{2\pi^2}{3} N(E_{\mathrm{F}})k_{\mathrm{B}}^2 T \left\{1-\frac{3\pi^2}{10}\left(\frac{T}{T_{\mathrm{F}}}\right)^2\right\}. \tag{8.20}$$

It is not an unusual mistake that the T^3-term in C_{el} is neglected in transition metals on the ground that $T_{\mathrm{F}} = E_{\mathrm{F}}/k_{\mathrm{B}}$ is of the order of 30000 K to 60000 K. A correct treatment has to consider whether $T \ll T', T''$. We may call T' and T'' effective degeneracy temperatures. They can be of the order of only 1000 K in transition metals, where this

T^3 term may thus be important in simulating an anomaly in the Debye term (4.70).

In ferromagnetic metals, one must introduce separate density-of-states functions, $N_+(E)$ and $N_-(E)$, for the two spin directions. Our previous relations remain valid if we put

$$2N(E) = N_+(E) + N_-(E). \tag{8.21}$$

However, we should note that the splitting of the two spin bands, and hence $N_+(E_F)$ and $N_-(E_F)$, varies with the temperature.

Example: C_{el} for a realistic density of states. In this example we shall get an idea about the magnitude of the terms in (8.16) in a real metal. Figure 8.2 shows the theoretically calculated $N(E)$ of Pd (Mueller et al. 1970). The dashed curve is a second-order polynomial in $(E - E_F)$ fitted to the region around the Fermi level, cf. (8.8). It gives $T_v' = 1080$ K and $T_v'' = 960$ K. For the expansion (8.16) to be at all relevant it is necessary that the electronic heat capacity does not contain significant contributions from energies E where the second-order polynomial fails to reproduce $N(E)$ (e.g., to the left of the peak denoted A in fig. 8.2). In ch. 15, §5 it is shown that C_{el} probes $N(E)$ within an approximate

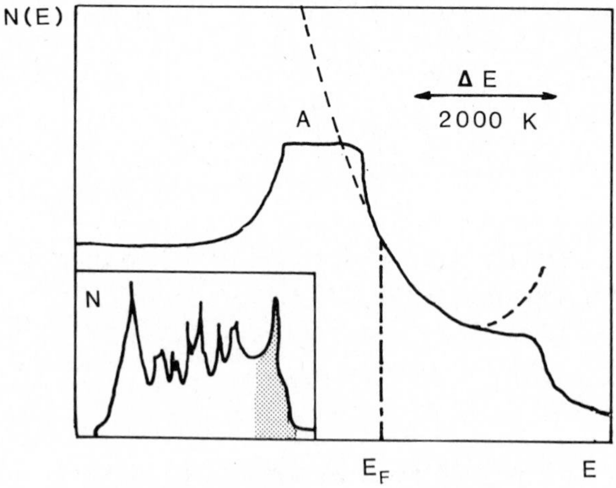

Fig. 8.2. The inset shows the full calculated density of states $N(E)$ for d-electrons in Pd (Mueller et al. 1970). The shaded area is $N(E)$ near the Fermi level E_F, as shown in the main figure. The dashed curve is a second-order polynomial in $E - E_F$, made to give a good fit to $N(E)$ near E_F. The arrow marks an energy width ΔE corresponding to $\Delta E/k_B = 2000$ K.

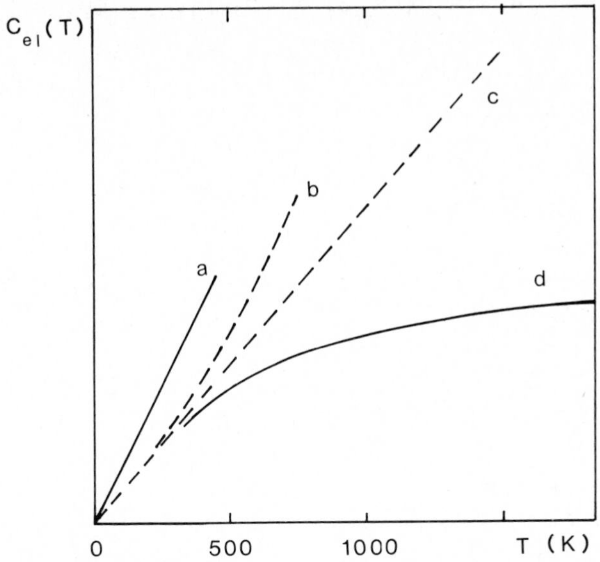

Fig. 8.3. The heat capacity $C_{el}(T)$, in arbitrary units, calculated from the density of states in fig. 8.2. The curve a is $(2\pi^2/3)N(E_F)k_B^2(1+\lambda)T$ with $\lambda = 0.7$. Curves b, c and d are based on the band-structure $N(E)$ only, i.e. with no corrections due to electron–phonon interactions. Curve b results from the dashed $N(E)$ in fig. 8.2 (the second-order polynomial), curve c is curve a without the factor $(1+\lambda)$ (i.e. a constant $N(E)$) and curve d contains the full $N(E)$, using eq. (8.13). From calculations by Thiessen (1985, unpublished).

energy interval $\pm 5k_BT$ around $\mu(T)$. The polynomial expansion (8.8) therefore gives a poor account of C_{el} when T is larger than ~ 500 K. Figure 8.3 shows C_{el} in different approximations; (8.1), according to (8.16) and according to a numerical calculation based on the full $N(E)$. Similar calculations by Shimizu (1981) show that the use of $N(E_F)$, instead of the full $N(E)$, may overestimate (Nb, V) or underestimate (Zr) C_{el} at 1000 K by a factor of two. Finally, we note that the measured C_{el} also contains many-body corrections (§3.5, 3.6).

2.3. *The entropy*

A useful expression for the entropy is

$$S_{el} = -2k_B \int_{-\infty}^{\infty} \{f(E)[\ln f(E)] + [1-f(E)]\ln[1-f(E)]\}\, N(E)\mathrm{d}E. \tag{8.22}$$

The parenthesis {...} in (8.22) is an even function of $E-\mu(T)$ and is sharply peaked at $\mu(T)$. When the density of states varies slowly with E near the Fermi level, $N(E)$ can be taken outside the integral, as a constant $N(E_F)$. We get

$$S_{el} = \frac{2\pi^2}{3} N(E_F) k_B^2 T, \tag{8.23}$$

Thus $S_{el} = C_{el}$, a result which follows immediately from the general thermodynamic relation

$$S_{el} = \int_0^T \frac{C_{el}(T')}{T'} \, dT', \tag{8.24}$$

when $C_{el}(T')$ is linear in T'.

3. The electronic entropy and heat capacity in real metals

3.1. Introduction

The Sommerfeld electron theory of metals leading to expressions such as (8.16) for the heat capacity C_{el} and (8.22) for the entropy S_{el} is not in quantitatively good agreement with experiment, even if one uses an electron density of states $N(E_F)$ determined from an accurate band calculation. The main reason is that the Sommerfeld model neglects important electron-phonon many-body corrections. Their existence was realised by Buckingham (1951) and Buckingham and Schafroth (1954) but their magnitude was not known until much later (e.g. Ashcroft and Wilkins 1965, Allen and Cohen 1969). The theory of electron–phonon many-body corrections to the electron thermal properties has been reviewed by Grimvall (1976, 1981).

3.2. A general formula for the entropy

We shall be mainly concerned with electron–phonon effects but the formalism allows for some interesting connections with the Sommerfeld theory and with the effect of lattice defects on the thermal properties of electrons. A key result is the entropy expression (Éliashberg 1962, Grimvall 1969b, 1981)

$$S_{\mathrm{el}} = 2k_{\mathrm{B}} \frac{V}{(2\pi)^3} \frac{1}{2\mathrm{i}\pi k_{\mathrm{B}} T} \int \mathrm{d}^3 \boldsymbol{k} \int_{-\infty}^{\infty} \omega \left(-\frac{\partial f}{\partial \omega} \right) [\ln G_{\mathrm{R}}(\omega, \boldsymbol{k}; T) - \ln G_{\mathrm{A}}(\omega, \boldsymbol{k}; T)] \mathrm{d}\omega, \tag{8.25}$$

where G_{R} and G_{A} are "retarded" and "advanced" Green functions, defined as

$$G_{\mathrm{R}}(\omega, \boldsymbol{k}; T) = [\omega - \varepsilon(\boldsymbol{k}) - \mathrm{Re}\, M_{\mathrm{ep}}(\omega, \boldsymbol{k}; T) - \mathrm{i}\,\mathrm{Im}\, M(\omega, \boldsymbol{k}; T)]^{-1} \tag{8.26}$$

and

$$G_{\mathrm{R}}(-\omega, \boldsymbol{k}; T) = [-\omega - \varepsilon(\boldsymbol{k}) + \mathrm{Re}\, M(\omega, \boldsymbol{k}; T) - \mathrm{i}\,\mathrm{Im} M(\omega, \boldsymbol{k}; T)]^{-1}, \tag{8.27}$$

and G_{A} is the complex conjugate of G_{R}, $G_{\mathrm{A}} = G_{\mathrm{R}}^*$. Here $\varepsilon(\boldsymbol{k}) = E_{\boldsymbol{k}} - E_{\mathrm{F}}$ is an electron band energy for a state of wave vector $\boldsymbol{k}$ and $\mathrm{Re}\, M$ and $\mathrm{Im}\, M$ are the real and imaginary parts of the electron self-energy. Eventually we shall let M be the electron–phonon self-energy M_{ep}, but we first check that eq. (8.25) contains the Sommerfeld result as a special case. When many-body corrections are negligible, $\mathrm{Re}\, M = 0$. Further, $\mathrm{Im}\, M = -\delta \mathrm{sgn}\,(\omega)$, i.e. $\mathrm{Im}\, M$ equals an infinitesimal negative quantity $-\delta$ when $\omega > 0$ and an infinitesimal positive quantity, δ, when $\omega < 0$. The Fermi level corresponds to $\omega = 0$. Physically, this form of $\mathrm{Im}\, M$ assures that when an electron is excited above E_{F}, and hence a "hole" is created for a state with $E(k) < E_{\mathrm{F}}$, the electron will eventually decay to a state of lower energy and the hole will be filled by an electron decaying into it. Mathematically, the form of $\mathrm{Im}\, M$ says whether the poles of the integrand in (8.25) lie in the upper or the lower complex ω-plane.

The entropy is real (i.e. with no imaginary part), like all physically measurable quantities, but the integrand in (8.25) is complex. We can consider the integrations as being performed along the real axis in the complex plane of the integration variable. The analytic properties of the self-energy M are such that the resulting S_{el} is real. From the theory of complex analysis, it is known that

$$\int \frac{F(z)}{z - a \pm \mathrm{i}\delta} \mathrm{d}z = \mathrm{P} \int \frac{F(z)}{z - a} \mathrm{d}z \mp \mathrm{i}\pi F(a). \tag{8.28}$$

P denotes the principal value of the integral, and $F(z)$ is a function with no poles in the complex z-plane. We now rewrite (8.25), using (B.5), and integrate over $\varepsilon_{\boldsymbol{k}}$. This picks up terms corresponding to the imaginary part in (8.28), and gives

$$S_{\mathrm{el}} = \frac{2k_{\mathrm{B}}}{2\mathrm{i}\pi k_{\mathrm{B}}T} N(E_{\mathrm{F}}) \int (2\mathrm{i}\pi\omega)\omega(-\partial f/\partial\omega)\mathrm{d}\omega. \tag{8.29}$$

We have taken an isotropic and constant $N(E) = N(E_{\mathrm{F}})$. Equation (8.29) now immediately gives the desired result

$$S_{\mathrm{el}} = \frac{2\pi^2}{3} N(E_{\mathrm{F}})k_{\mathrm{B}}^2 T. \tag{8.30}$$

3.3. *Effects of electron scattering*

In the steps leading to eq. (8.29) it was assumed that $\operatorname{Im} M$ is infinitesimal. In a real solid, scattering against phonons or impurities gives $\operatorname{Im} M$ a finite value. It is not difficult to show (Thiessen 1986) that we can still use expression (8.22) for the entropy if only the density of states $N(E)$ is replaced by a "smoothed" function $N^*(E)$ defined as

$$N^*(E) = \int_{-\infty}^{\infty} N(\varepsilon) \frac{(\Gamma/\pi)}{(E-\varepsilon)^2 + \Gamma^2} \mathrm{d}\varepsilon. \tag{8.31}$$

When $\Gamma \to 0$, the Lorentz function in the integrand becomes a delta function, $\delta(E-\varepsilon)$, and $N^*(E) = N(E)$. If we take $N(E) = N(E_{\mathrm{F}}) + (E-E_{\mathrm{F}})N'$ (8.31) diverges, but in real cases it is only when $N(E)$ has a strong non-linear energy dependence near E_{F} that the finite lifetime of the electrons affects the entropy. Electron states within several $k_{\mathrm{B}}T$ around the Fermi level contribute to the entropy, so the effect of a finite Γ is negligible if $\Gamma \ll k_{\mathrm{B}}T$. For the phonon-limited lifetime, one has (Grimvall 1981) $\Gamma = \pi\lambda k_{\mathrm{B}}T$ at high T where λ is introduced in eq. (8.33). Since λ is of the order of unity, the phonon-limited Γ could have a significant influence on S_{el}. However, calculations on realistic $N(E)$ for transition metals (Thiessen 1986) show that the effect is usually small ($< 10\%$, any sign). In the case of impurity scattering it is convenient to relate $\Gamma = \hbar/2\tau$ to the electron lifetime as it enters the electrical conductivity, $\sigma = ne^2\tau/m_{\mathrm{b}}$ (eq. 11.2). It follows that when the electrical resistance due to impurities is less than that due to electron–phonon scattering, as is the case with dilute impurities, the influence of impurity scattering on S_{el} is negligible.

3.4. Electron–phonon many-body corrections to the electronic entropy

We want to evaluate the integral expression (8.25) for the entropy when M is the electron–phonon self-energy M_{ep} (for details, see Grimvall (1981)). For simplicity we first assume that the electron density of states is isotropic and constant, $N(E_F)$, in the vicinity of the Fermi level. M_{ep} varies only weakly with the magnitude of the electron wave vector $\boldsymbol{k}$, but may have a significant directional dependence. We shall assume an isotropic $M_{ep} = M_{ep}(\omega, k_F; T)$. Here ω is an energy defined relative to the chemical potential μ. The integration over $\varepsilon_{\boldsymbol{k}}$ is easily performed, as in (8.29), which yields

$$S_{el} = \frac{N(E_F)k_B}{(k_B T)^2} \times \int_0^\infty \omega[\cosh(\omega/2k_B T)]^{-2}[\omega - \operatorname{Re} M_{ep}(\omega, k_F; T)]\,d\omega. \tag{8.32}$$

This result holds for all temperatures. In the low temperature limit ($T \ll \theta_D$ where θ_D is a Debye temperature) the integral (8.32) picks up M_{ep} very close to the Fermi level, i.e. for small ω. There we expand $M_{ep}(\omega, k_F; T \ll \theta_D)$ as

$$M_{ep} = -\lambda\omega. \tag{8.33}$$

The resulting integral in (8.32) has the same form as in the Sommerfeld model, apart from a factor $1+\lambda$, and the final low temperature result is

$$S_{el} = (2\pi^2/3)N(E_F)(1+\lambda)k_B^2 T. \tag{8.34}$$

The high temperature limit of (8.32) agrees exactly with the Sommerfeld model, since M_{ep} goes to zero as $(\theta_D/T)^2$ for $T \gtrsim \theta_D$. At intermediate temperatures, one has to perform the integration in (8.32) numerically. It is convenient to split S_{el} into two parts; S_b which is the Sommerfeld (or electron-band theory) result and S_{ep} which is the correction caused by electron–phonon many-body interactions:

$$S_{el} = S_b + S_{ep}. \tag{8.35}$$

In the low temperature limit,

$$S_{ep} = \lambda S_b. \tag{8.36}$$

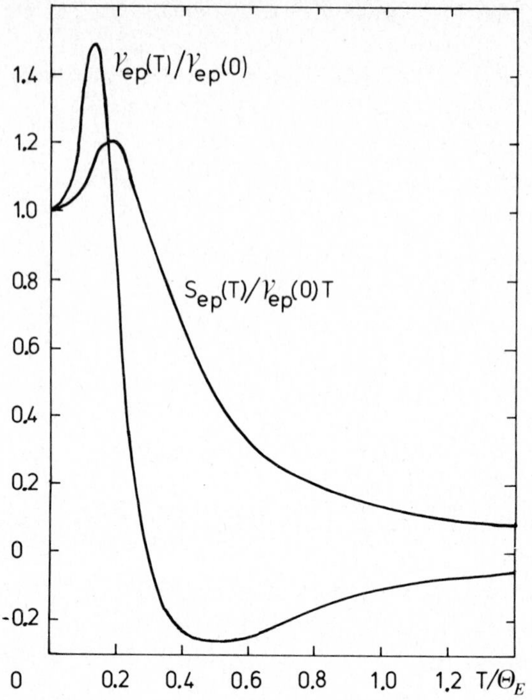

Fig. 8.4. The temperature dependence of the electron–phonon renormalisation contribution to the electronic heat capacity, $\gamma_{ep}(T)$, and to the electronic entropy, $S_{ep}(T)$. From Grimvall (1981).

If M_{ep} is calculated with an Einstein model for the lattice vibrations S_{ep} can be expressed as a universal function $S_{ep}(T)/\gamma_{ep}(0)T$; see fig. 8.4.

3.5. Electron–phonon many-body corrections to the electronic heat capacity

The electronic heat capacity (at constant volume) is obtained from

$$C_{el} = T(\partial S_{el}/\partial T)_V. \tag{8.37}$$

Equation (8.32) gives the low temperature (i.e. $T \ll \theta_D$) result

$$C_{el} = \gamma_b(1+\lambda)T = (2\pi^2/3)N(E_F)(1+\lambda)k_B^2 T. \tag{8.38}$$

We define a "thermal" effective electron mass m_{th} by

$$m_{th} = (\gamma/\gamma_{fe})m, \tag{8.39}$$

where γ is the measured coefficient in $C_{el} = \gamma T$ and γ_{fe} is the coefficient resulting from the free-electron version of the Sommerfeld model. With the electron–phonon many-body correction written explicitly we have

$$m_{th} = (\gamma_b/\gamma_{fe})(1+\lambda)m = m_b(1+\lambda). \quad (8.40)$$

In analogy to our treatment of the entropy, the electronic heat capacity is split into two parts;

$$C_{el} = [\gamma_b + \gamma_{ep}(T)]T, \quad (8.41)$$

where $\gamma_{ep}(T=0) = \lambda(T=0)\gamma_b$. At high temperatures, γ_{ep} tends to zero. There is no temperature dependence in γ_b since we have assumed that $N(E)$ is a constant, within the energy interval probed by the heat capacity. Figure 8.4 shows $\gamma_{ep}(T)$ in an Einstein phonon model. As a rough rule of thumb, we can take $\gamma_{ep}(T)/\gamma_b = \lambda$ for $T < \theta_D/4$ and zero for $T > \theta_D/3$, where θ_D is some characteristic Debye temperature. The temperature dependence of $\gamma_{ep}(T)$ and $S_{ep}(T)$ given in fig. 8.4 is not much altered if one uses a true phonon spectrum instead of the Einstein model.

3.6. Other many-body corrections

Electron–electron interactions: In a uniform electron gas, the electron–electron many-body corrections to the thermal electron mass are small. In a free-electron-like metal the corrections are at most a few percent. In transition metals there are important electron–electron many-body terms but to a large extent they are folded into the single-particle density of states $N(E)$ obtained in a band structure calculation. The remaining correction to the thermal mass probably is only a few percent and of uncertain sign. Lacking more detailed information it is therefore best to neglect these effects.

Electron–paramagnon interactions: In metals which are close to a magnetic instability, there are electron–paramagnon many-body corrections. We can write (Gladstone et al. 1969, Burnell et al. 1982, Leavens and MacDonald 1983)

$$m_{th} = m_b(1+\lambda_{el-ph}+\lambda_{sp}), \quad (8.42)$$

where λ_{sp} refers to spin fluctuations (i.e. paramagnons). In the Stoner model (ch. 15 §5), ferromagnetism arises when the density of states $N(E_F)$ is so large that $IN(E_F) > 1$. Here I is an electron interaction parameter. It is difficult to calculate λ_{sp} accurately (Daams et al. 1981, Leavens and MacDonald 1983). Among several proposed expressions we quote that of Doniach and Engelsberg (1966);

$$\lambda_{sp} = 3IN(E_F)\ln\left\{1+\frac{vIN(E_F)}{12[1-IN(E_F)]}\right\}. \tag{8.43}$$

Here v is a parameter roughly of the order of 1/2. In simple metals, λ_{sp} is negligible, and also for most transition metals it is < 0.05. However, there are metals ($LuCo_2$; Ikeda and Gschneidner (1980), MnSi; Taillefer et al. (1985)) with λ_{sp} of the order of 4, i.e. a larger enhancement than the highest known λ_{el-ph}. Very little seems to be known about the temperature dependence of λ_{sp}. Naively, one might expect that λ_{sp} is almost independent of T since the temperature scale is set by $E_F/k_B \sim 10^4$ K. This argument is as misleading as in the case of C_{el} (§2.2) and rapid variations of $N(E)$ with E should be accounted for.

Electron–magnon interactions: In magnetically ordered materials, there are electron–magnon many-body corrections. The magnitude of their influence on m_{th} is not very well known, but it may be comparable to λ_{el-ph} in Ni and Co (Phillips 1967, Batallan et al. 1975) and even be the dominating enhancement in rare earths (Cole and Turner 1967, Nakajima 1967, Kim 1968).

4. The electron density of states in real metals

Aluminium is a good example of a free-electron-like metal. Its density of states $N(E)$ is shown in fig. 8.5, together with $N(E)$ of a typical transition metal, vanadium. The Fermi level E_F is determined from (8.5) at $T = 0$, i.e. from

$$2\int_0^{E_F} N(E)\mathrm{d}E = N_e. \tag{8.44}$$

The bottom of the conduction band has been put at $E = 0$. In a transition metal, $N(E)$ is dominated by the d-electrons. Often, E_F is obtained with $N(E)$ replaced by $N_d(E)$ and N_e replaced by N_d, where subscript d refers to the d-electrons.

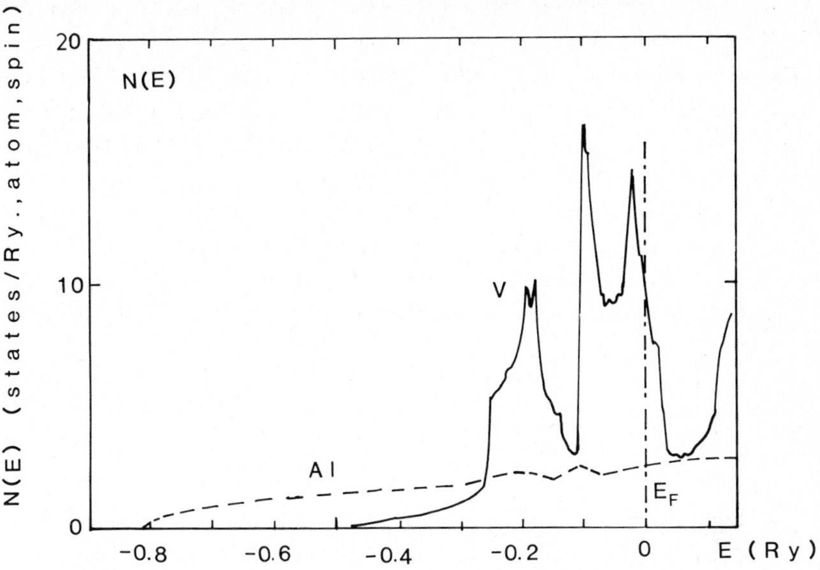

Fig. 8.5. Typical shapes of $N(E)$ curves for a free-electron-like metal (Al) and a transition metal (V) with $N(E)$ dominated by the d-band. Based on Moruzzi et al. (1978).

Now, suppose that aluminium is alloyed with a metal of a different valency. This alters the average N_e. In the rigid-band model, one assumes that $N(E)$ retains the same form as for the host (Al) but E_F, and then $N(E_F)$, are determined by the new N_e. Similarly, if vanadium is alloyed by a metal with a different number of d-electrons, N_d is changed. In a rigid-band model, one keeps $N(E)$ of V but alters E_F and $N(E_F)$ in accordance with N_d of the alloy. Although the rigid-band model may show the correct trend for $N(E_F)$ (Inoue and Shimizu 1976), it is a crude approximation (Sellmeyer 1978), Faulkner (1982). The shift $N_{alloy}(E_F) - N_{host}(E_F)$ may come out wrong by a factor of two or more. In ch. 1 §2.3 we accounted for the trend in crystal structures of transition metals on the basis of a rigid band model. Then we were interested in an integrated property of $N(E)$. The single value $N(E_F)$ may be more in error.

In the low temperature form of the heat capacity, $C = \gamma T + (12\pi^4/5)Nk_B(T/\Theta_D)^3$, the phonon term varies slowly with the composition of an alloy (at most a few per cent per at.% change in the composition), while the electronic term may vary much more. For instance, γ increases by 50% when going from $Cu_{0.66}Al_{0.34}$ to

$Cu_{0.69}Al_{0.31}$ and $\theta_D(-3)$ changes by less than 1% (Mountfield and Rayne 1984). In this case, we have an almost ordered Al_2Cu phase. Contrasting this, in a non-dilute random solid solution the alloying elements scatter the electrons, and hence give $N(E)$ a smooth shape (§3.3). Then, C_{el} does not depend strongly on the alloy composition. Defects like vacancies, dislocations, grain boundaries etc. involve so few atomic sites that their effect on $N(E)$ can be neglected, when it comes to a calculation of thermodynamic properties of the electrons.

The pressure (volume) dependence of $N(E)$ is discussed in ch. 10. Electron properties in very small particles are reviewed by Perenboom et al. (1981).

5. Band structure calculations

It is beyond the scope of this book to review the large field of band structure calculations in metals. Moruzzi et al. (1978) calculated $N(E)$ for all metals with atomic number $Z < 49$. In the Landolt–Börnstein series, Cracknell (1984) presents graphs of $N(E)$ for metallic elements and Sellmeyer (1981) gives the same information for ordered compounds and disordered alloys.

CHAPTER 9

HEAT CAPACITY OF REAL SOLIDS

1. Introduction

In previous chapters we have treated various contributions to the heat capacity of solids. The purpose of this chapter is to give a brief summary and to indicate the relative magnitude of different terms in the total heat capacity. Our example is that of a typical transition metal. In non-transition metals, the electronic terms are much smaller, and in insulators they are, of course, absent. Our approach is only qualitative, and neglects some effects (e.g., order–disorder phenomena, including magnetic excitations, are not considered). Experimental values of the heat capacity of materials are tabulated or plotted in several reference works, e.g., JANAF Thermochemical Tables (1971) and its supplements, Hultgren et al. (1973a, b) and Touloukian and Buyco (1970).

2. An example

Figure 9.1 shows, qualitatively, the total heat capacity C_p of a typical transition metal. The various contributions are now briefly discussed. The relative magnitude of these, as they are drawn in the figure, only serves to give a rough idea, and the same holds for the temperature scale.

2.1. Low temperatures

Contribution A. At low temperatures, C_p is dominated by the electronic term $C_{el} = \gamma_b(1+\lambda)T$ (ch. 8 §3). The part $\gamma_b T$ is the usual heat capacity, as treated in textbooks and with $\gamma_b = (2\pi^2/3)k_B^2 N(E_F)$ obtained from an accurate band-structure calculation. The factor $1+\lambda$ is an enhancement due to electron–phonon interactions. The magnitude of λ varies with the material, from ~ 0.1 to ~ 3, with typical values of about 0.4.

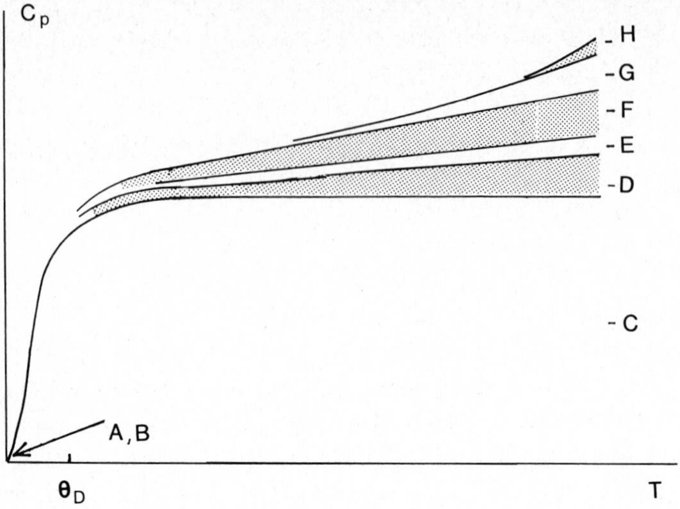

Fig. 9.1. Various contributions to the heat capacity. See text for details.

Contribution B. The phonon contribution to the heat capacity, C_{vib}, is given by the usual Debye T^3-law, $C_p = (12\pi^4/5)Nk_B(T/\Theta_D)^3$ for all crystalline solids (ch. 4 §7).

Other contributions. At very low temperatures (typically $T < 0.01\Theta_D$ or below) there may be contributions from, e.g., superconductivity or nuclear spins. Such effects have not been treated in this book. Very precise measurements may reveal a heat capacity from localised vibrations of heavy impurities (ch. 6 §4), the vibration of dislocations (ch. 6 §6) or surface effects in very small or fine-grained samples (ch. 6 §3). Amorphous solids have a small vibrational term which is linear in T, probably coming from two-level tunnelling systems (ch. 7 §2). None of these corrections to C_p are of practical importance.

2.2. *High temperatures*

Contribution C. The major part of C_p is given by the harmonic vibrational heat capacity (ch. 4 §7). It approaches the asymptotic value $3k_B$ per atom at high T. At intermediate temperatures, the term is well described by a Debye or an Einstein model (fig. 4.15).

Contribution D. This is the $C_p - C_V$ term (ch. 5 §4, ch. 10 §§4, 11). It is usually dominated by a vibrational part, described by the quasiharmonic model and linear in T, to leading order. There are also small (often negligible) parts in $C_p - C_V$ from higher-order anharmonicity (ch. 10 §7) and of electronic origin (ch. 10 §8).

Contribution E. The explicit anharmonic contributions to C_V are linear in T, to leading order (ch. 5 §4). These anharmonic terms can have any sign. In non-transition metals and in alkali halides, the contribution E is about 1/3 to 1/10 of the contribution D, but in Ge and Si, the (positive) contribution E is larger than D (Leadbetter 1968, Leadbetter and Settatree 1969, Leadbetter et al. 1969).

Contribution F. The electronic heat capacity is given by $C_{el} = \gamma_b T$. There is no enhancement factor $1+\lambda$ at high temperatures, so low-temperature data for C_{el} cannot be directly extrapolated to high T.

Contribution G. The term C_{el} in F, which is linear in T, requires that the electron density of states $N(E)$ is well approximated by the Fermi level value $N(E_F)$. In some transition metals, this is not the case (fig. 8.2). Then there is a correction to $C_{el} = \gamma_b T$, which can have any sign and can be as large as 50% (or more) of C_{el} at $T = T_m$.

Contribution H. The heat capacity from vacancies varies exponentially in T_m/T (ch. 7 §2.1). In fig. 9.1, which represents a metal, the vacancy part is somewhat exaggerated, but in certain non-metals (e.g., AgBr; Christy and Lawson 1951) lattice defects give a significant contribution to C_p near T_m.

CHAPTER 10

THERMAL EXPANSION AND RELATED PROPERTIES

1. Introduction

Most materials increase their volume as the temperature T is raised. Since also the thermal displacement of atoms increases, one might think that the atoms "push" their neighbours apart. However, this is a misleading argument. A crystal with perfectly harmonic lattice vibrations may show no thermal expansion at all. Many solids, for instance silicon and germanium and some alkali halides, shrink with increasing T, at low temperatures. Some solids with non-cubic lattice structures, e.g., zinc and uranium, shrink in one direction but expand in others so that there is a net volume increase. There are also materials, such as invar alloys, which have a very small or slightly negative coefficient of thermal expansion at ambient temperatures. Figure 10.1 exemplifies how the linear expansion coefficient α varies with T.

A correct approach to thermal expansion relies on basic thermodynamics, in particular the fact that the volume of a solid in thermal equilibrium is such as to minimise the free energy. A temperature dependence of the specimen volume V arises when there are contributions to the free energy which vary with both T and V (under restrictions such as a constant number of atoms and constant pressure). However, many important relations involving the expansion coefficient can be derived from macroscopic thermodynamics, without reference to any microscopic model. We shall begin this chapter with such relations and then continue with microscopic models for insulators, metals and magnetic materials.

The field of thermal expansion has been extensively reviewed by Barron et al. (1980), with emphasis on the low-temperature behaviour. Monographs by Yates (1972) and Krishnan et al. (1979) give a general introduction and a detailed survey of experimental data. See also Wallace (1972) for a more theoretical account.

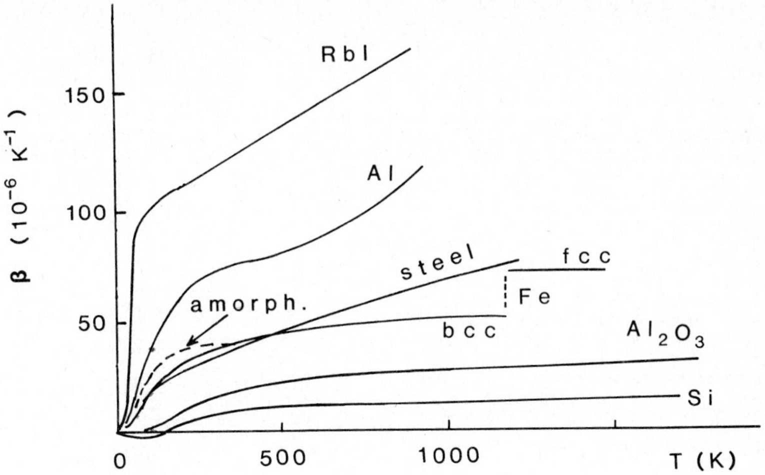

Fig. 10.1. The cubic expansion coefficient $\beta(T)$ of RbI, Al, α-Fe and γ-Fe (bcc and fcc), stainless steel, amorphous $Pd_{0.775}Si_{0.165}Cu_{0.06}$, Al_2O_3 and Si. Data from Touloukian et al. (1975, 1977) and, for Pd–Si–Cu, from Kaspers et al. (1983). The curves for RbI, Al and Si end at the melting temperature.

2. Equations-of-state based on macroscopic parameters

2.1. General aspects

We simplify the problem and assume an isotropic material or a single crystal with cubic lattice symmetry, so that the thermal expansion is isotropic. (Anisotropy is considered in §11.) We also assume that the material is under hydrostatic pressure (including atmospheric pressure, i.e. $p = 0$). The cubic (volume) expansion coefficient β is defined by

$$\beta = \frac{1}{V}\left(\frac{\partial V}{\partial T}\right)_p . \tag{10.1}$$

In an isotropic material, the linear expansion coefficient is

$$\alpha = \beta/3. \tag{10.2}$$

To evaluate the right-hand side of (10.1) we need an equation of state, i.e. a relation between p, V and T. For a comparison, consider the equation of state of an ideal gas:

$$pV = Nk_BT. \tag{10.3}$$

It yields an expansion coefficient

$$\beta = \frac{1}{V}\left(\frac{Nk_B}{P}\right) = \frac{1}{T}. \tag{10.4}$$

In contrast to this universal behaviour, β of different solids may even vary qualitatively. In many cases, however, $\beta(T)$ of a crystalline solid has a temperature dependence which is remarkably similar to that of the heat capacity. We shall see that in a simple model, $\beta(T)$ is in fact proportional to $C_V(T)$.

2.2. *Variables p, T*

Let $V(p, T)$ be the volume of a specimen and V_0 its volume at $T = 0$ and $p = 0$. We expand $V(p, T)$ in powers of p:

$$V = V_0\{1 + a_0(T) + a_1(T)p + a_2(T)p^2 + \ldots\}. \tag{10.5}$$

Then, keeping only the leading terms,

$$\beta = \frac{1}{V}\left(\frac{\partial V}{\partial T}\right)_p = \frac{1}{1+a_0}\left(\frac{\mathrm{d}a_0}{\mathrm{d}T}\right) \approx \frac{\mathrm{d}a_0}{\mathrm{d}T}. \tag{10.6}$$

By definition, $a_0(T = 0) = 0$. For most materials, $a_0 < 0.1$ at the melting temperature T_m. Hence it is a good approximation to put $1 + a_0 = 1$ in the denominator of (10.6), at least when $T \ll T_m$.

The compressibility $\kappa_T = (-1/V)(\partial V/\partial p)_T$ is usually taken at zero (atmospheric) pressure. Then (K is the bulk modulus)

$$\kappa_T = K_T^{-1} = -\frac{a_1}{1+a_0} \approx -a_1(T). \tag{10.7}$$

Most crystalline solids have $a_0(T) > 0$ and $a_2(T) > 0$. Since thermodynamics requires that $K > 0$, it always holds that $a_1 < 0$. Keeping only the first four terms in (10.5) we have

$$V/V_0 = 1 + a_0(T) + a_1(T)p + a_2(T)p^2. \tag{10.8}$$

If we consider a fixed temperature and regard $V(p=0) = (1+a_0)V_0$ as known, (10.8) is a two-parameter equation of state. However, there are other two-parameter relations between V and p at fixed T. For instance, we can assume that the isothermal bulk modulus K_T varies linearly with p;

$$K_T(p;T) = K_0(T) + K_1(T)p. \tag{10.9}$$

From the definition $K_T(p) = -V(\partial p/\partial V)_T$ we find Murnaghan's (1944) logarithmic equation of state,

$$\ln[V(p=0)/V] = \frac{1}{K_1}\ln[1+(K_1/K_0)p]. \tag{10.10}$$

At low pressures, the expressions (10.8) and (10.10) are of course equivalent. (Keep only the term linear in p and use the fact that $K^{-1} = -a_1$.) It has been empirically established (Anderson 1966b) that for many solids, Murnaghan's equation gives a better representation than the polynomial expression (10.8), even if p^3 and p^4 terms are added to the polynomial.

2.3. *Variables V, T*

In analogy to the approach in the preceding section we now expand the pressure in powers of the small quantity $(V_0 - V)/V$:

$$p = p_0(T) + p_1(T)\left(\frac{V_0 - V}{V}\right) + p_2(T)\left(\frac{V_0 - V}{V}\right)^2 + \dots. \tag{10.11}$$

If $V_0 - V$ from (10.5) is inserted in (10.11) and the coefficients of equal powers of p on the left- and right-hand sides are put equal, we obtain relations between a_0, a_1, a_2 and p_0, p_1, p_2. In particular, when only the lowest-order terms are kept,

$$\beta \approx \frac{\mathrm{d}(p_0/p_1)}{\mathrm{d}T}, \tag{10.12}$$

and

$$K = \kappa^{-1} \approx p_1. \tag{10.13}$$

Proceeding as for Murnaghan's equation above gives

$$p = \frac{K_0}{K_1}\left[\left(\frac{V(p=0)}{V}\right)^{K_1} - 1\right]. \tag{10.14}$$

This is just an alternative way of writing (10.10). The relation has also been obtained by Cook and Rogers (1963). Anderson (1966b) and Thomsen and Anderson (1969) have discussed (10.14) and related equations of state.

3. Coupled thermal conduction and thermal expansion. Thermoelastic effects

A material, subject to a temperature gradient, is normally in a state of spatially varying strain. The thermal expansion and the thermal conduction are coupled through the equation

$$\nabla(\kappa\nabla T) = \rho c_V(\partial T/\partial t) + 3\alpha T K_T \sum_{i=1}^{3} (\partial\varepsilon_i/\partial t). \tag{10.15}$$

This is the specialisation to isotropic materials of a general relation (e.g., Fung 1965) for the anisotropic case. The strains correspond to $i = x, y, z$, κ is the thermal conductivity, ρ the mass density, c_V the specific (i.e. per mass) heat capacity, and K_T the isothermal bulk modulus. If the sample is clamped so that the strain is zero, or if the coefficient of thermal expansion is zero, the last term in (10.15) vanishes. Then we recover Fourier's law of heat conduction, (12.8). If heat flow is prevented, i.e. $\boldsymbol{Q} = -\kappa(\nabla T) = 0$, we obtain

$$\left(\frac{\partial T}{\partial t}\right) = -\frac{3\alpha T K_T}{\rho c_V}\sum_{i=1}^{3}\left(\frac{\partial\varepsilon_{ii}}{\partial t}\right). \tag{10.16}$$

After integration,

$$\Delta T = -\frac{3\alpha T K_T}{\rho c_V}\sum_{i=1}^{3}\Delta\varepsilon_i = -\frac{3\alpha T K_T}{\rho c_V}\left(\frac{\Delta V}{V}\right). \tag{10.17}$$

This is Kelvin's formula for the thermoelastic effect, i.e. the change in temperature, ΔT, caused by changes in strain, $\Delta\varepsilon_i$. A uniaxial tension in the x-direction gives, by (3.25) and (3.36), $\varepsilon_1 = \sigma/E$ and

$\varepsilon_2 = \varepsilon_3 = -\sigma\nu/E$. Insertion in (10.17) yields

$$\frac{\Delta T}{T} = -\varepsilon_x \gamma_G (1-2\nu), \tag{10.18}$$

where γ_G is the Grüneisen parameter, (10.36). For most materials, $\gamma_G(1-2\nu) \approx 1$. A 1% strain of a wire thus gives a temperature decrease of about 1%.

Example: Direct measurement of the Grüneisen parameter. The relation between ΔT and γ_G facilitates a direct measurement of γ_G instead of the indirect evaluation through β, C_V and K_T (Caglioti 1982, Bottani et al. 1978). A plot of $-T^{-1}(\mathrm{d}T/\mathrm{d}t)$ versus $\mathrm{d}\varepsilon/\mathrm{d}t$ has a slope $\gamma_G(1-2\nu)$ for small times t. Anelastic effects give rise to a non-linear behaviour at high strains (i.e. after long times). Figure 10.2 shows experimental data for Al, from Bottani et al. (1978). The slope $\gamma_G(1-2\nu)$, for the elemental metals, lies within the shaded area (data from Gschneidner 1964). Dato and Köhler (1984) and Baier and Köhler (1986) have measured the thermoelastic behaviour of the trigonal semiconductors Bi_2Te_3 and Te.

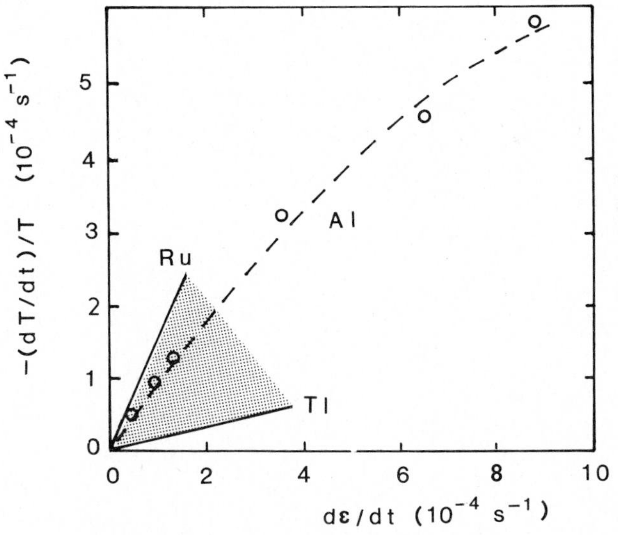

Fig. 10.2. The temperature change $\mathrm{d}T/\mathrm{d}t$ as a metal wire is strained at a rate $\mathrm{d}\varepsilon/\mathrm{d}t$. Data for Al (circles plus dashed line as a guide to the eye) from Bottani et al. (1978). All metallic elements start with a slope falling in the shaded area, with Ru and Tl as extreme cases.

4. Some important thermodynamic relations

4.1. Definitions

For later reference we define some important thermodynamic quantities and express them as derivatives of thermodynamic functions. The heat capacities at constant volume and at constant pressure are defined by

$$C_V = \left(\frac{\partial E}{\partial T}\right)_V = T\left(\frac{\partial S}{\partial T}\right)_V, \tag{10.19}$$

$$C_p = \left(\frac{\partial H}{\partial T}\right)_p = T\left(\frac{\partial S}{\partial T}\right)_p. \tag{10.20}$$

The isothermal and the adiabatic (= isentropic) compressibilities κ and bulk moduli K are defined by

$$\kappa_T = (K_T)^{-1} = -\frac{1}{V}\left(\frac{\partial V}{\partial p}\right)_T = \frac{1}{V}\left(\frac{\partial^2 F}{\partial V^2}\right)_T^{-1}, \tag{10.21}$$

$$\kappa_S = (K_S)^{-1} = -\frac{1}{V}\left(\frac{\partial V}{\partial p}\right)_S = \frac{1}{V}\left(\frac{\partial^2 E}{\partial V^2}\right)_S^{-1}. \tag{10.22}$$

The coefficient of thermal expansion can be expressed in the Helmholtz free energy F, or the entropy S,

$$\begin{aligned}\beta &= \frac{1}{V}\left(\frac{\partial V}{\partial T}\right)_p = -\frac{1}{V}\frac{(\partial p/\partial T)_V}{(\partial p/\partial V)_T} = \frac{1}{K_T}\left(\frac{\partial p}{\partial T}\right)_V \\ &= -\frac{1}{K_T}\left(\frac{\partial^2 F}{\partial T \partial V}\right) = \frac{1}{K_T}\left(\frac{\partial S}{\partial V}\right)_T,\end{aligned} \tag{10.23}$$

where we have used

$$p = -\left(\frac{\partial F}{\partial V}\right)_T. \tag{10.24}$$

There is a simple expression for βK_T:

$$\beta K_T = -\left(\frac{\partial V}{\partial T}\right)_p\left(\frac{\partial p}{\partial V}\right)_T = \left(\frac{\partial p}{\partial T}\right)_V = \left(\frac{\partial S}{\partial V}\right)_T. \tag{10.25}$$

The thermodynamic Grüneisen parameter γ_G is defined by

$$\gamma_G = \frac{\beta V K_T}{C_V}. \tag{10.26}$$

This is a very important quantity, which can also be written

$$\gamma_G = -\frac{V(\partial V/\partial T)_p}{C_V(\partial V/\partial p)_T} = \frac{V(\partial p/\partial T)_V}{C_V} = \frac{V(\partial S/\partial V)_T}{C_V}. \tag{10.27}$$

The Grüneisen parameter γ_G is sometimes called the Grüneisen constant but it may vary by a factor of two or more as a function of T and should rather be called the Grüneisen function $\gamma_G(T, V)$. Still γ_G varies much less than C_V and β as a function of T.

4.2. $C_p - C_V$ and related quantities

Using macroscopic thermodynamics, in particular Maxwell relations, one can derive a number of useful relations between C_p, C_V, K_T, K_S, κ_T, κ_S and β. Most of those quoted below, and several others, are given by Wallace (1972). In any relevant textbook one finds

$$C_p - C_V = VT\beta^2 K_T, \tag{10.28}$$

which may be rewritten as

$$C_p - C_V = \frac{V\beta^2 K_T}{C_p^2} C_p^2 T = AC_p^2 T. \tag{10.29}$$

The merit of this formulation is that the parameter A is often approximately constant, over a wide range of temperatures $T > \Theta_D$. Therefore, $AC_p^2 T$ can be used to extrapolate $C_p - C_V$. When A is regarded as a constant, (10.29) is known as the Nernst–Lindemann relation. Another relation of practical importance is

$$C_p = C_V\left(1 + \frac{\gamma_G^2 T C_V}{VK_T}\right) = C_V(1 + \beta\gamma_G T). \tag{10.30}$$

The compressibilities at constant temperature and at constant entropy are related by

$$\kappa_T - \kappa_S = \frac{VT\beta^2}{C_p}, \tag{10.31}$$

The corresponding bulk moduli obey

$$K_T - K_S = K_T K_S(\kappa_S - \kappa_T), \tag{10.32}$$

and

$$K_T = K_S\left(1 - \frac{\gamma_G^2 T C_V^2}{V C_p K_T}\right). \tag{10.33}$$

The relations (10.30) and (10.33) imply that

$$C_p > C_V; \qquad K_S > K_T. \tag{10.34}$$

From (10.30) and (10.33) simple algebra leads to

$$\frac{C_p}{C_V} = \frac{K_S}{K_T} = \frac{\kappa_T}{\kappa_S}. \tag{10.35}$$

The thermodynamic Grüneisen parameter may be written

$$\gamma_G = \frac{\beta V K_T}{C_V} = \frac{\beta V K_S}{C_p}. \tag{10.36}$$

Example: Estimation of $C_p - C_V$ in Al. We determine the Nernst–Lindemann parameter A from a fit to experimental data, $C_p - C_V = AC_p^2 T$, at $T = 400$ K. An extrapolation to 800 K, with a constant A, leads to an overestimation of $C_p - C_V$ by 14% (Brooks and Bingham 1968), compared with the measured value. On the other hand, if we assume a linear temperature dependence, $C_p - C_V = BT$, and determine B from a fit at 400 K, we will underestimate $C_p - C_V$ by 23% at 800 K.

5. Thermodynamic properties reduced to fixed volume and fixed pressure

5.1. Reduction to fixed volume

Experiments are usually carried out at constant (i.e. ambient) pressure, while theoretical calculations are more conveniently performed at constant specimen volume. However, there is sometimes confusion about what is meant by "constant volume". In the heat capacity $C_V(T)$ one takes the ratio $\Delta E/\Delta T$ of infinitesimal quantities ΔE and ΔT at a certain temperature T_1 while the volume V is kept constant. If the ratio is taken at a different temperature T_2, the volume is again kept constant but normally $V(T_1) \neq V(T_2)$. One therefore must distinguish between $C_{V_0}(T)$, which is C_V reduced to a fixed volume V_0, and $C_V(T) = T(\partial S/\partial T)_V$. Using

$$\left(\frac{\partial C_V}{\partial V}\right)_T = T\left(\frac{\partial^2 S}{\partial T \partial V}\right) = T\left(\frac{\partial^2 p}{\partial T^2}\right)_V \tag{10.37}$$

gives, to lowest order in $(V-V_0)/V_0$ (cf. Wallace 1972),

$$C_V(T) - C_{V_0}(T) = (V-V_0)T[K_T(\partial\beta/\partial T)_p + 2\beta(\partial K_T/\partial T)_p + \beta^2 K_T(\partial K_T/\partial p)_T]. \tag{10.38}$$

Calculations by Wallace (1972) show that there is a strong cancellation among the terms on the right-hand side of (10.38). For instance, at $T = 2\Theta_D$, C_{V_0} is larger than C_V by 2% in KCl and the difference changes sign at $\sim 1.4\Theta_D$.

The reduction of the energy and the entropy to fixed volume is given by

$$E(T, V) - E(T_0, V_0) = \int_{T_0}^{T} C_{V_0}(T')\mathrm{d}T' + \int_{V_0}^{V} T^2 \frac{\partial}{\partial T}\left[\frac{p(T, V')}{T}\right]\mathrm{d}V', \tag{10.39}$$

$$S(T, V) - S(T_0, V_0) = \int_{T_0}^{T} \frac{C_{V_0}(T')}{T'}\mathrm{d}T' + \int_{V_0}^{V} \frac{\partial p(T, V')}{\partial T}\mathrm{d}V'. \tag{10.40}$$

To lowest order in $(V-V_0)/V_0$ we may write $S(T,V)-S(T,V_0)$ directly from (10.27) as

$$S(T,V)-S(T,V_0)=(V-V_0)\gamma_G C_V/V_0=(V-V_0)\beta K_T. \tag{10.41}$$

At room temperatures, $S\sim 2Nk_B$ (very roughly). Then, with $\gamma_G\sim 1.5$ and $C_V\sim 3Nk_B$, we obtain, as a crude estimate, $[S(V)-S(V_0)]/S(V_0)\sim 2(V-V_0)/V_0$. Thus, the difference between room temperature values of $S(V)$ and $S(V_0)$ is often $\sim 1\%$ if $V-V_0$ is due to thermal expansion.

The expansion coefficient depends on the volume as

$$\left(\frac{\partial\beta(T,V)}{\partial V}\right)_T=\frac{1}{K_T}\left(\frac{\partial^2 S(T,V)}{\partial V^2}\right)-\frac{1}{K_T^2}\left(\frac{\partial K_T}{\partial V}\right)_T\left(\frac{\partial S}{\partial V}\right)_T. \tag{10.42}$$

To lowest order in $V-V_0$ we have

$$\beta(V)-\beta(V_0)=\left(\frac{V-V_0}{V_0}\right)\frac{1}{K_T}\left(\frac{\partial K_T}{\partial T}\right)_p. \tag{10.43}$$

The temperature dependence of K_T (and other elastic constants) was discussed in ch. 3 §9.

Example: $S(T,V)-S(T,V_0)$ in the quasi-harmonic approximation. Let $T>\Theta_D(0)$ so that

$$S\approx 3Nk_B\{4/3+\ln[T/\Theta_D(0)]\}, \tag{10.44}$$

and $C_V\approx 3Nk_B$. Then, by (10.41),

$$\begin{aligned}S(T,V)-S(T,V_0)&=3Nk_B\ln[\Theta_D(0;V_0)/\Theta_D(0;V)]\\&=[(V-V_0)/V_0]\gamma_G C_V=[(V-V_0)/V_0]3Nk_B\gamma_G,\end{aligned} \tag{10.45}$$

i.e. for small shifts in $\Theta_D(0)$

$$\frac{\Theta_D(0;V_0)-\Theta_D(0;V)}{\Theta_D(0;V_0)}=\frac{V-V_0}{V_0}\gamma_G. \tag{10.46}$$

Within the quasi-harmonic approximation, and at high temperatures, we thus have $\gamma_G=\gamma(0)$.

5.2. Reduction to fixed pressure

In analogy to the considerations in the preceding section, one should distinguish between the heat capacity at constant pressure, C_p, and the heat capacity reduced to a fixed pressure, C_{p_0}. General relations are

$$\left(\frac{\partial C_p}{\partial p}\right)_T = -T\left(\frac{\partial^2 V}{\partial T^2}\right)_p, \tag{10.47}$$

$$H(T,p)-H(T_0,p_0) = \int_{T_0}^{T} C_p(T',p_0)\,\mathrm{d}T' - \int_{p_0}^{p} T^2\left(\frac{\partial}{\partial T}\left[\frac{V(T,p')}{T}\right]\right)\mathrm{d}p', \tag{10.48}$$

$$S(T,p)-S(T_0,p_0) = \int_{T_0}^{T} C_p(T',p_0)\frac{\mathrm{d}T'}{T'} - \int_{p_0}^{p}\left(\frac{\partial V(T,p')}{\partial T}\right)\mathrm{d}p'. \tag{10.49}$$

Keeping only terms linear in $p-p_0$ gives

$$C_p(T)-C_{p_0}(T) = -(p-p_0)VT(\partial\beta/\partial T), \tag{10.50}$$

$$S(T,p)-S(T,p_0) = -(p-p_0)(\partial V/\partial T) = -(p-p_0)\beta V. \tag{10.51}$$

Example: An Einstein model result for $C_p-C_{p_0}$. If the expansion coefficient β is described by the quasi-harmonic model and an Einstein spectrum, then $\beta(T) = \beta_0\tilde{C}_{\mathrm{E}}(T/\Theta_{\mathrm{E}})$, (15.2). Here β_0 is the value of β when $T \gg \Theta_{\mathrm{E}}$ and $\tilde{C}_{\mathrm{E}}$ is normalised to 1 for $T \gg \Theta_{\mathrm{E}}$. One then finds that $T(\mathrm{d}\beta/\mathrm{d}T)$ has its maximum value $0.66\beta_0$ for $T = 0.28\Theta_{\mathrm{E}}$. In this model, $(\mathrm{d}\beta/\mathrm{d}T) \approx 0$ when $T \gg \Theta_{\mathrm{D}}$. Thus we entirely miss the anharmonic contribution to $(\mathrm{d}\beta/\mathrm{d}T)$ which may give $T(\mathrm{d}\beta/\mathrm{d}T) \sim \beta$ (cf. fig. 10.1) near the melting temperature T_{m}.

6. Microscopic models for thermal expansion

In previous sections of this chapter we have dealt with general macroscopic thermodynamic relations, which require no knowledge of the underlying microscopic mechanisms. Now, assume that the total free energy, and thus the entropy, is the sum of independent contributions labelled by r. These terms may originate from atomic vibrations, electronic and magnetic excitations, etc. The relation (10.23) can be generalised as

$$\beta = \sum_r \beta_r = \frac{1}{K_T} \sum_r \left(\frac{\partial S_r}{\partial V}\right)_T. \tag{10.52}$$

Each contribution to the entropy may be associated with a Grüneisen parameter $(\gamma_G)_r$ through

$$(\gamma_G)_r = \frac{V}{(C_V)_r} \left(\frac{\partial S_r}{\partial V}\right)_T. \tag{10.53}$$

The total Grüneisen parameter γ_G is a weighted average of $(\gamma_G)_r$,

$$\gamma_G = \frac{\sum_r (\gamma_G)_r (C_V)_r}{\sum_r (C_V)_r}. \tag{10.54}$$

From (10.53) and (10.54) we get

$$\gamma_G = \frac{\sum_r [V/(C_V)_r](\partial S_r/\partial V)_T (C_V)_r}{\sum_r (C_V)_r} = \frac{V(\partial S/\partial V)_T}{C_V}, \tag{10.55}$$

i.e. we recover (10.27), as expected. In the following sections we discuss various contributions to β_r and $(C_V)_r$. We shall see that the dimensionless parameter $(\gamma_G)_r$ in (10.54) is often of the order of 2. Then the expansion coefficient has contributions from those excitations (vibrational, electronic, magnetic, etc.) which contribute to a significant fraction of the heat capacity at the temperature of interest.

7. Phonon contribution to the thermal expansion

7.1. The quasi-harmonic approximation

It is often stated that an insulator with harmonic lattice vibrations has no thermal expansion. However, nearest-neighbour interactions in the form of ideal springs (Hooke's law) yield a negative expansion coefficient (Barron 1957). The reason is that the restoring forces perpendicular to the direction of an atomic displacement do depend on the strain state. We thus see that there is no *a priori* reason why a solid should *expand* on heating, although this is usually observed.

In the quasiharmonic approximation, each phonon mode $(\boldsymbol{q}, \lambda)$ contributes to the entropy an amount (4.44)

$$S(\boldsymbol{q}, \lambda) = k_B\{(x/2)\coth(x/2) - \ln[2\sinh(x/2)]\}, \tag{10.56}$$

where $x = \hbar\omega(\boldsymbol{q}, \lambda)/k_B T$. The frequency $\omega(\boldsymbol{q}, \lambda; V)$ is volume dependent:

$$\left(\frac{\partial \ln \omega(\boldsymbol{q}, \lambda; V)}{\partial \ln V}\right) = -\gamma(\boldsymbol{q}, \lambda). \tag{10.57}$$

Since $\beta = (1/K_T)(\partial S/\partial V)_T$, we are interested in

$$\left(\frac{\partial S(\boldsymbol{q}, \lambda)}{\partial V}\right)_T = \left(\frac{\partial S}{\partial x}\right)\left(\frac{\partial x}{\partial V}\right) = \frac{\hbar}{k_B T}\left(\frac{\partial \omega}{\partial V}\right)\left(\frac{\partial S}{\partial x}\right). \tag{10.58}$$

Compare this with the expression for $C_V(\boldsymbol{q}, \lambda)$:

$$C_V(\boldsymbol{q}, \lambda) = T\left(\frac{\partial S}{\partial T}\right)_V = T\left(\frac{\partial S}{\partial x}\right)\left(\frac{\partial x}{\partial T}\right) = -\left(\frac{\hbar\omega}{k_B T}\right)\left(\frac{\partial S}{\partial x}\right) \tag{10.59}$$

Combining (10.57)–(10.59) gives

$$\left(\frac{\partial S(\boldsymbol{q}, \lambda)}{\partial V}\right)_T = [\gamma(\boldsymbol{q}, \lambda)/V]C_V(\boldsymbol{q}, \lambda). \tag{10.60}$$

By (10.52), the coefficient of thermal expansion due to all phonon modes, is

$$\beta = \frac{1}{VK_T}\sum_{\boldsymbol{q}\lambda} C_V(\boldsymbol{q}, \lambda)\gamma(\boldsymbol{q}, \lambda). \tag{10.61}$$

Since C_V for phonons at low temperatures has the temperature dependence $A_1T^3 + A_2T^5 + \ldots$, the same powers of T appear in the phonon contribution to $\beta(T)$ at low T.

We can express β in terms of the Grüneisen parameter γ_G, eq. (10.36), as

$$\beta = \frac{C_V\gamma_G}{VK_T}, \tag{10.62}$$

giving

$$\gamma_{\rm G} = \frac{\sum_{q\lambda} \gamma(\boldsymbol{q},\lambda) C_V(\boldsymbol{q},\lambda)}{\sum_{q\lambda} C_V(\boldsymbol{q},\lambda)}. \tag{10.63}$$

At low temperatures, $C_V(\boldsymbol{q},\lambda)$ is appreciable only for those phonon modes which have $\hbar\omega(\boldsymbol{q},\lambda)/k_{\rm B}T \lesssim 1$, while at high temperatures all C_V are close to the Dulong–Petit value $k_{\rm B}$. Since $\gamma_{\rm G}(\boldsymbol{q},\lambda)$ may vary by a factor of two or more between low-frequency transverse modes and high-frequency longitudinal modes, for example, the weighted average $\gamma_{\rm G}$ in (10.63) may show a considerable temperature dependence in the range from 0 K to $\Theta_{\rm D}$.

We now turn to the volume change $\Delta V = V(T) - V(0)$. It is obtained as

$$\Delta V = \int_0^T V(T')\beta(T')\,{\rm d}T'. \tag{10.64}$$

If $\gamma_{\rm G}$ is approximated by a temperature-independent constant, and if we neglect the temperature dependence of K_T and other anharmonic corrections, (10.62) yields

$$\Delta V = \frac{\gamma_{\rm G}}{K_T}\int_0^T C_V(T')\,{\rm d}T' = \frac{\gamma_{\rm G}}{K_T}[E_{\rm har}(T) - E_{\rm har}(0)]. \tag{10.65}$$

$E_{\rm har}(0)$ is the zero-point vibrational energy. When one lacks detailed information about the temperature dependence of $\gamma_{\rm G}(T)$ it may be a useful approximation to take ΔV proportional to the thermal energy, as in (10.65). This gives essentially the Mie–Grüneisen equation-of-state at zero pressure. With an Einstein approximation for $E_{\rm har}(T)$,

$$V(T) - V(0) = \frac{3N\gamma_{\rm G}k_{\rm B}T}{K_T}\,\frac{\Theta_{\rm E}/T}{\exp(\Theta_{\rm E}/T) - 1}. \tag{10.66}$$

Thomsen and Anderson (1969) have discussed the partial lack of consistency between the Mie–Grüneisen equation and other equations of state like the Murnaghan equation, and warn against the simultaneous use of equations of state which are not mutually consistent.

7.2. *Higher-order anharmonicity*

We noted in ch. 5 §5 that the third- and fourth-order anharmonic effects are correctly accounted for if one inserts, in the harmonic expression for the entropy, the shifted frequencies

$$\omega(\boldsymbol{q},\lambda) = \omega_0(\boldsymbol{q},\lambda)+\Delta_2(\boldsymbol{q},\lambda)+\Delta_3(\boldsymbol{q},\lambda)+\Delta_4(\boldsymbol{q},\lambda). \tag{10.67}$$

In the previous section we only considered the quasiharmonic approximation, i.e. $\omega(\boldsymbol{q},\lambda) = \omega_0+\Delta_2$. These results are easily extended to include Δ_3 and Δ_4. We define generalised Grüneisen parameters γ_2 and γ_{34} by

$$\gamma(\boldsymbol{q},\lambda) = \gamma_2(\boldsymbol{q},\lambda)+\gamma_{34}(\boldsymbol{q},\lambda), \tag{10.68}$$

with (assuming $\Delta_2, \Delta_3, \Delta_4 \ll \omega_0$)

$$\gamma_2 = -\left(\frac{\partial \ln(\omega_0+\Delta_2)}{\partial \ln V}\right), \tag{10.69}$$

and

$$\gamma_{34} = -\left(\frac{\partial \ln(\omega_0+\Delta_3+\Delta_4)}{\partial \ln V}\right), \tag{10.70}$$

where Δ_2 and $(\Delta_3+\Delta_4)$ may be of the same order of magnitude. However, Δ_2 is directly proportional to the volume change while $(\Delta_3+\Delta_4)$ is likely to vary much more slowly with V. We thus expect that $\gamma_{34} \ll \gamma_2$.

From the definition of γ_G, (10.27), we have

$$\gamma_G = \frac{(\partial S/\partial \ln V)_T}{C_V} = \frac{(\partial S/\partial \ln V)_T}{(\partial S/\partial \ln T)_V}. \tag{10.71}$$

It is convenient to introduce isothermal and isochoric (i.e. "constant volume") Grüneisen parameters (Varley 1956, Barron et al. 1980):

$$\gamma_T = -(\partial \ln \Theta^S/\partial \ln V)_T, \tag{10.72}$$

$$\gamma_V = -(\partial \ln \Theta^S/\partial \ln T)_V, \tag{10.73}$$

where $\Theta^S(T)$ is the entropy Debye temperature, which is here assumed to include $\Delta_2+\Delta_3+\Delta_4$. After a few manipulations one obtains

$$\gamma_G = \gamma_T/(1+\gamma_V). \tag{10.74}$$

At low temperatures, $\gamma_V \neq 0$ because the phonon frequency spectrum is not a Debye spectrum, and this gives a temperature dependent $\Theta^S(T)$, eq. (4.125). When $T > \Theta^S/2$ this effect is of little importance. Instead, the explicit anharmonicity gives a non-zero γ_V through the frequency shifts $\Delta_3+\Delta_4$. Then, $\gamma_V \approx (\Delta_3+\Delta_4)/\omega_0$, which is usually less than 5% at $T = T_m$. Further, $\gamma_T \approx \gamma_2+\gamma_{34} \approx \gamma_2$. Therefore, the Grüneisen parameter γ_G derived from macroscopic thermodynamic quantities ($\gamma_G = \beta V K_S/C_p$) gives information about γ_2, i.e. the quasiharmonic shift. It is then assumed that β and C_p do not have significant contributions of non-vibrational origin.

Leadbetter (1968) has discussed the separation of the quasiharmonic and the explicitly anharmonic parts in S, C_V and β, with an application to Al and Pb, and a similar analysis was performed for Ge (Leadbetter and Settatree 1969), NaCl and KBr (Leadbetter et al. 1969). Rosén and Grimvall (1983) considered non-transition metals and Vieira and Hortal (1971) analysed KF.

7.3. *High-temperature expansion of γ_G in terms of $\gamma(n)$*

We consider γ_G within the quasiharmonic approximation. From (10.63) and the high-temperature expansion (4.74) of C_V we obtain, after a rearrangement of terms, (Barron 1957, Barron et al. 1964)

$$\begin{aligned}\gamma_G(T) = {} & \gamma(0) + (1/12)[\hbar\omega(2)/k_BT]^2[\gamma(0)-\gamma(2)] \\ & + (1/240)[\gamma(4)-\gamma(0)][\hbar\omega(4)/k_BT]^4 - (1/144)[\gamma(2)-\gamma(0)] \\ & \times [\hbar\omega(2)/k_BT]^4 + \cdots .\end{aligned} \tag{10.75}$$

When $T \gg \Theta_D, \gamma_G(T) \to \gamma(0)$, as we have noted in §7.2. A plot of $\gamma(T)$ versus $1/T^2$ yields $\gamma(0)$ as the intercept at $1/T^2 = 0$. Such an analysis makes use of $\gamma_G(T)$ at intermediate temperatures ($T \sim \Theta_D/2$) where anharmonic effects are small. Therefore this determination of $\gamma(0)$ is quite accurate. After $\gamma(0)$ has been obtained one may plot $[\gamma(0)-\gamma_G(T)]T^2$ versus $1/T^2$ to give $\gamma(2)$ and $\gamma(4)$. One can prove that (cf. (4.146))

$$\gamma(n) = \frac{\int \gamma_G(T) C_{har}(T) T^{n-1} dT}{\int C_{har}(T) T^{n-1} dT}. \tag{10.76}$$

The integrals converge for $-3 < n < 0$. Thus, an analysis of $C_V(T)$ may give good estimations of $\gamma(-2)$, $\gamma(-1)$, $\gamma(0)$, $\gamma(2)$ and $\gamma(4)$. The pressure dependence of the elastic constants yields $\gamma(-3)$. Values of $\gamma(n)$ obtained in this way are given in fig. 5.1.

8. Electronic contribution to the thermal expansion

The electronic contribution to the coefficient of thermal expánsion, β_{el}, is (Varley 1956)

$$\beta_{el} = \frac{(C_V)_{el}(\gamma_G)_{el}}{VK_T}. \tag{10.77}$$

VK_T refers to the real solid (i.e. with all contributions to K_T included) but $(C_V)_{el}$ and $(\gamma_G)_{el}$ contain the electronic contribution only:

$$(\gamma_G)_{el} = \frac{V}{(C_V)_{el}}(\partial S_{el}/\partial V)_T. \tag{10.78}$$

Many-body electron-phonon interactions give to $S_{el}(T)$ a complicated temperature dependence at $T \sim \Theta_D/5$, but in the limit of low and high temperatures ($T < \Theta_D/10$ and $T > \Theta_D/2$) the theory is simple. We can write (ch. 8 §3)

$$S_{el} = (2\pi^2/3)N(E_F)(1+\lambda)k_B^2 T, \qquad \text{low } T; \tag{10.79}$$

$$S_{el} = (2\pi^2/3)N(E_F)k_B^2 T, \qquad \text{high } T. \tag{10.80}$$

These expressions require that the electron density of states $N(E)$ varies slowly with the energy E in the vicinity of the Fermi level, so that it suffices to consider the value $N(E_F)$ at the Fermi level. We have

$$(\gamma_G)_{el} = \frac{\mathrm{d}\ln N(E_F)}{\mathrm{d}\ln V} + \frac{\mathrm{d}\ln(1+\lambda)}{\mathrm{d}\ln V}, \qquad \text{low } T; \tag{10.81}$$

$$(\gamma_G)_{el} = \frac{\mathrm{d}\ln N(E_F)}{\mathrm{d}\ln V}, \qquad \text{high } T. \tag{10.82}$$

In a free-electron system, (B.7), $N(E_F) \sim V\sqrt{E_F} \sim V^{2/3}$. Then,

$$\frac{\mathrm{d}\ln N(E_F)}{\mathrm{d}\ln V} = \frac{2}{3}. \tag{10.83}$$

In transition metals we may assume that the width of the d-band varies as $V^{-5/3}$ (Heine 1967) and that the "shape" of $N_d(E)$ does not depend on V (Lang and Ehrenreich 1968). Since the area under $N_d(E)$ is invariant (5 electrons per atom and spin), it follows that $N_d(E_F) \sim 1/W_d \sim V^{5/3}$. We get (Shimizu 1974, Fletcher 1978)

$$\frac{\mathrm{d}\ln N(E_F)}{\mathrm{d}\ln V} = \frac{5}{3}. \tag{10.84}$$

The quantity $\mathrm{d}\ln(1+\lambda)/\mathrm{d}\ln V$ has been reviewed by Grimvall (1981). A major contribution to the volume dependence of λ comes from an average over the phonon frequencies; $\lambda \sim 1/\langle\omega^2\rangle$. If we only include that effect,

$$\frac{\mathrm{d}\ln(1+\lambda)}{\mathrm{d}\ln V} = \frac{\lambda}{1+\lambda}\frac{\mathrm{d}\ln\lambda}{\mathrm{d}\ln V} = \frac{2\lambda}{1+\lambda}\gamma(-2). \tag{10.85}$$

Usually $2\lambda/(1+\lambda)$ lies in the interval 1/3 to 3/2, with a typical value of 1/2 for transition metals. An approximate value of $\gamma(-2)$ is 1.5 ± 1. We conclude that at low temperatures, the electron–phonon many-body interactions give a significant contribution to the electronic Grüneisen parameter $(\gamma_G)_{el}$. Hence, an account of the thermal expansion at very low temperatures ($T < \Theta_D/20$), where $(\beta)_{el} \sim T$ dominates over the phonon part $(\beta)_{ph} \sim T^3$, cannot be based on a discussion of the electron-band density of states $N(E_F)$ alone. At high temperatures ($T > \Theta_D/2$) a description of $(\beta)_{el}$ in terms of $N(E_F)$ is sufficient, but then the total expansion coefficient is dominated by the phonon part $(\beta)_{ph}$.

Barron et al. (1980) reviewed experimental values of $(\gamma_G)_{el}$. Most of them lie in the interval 1.5 ± 0.5, but they are occasionally negative (e.g., Sr, Cr) or very large ($(\gamma_G)_{el} = 3.8$ for Re). Fletcher and Yahaya (1979) obtained $(\gamma_G)_{el}$ from band structure theory for 22 transition metals. See also Sundqvist et al. (1985) for theoretical calculations on the electron band structure under pressure. In Zr–Nb–Mo–Re alloys, with e/a (d-electrons per atom) ranging from 4.3 to 6.3, the experimental $(\gamma_G)_{el}$ varied smoothly between slightly negative values and about +2.0 (Ohta and Shimizu 1982).

Example: Comparison between $(\beta)_{el}$ and $(\beta)_{ph}$ in metals. As a model example, consider a solid described by a Debye phonon spectrum with $(\gamma_G)_{ph} = 2$. We take $\lambda = 2/5$, $\mathrm{d}\ln(1+\lambda)/\mathrm{d}\ln V = [2\lambda/(1+\lambda)](\gamma_G)_{ph} = 8/7$

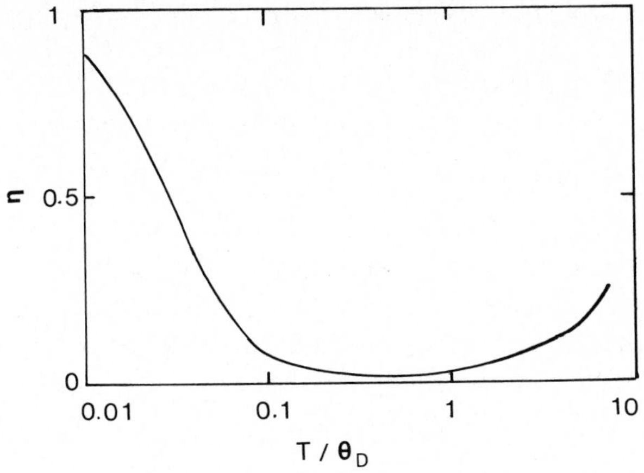

Fig. 10.3. Relative contribution, η, of conduction electrons to the total expansion coefficient of a typical transition metal.

and $\mathrm{d}\ln N(E_{\mathrm{F}})/\mathrm{d}\ln V = 5/3$. Further, let $(C_V)_{\mathrm{el}} = Nk_{\mathrm{B}}T/T_{\mathrm{m}}$ when $T > \Theta_{\mathrm{D}}$ and assume a melting temperature $T_{\mathrm{m}} = 8\Theta_{\mathrm{D}}$. These numbers are typical of a transition metal or alloy. We give VK_T an arbitrary fixed value, neglect anharmonic effects in $(C_V)_{\mathrm{ph}}$ and consider only the relative importance, η, of $(\beta)_{\mathrm{el}}$ and $(\beta)_{\mathrm{ph}}$. Figure 10.3 shows the result.

9. Magnetic contribution to the thermal expansion

The magnetic contribution to the thermal expansion can be written

$$\beta_{\mathrm{magn}} = \frac{C_{\mathrm{magn}}(\gamma_{\mathrm{G}})_{\mathrm{magn}}}{VK_T}. \tag{10.86}$$

Here C_{magn} is the magnetic part of the heat capacity and

$$(\gamma_{\mathrm{G}})_{\mathrm{magn}} = \frac{V}{C_{\mathrm{magn}}(T)}\left(\frac{\partial S_{\mathrm{magn}}}{\partial V}\right)_T. \tag{10.87}$$

In a ferromagnetic insulator at low temperatures, the magnetic part of the heat capacity is described by magnons, or spin waves, of energy $\omega_{\mathrm{m}}(\boldsymbol{q}, i)$; ch. 7 §4. The index i distinguishes between acoustic and optical branches and $\boldsymbol{q}$ is a wave vector in the first Brillouin zone of the lattice defined by the magnetic atoms. Magnons have thermal properties which are closely similar to those of phonons. From the general entropy expression (4.43), $S(\boldsymbol{q},i) = k_B[(1+n)\ln(1+n) - n\ln(n)]$, where $n = \{\exp[\hbar\omega_m(\boldsymbol{q},i)/k_BT] - 1\}^{-1}$, we can define a parameter

$$[\gamma_G(\boldsymbol{q},i)]_{magn} = -\{\partial\ln[\omega_m(\boldsymbol{q},i)]/\partial\ln V\}_T. \tag{10.88}$$

This is in direct analogy to the mode Grüneisen parameters of phonons. The volume dependence of ω_m depends on how the exchange interaction between adjacent spins varies with the distance between them. However, it may be of more practical use to relate $(\gamma_G)_{magn}$ to the volume dependence of the Curie temperature T_c. Within a simple model we have (ch. 7 §4)

$$\hbar\omega = Da^2q^2; \qquad D = k_BT_c/(1+S). \tag{10.89a, b}$$

S is the magnitude of the spin on a lattice site and a is a lattice parameter. Then,

$$[\gamma_G(\boldsymbol{q},i)]_{magn} = -(\partial\ln D/\partial\ln V)_T = -(\partial\ln T_c/\partial\ln V), \tag{10.90}$$

i.e. a common parameter $(\gamma_G)_{magn}$ for all $\boldsymbol{q}$. (The contributions from a^2 and q^2 of (10.89) cancel in (10.90).) The expression (7.18) for C_{magn} now yields

$$\beta_{magn} = \frac{0.113Nk_B}{VK_T}\left(\frac{T(S+1)}{T_c}\right)^{3/2}\left(-\frac{\partial\ln T_c}{\partial\ln V}\right). \tag{10.91}$$

The spin wave contribution to the thermal expansion can be observed only at very low temperatures (Lord 1967). At high temperatures, the spin wave description is inadequate. We should get β from $(\partial S_{magn}/\partial V)$, but we have no simple and realistic model to evaluate this quantity. For the same reason, we cannot in this book discuss the very intricate question of β_{magn} in iron. For a review see Steinemann (1978, 1979). These effects are especially relevant in invar alloys.

10. Vacancy contribution to the thermal expansion

Let the formation volume of a vacancy be V_{vac} and the atomic volume be Ω_a. In thermal equilibrium, the crystal volume $V(T)$ is

$$V(T) = V(0) + V_{\mathrm{vac}}[V(0)/\Omega_{\mathrm{a}}]\exp(S_{\mathrm{vac}}/k_{\mathrm{B}})\exp(-E_{\mathrm{vac}}/k_{\mathrm{B}}T). \tag{10.92}$$

The term $V(0)/\Omega_{\mathrm{a}}$ is the number of atoms in the solid, and the two exponential factors give the equilibrium concentration c_{vac} of vacancies. Then the vacancies contribute to the expansion coefficient an amount

$$\begin{aligned}\beta_{\mathrm{vac}} &= (1/V)(\partial V/\partial T)\\ &= (E_{\mathrm{vac}}/k_{\mathrm{B}}T^2)(V_{\mathrm{vac}}/\Omega_{\mathrm{a}})\exp(S_{\mathrm{vac}}/k_{\mathrm{B}})\exp(-E_{\mathrm{vac}}/k_{\mathrm{B}}T)\\ &= (E_{\mathrm{vac}}/k_{\mathrm{B}}T^2)(V_{\mathrm{vac}}/\Omega_{\mathrm{a}})c_{\mathrm{vac}}.\end{aligned} \tag{10.93}$$

The vacancy concentration is appreciable only near the melting temperature T_{m}. Let us take $E_{\mathrm{vac}}/k_{\mathrm{B}}T_{\mathrm{m}} \sim 8$ (table 2.1), $c_{\mathrm{vac}}(T_{\mathrm{m}}) \sim 0.002$ and $V_{\mathrm{vac}}/\Omega_{\mathrm{a}} \sim 1/2$. Then, $T\beta_{\mathrm{vac}} \sim 0.01$. This can be compared with the empirical rule $T_{\mathrm{m}}\beta_{\mathrm{tot}} \sim 0.06$ (ch. 15 §3). It is obvious that vacancies may give a significant contribution to the expansion coefficient at high temperatures, but because of the sensitivity of the exponential factor to small changes in E_{vac} it is difficult to give an accurate value for β_{vac}. See, for example, Kraftmakher (1972, 1978).

11. Thermal expansion of anisotropic solids

11.1. Introduction

Quite generally, the expansion coefficient, α, is a symmetric tensor of rank two. It is related to the symmetric strain tensor ε by

$$\begin{pmatrix}\varepsilon_{11} & \varepsilon_{12} & \varepsilon_{13}\\ \varepsilon_{12} & \varepsilon_{22} & \varepsilon_{23}\\ \varepsilon_{13} & \varepsilon_{23} & \varepsilon_{33}\end{pmatrix} = \begin{pmatrix}\alpha_{11} & \alpha_{12} & \alpha_{13}\\ \alpha_{12} & \alpha_{22} & \alpha_{23}\\ \alpha_{13} & \alpha_{23} & \alpha_{33}\end{pmatrix}\Delta T, \tag{10.94}$$

where ε_{ij} are the strains caused by a temperature increment ΔT. The cubic expansion coefficient, β, is obtained through the relation $\Delta V/V = \varepsilon_{11} + \varepsilon_{22} + \varepsilon_{33}$. One has

$$\beta = \alpha_{11} + \alpha_{22} + \alpha_{33}. \tag{10.95}$$

A general definition of α_{ij} is

$$\alpha_{ij} = \left(\frac{\partial \varepsilon_{ij}}{\partial T}\right)_{\sigma}. \tag{10.96}$$

The derivative is taken with all components σ_{ij} of the stress tensor held constant. Usually, σ refers to constant (ambient) pressure p, and we write

$$(\alpha_{ij})_p = \left(\frac{\partial \varepsilon_{ij}}{\partial T}\right)_p. \tag{10.97}$$

Table 10.1 gives the independent α_{ij} in different crystal symmetries. It also gives α_{ij} expressed as α_μ, with only one index $\mu = 1-6$, using Voigt's contraction scheme (table 3.1).

Table 10.1
Non-vanishing α_{ij} and α_μ in different crystal symmetries

Symmetry	Tensor components α_{ij}	In Voigt's notation
Cubic	$\alpha_{11} = \alpha_{22} = \alpha_{33} = \alpha = \beta/3$	$\alpha_1 = \alpha_2 = \alpha_3 = \alpha$
Hexagonal, Trigonal, Tetragonal	$\alpha_{11} = \alpha_{22} = \alpha_\perp;\ \alpha_{33} = \alpha_{\parallel}$	$\alpha_1 = \alpha_2 = \alpha_\perp,\ \alpha_3 = \alpha_{\parallel}$
Orthorhombic	$\alpha_{11},\ \alpha_{22},\ \alpha_{33}$	$\alpha_1,\ \alpha_2,\ \alpha_3$
Monoclinic	$\alpha_{11},\ \alpha_{22},\ \alpha_{33},\ \alpha_{13}$	$\alpha_1,\ \alpha_2,\ \alpha_3,\ \alpha_5/2$
Triclinic	$\alpha_{11},\ \alpha_{22},\ \alpha_{33},\ \alpha_{23},\ \alpha_{13},\ \alpha_{12}$	$\alpha_1,\ \alpha_2,\ \alpha_3,\ \alpha_4/2,\ \alpha_5/2,\ \alpha_6/2$

The linear expansion coefficient in a certain crystallographic direction [hkl] is

$$\alpha[hkl] = n_1^2\alpha_{11} + n_2^2\alpha_{22} + n_3^2\alpha_{33}. \tag{10.98}$$

Here n_1, n_2 and n_3 are the direction cosines of [hkl]. In axial crystals (hexagonal, trigonal and tetragonal lattices),

$$\begin{aligned}\alpha[hkl] &= (n_1^2+n_2^2)\alpha_{11} + n_3^2\alpha_{33} = \alpha_\perp \sin^2\theta + \alpha_{\parallel}\cos^2\theta\\ &= \alpha_\perp + (\alpha_{\parallel} - \alpha_\perp)\cos^2\theta,\end{aligned} \tag{10.99}$$

where θ is the angle between [hkl] and the c-axis, and we have used the result that $n_1^2 + n_2^2 + n_3^2 = 1$.

Example: Thermal expansion in some axial crystals. Table 10.2 gives $\alpha_\perp$ and $\alpha_{\parallel}$ in two metallic and two non-metallic crystals; the data are from

Table 10.2
Components of the thermal expansion coefficient in some axial crystals (at 293 K)

	Mg	Zn	RuO_2	Graphite
$\alpha_\perp[10^{-6}\ K^{-1}]$	24	14	7	~ 1
$\alpha_\parallel[10^{-6}\ K^{-1}]$	26	65	−2	~ 20

Touloukian et al. (1975, 1977). The cubic expansion coefficient is

$$\beta = 2\alpha_\perp + \alpha_\parallel. \tag{10.100}$$

Note that magnesium is almost isotropic in its expansion properties.

11.2. Grüneisen parameters in non-cubic lattices. General relations

In an anisotropic solid, Grüneisen parameters $\gamma_{G,i}$ are defined as a generalisation of the relation $\gamma = (V/C_V)(\partial S/\partial V)_T$ for the isotropic case. We have (see, for example, Wallace (1972), Barron et al. (1980) and eq. (5.40))

$$\gamma_{G,i} = \frac{1}{C_\varepsilon}\left(\frac{\partial S}{\partial \varepsilon_i}\right)_{T,\varepsilon_i'}, \tag{10.101}$$

where $C_\varepsilon(T)$ is the heat capacity at constant strain ε, a quantity which is further discussed in §11.4. The index ε_i' in the derivative means that all strain components except ε_i are held constant.

The generalisations of the relations $\gamma_G = V\beta K_T/C_V = V\beta K_S/C_p$ are, with Voigt's notation for α_μ,

$$\gamma_{G,i} = \frac{V}{C_\varepsilon}\sum_{j=1}^{6} \alpha_j (c_{ij})_T, \tag{10.102}$$

$$\gamma_{G,i} = \frac{V}{C_\sigma}\sum_{j=1}^{6} \alpha_j (c_{ij})_S. \tag{10.103}$$

There are six components of the Grüneisen parameter, $\gamma_{G,i} (i = 1-6)$; C_σ is the heat capacity at constant stress σ and c_{ij} are elastic stiffness coefficients. The "inverse" relations to (10.102) and (10.103) are

$$\alpha_i = (C_\varepsilon/V)\sum_{j=1}^{6} (s_{ij})_T \gamma_{G,j}, \tag{10.104}$$

$$\alpha_i = (C_\sigma/V) \sum_{j=1}^{6} (s_{ij})_S \gamma_{G,j}. \tag{10.105}$$

As an illustration we check that (10.102) and (10.105) contain the well-known relations for a cubic lattice. Then, $\alpha_1 = \alpha_2 = \alpha_3 = \alpha$, while $\alpha_i = 0$ when $i = 4, 5$ and 6. Further, $C_\varepsilon = C_V$. Then

$$\gamma_{G,i} = (\alpha V/C_V) \sum_{j=1}^{6} (c_{ij})_T = 3\alpha V K_T/C_V = \beta V K_T/C_V. \tag{10.106}$$

Here we have used the result that $K_T = (1/3)\Sigma(c_{ij})_T$; from (3.90). Similarly, from (10.105), when $\gamma_{G,i} = \gamma_G$,

$$\alpha_i = (C_V\gamma_G/V) \sum_{j=1}^{6} (s_{ij})_T = (C_V\gamma_G/3V)\kappa_T. \tag{10.107}$$

In this context we also quote the following result (e.g., Musgrave 1970):

$$(s_{ij})_T - (s_{ij})_S = \frac{VT\alpha_i\alpha_j}{C_\sigma}. \tag{10.108}$$

When all $\alpha_i > 0$, which is often the case,

$$(c_{ij})_S > (c_{ij})_T, \tag{10.109}$$

$$(s_{ij})_T > (s_{ij})_S. \tag{10.110}$$

Again, we check that this reduces to the correct result, (10.31), for cubic lattices. Then $\kappa = 3(s_{11} + 2s_{12})$ and hence, by (10.109), $\kappa_T - \kappa_S = 3VT(\alpha^2 + 2\alpha^2)/C_p = VT\beta^2/C_p$.

11.3. Grüneisen parameters in hexagonal lattices

An important special case is that of hexagonal lattice symmetry. (The same relations hold for trigonal and tetragonal lattices.) There are only two independent α_{ij}; $\alpha_\perp$ and $\alpha_{\|}$. Further, there are two independent Grüneisen parameters; $(\gamma_G)_\perp$ and $(\gamma_G)_{\|}$. Performing the summations in (10.102) and (10.103) we obtain, for a stress corresponding to hydrostatic pressure p,

$$(\gamma_G)_\perp = (V/C_p)\{\alpha_\perp[(c_{11})_S + (c_{12})_S] + \alpha_{||}(c_{13})_S\}, \tag{10.111}$$

$$(\gamma_G)_{||} = (V/C_p)\{2\alpha_\perp(c_{13})_S + \alpha_{||}(c_{33})_S\}. \tag{10.112}$$

The "inversion" of these relations gives

$$\alpha_\perp = (C_p/V)\{[(s_{11})_S + (s_{12})_S](\gamma_G)_\perp + (s_{13})_S(\gamma_G)_{||}\}, \tag{10.113}$$

$$\alpha_{||} = (C_p/V)\{2(s_{13})_S(\gamma_G)_\perp + (s_{33})_S(\gamma_G)_{||}\}. \tag{10.114}$$

We may now write the cubic expansion coefficient as

$$\beta = 2\alpha_\perp + \alpha_{||} = C_p\gamma_{G,\mathrm{hex}}/VK_S, \tag{10.115}$$

with

$$\gamma_{G,\mathrm{hex}} = [2(\kappa_\perp)_S(\gamma_G)_\perp + (\kappa_{||})_S(\gamma_G)_{||}]/\kappa_S, \tag{10.116}$$

i.e. $\gamma_{G,\mathrm{hex}}$ is a weighted average of $\gamma_\perp$ and $\gamma_{||}$. The quantities $\kappa_\perp$ and $\kappa_{||}$ are properly defined compressibilities. Analogous relations for orthorhombic crystals are found in Barron et al. (1980).

Let the hexagonal unit cell have the conventional dimensions a and c. Then $d\varepsilon_1 = d\varepsilon_2 = da/a$ and $d\varepsilon_3 = dc/c$. Hence we can write

$$\alpha_\perp = (\partial \ln a/\partial T)_p, \tag{10.117}$$

$$\alpha_{||} = (\partial \ln c/\partial T)_p. \tag{10.118}$$

Further,

$$(\gamma_G)_\perp = (1/2C_\varepsilon)(\partial S/\ln a)_{T,c}, \tag{10.119}$$

$$(\gamma_G)_{||} = (1/C_\varepsilon)(\partial S/\partial \ln c)_{T,a}. \tag{10.120}$$

Munn (1969) has discussed the thermal expansion of Zn, Cd, Mg, Sn, In, Bi and Sb, which all have axial lattice structures. There is an empirical correlation that $(\gamma_G)_\perp > (\gamma_G)_{||}$ if $(c/a) > (c/a)_{\mathrm{ideal}}$ in the hcp lattices.

11.4. The generalisation of C_p–C_V

For isotropic materials, or crystals of cubic symmetry,

$$C_p - C_V = VT\beta^2 K_T. \tag{10.121}$$

All the quantities on the right are scalars, also for an anisotropic solid; C_p is the heat capacity measured with the stresses $\sigma_i = -p$ $(i = 1,2,3)$ and $\sigma_i = 0$ $(i = 4,5,6)$. Sometimes, (10.121) is used to find C_V from the measured C_p. However, this C_V is *not* the heat capacity at constant dimensions of the crystallographic unit cell. Instead it refers to constant volume *and* isotropic stress, i.e. there will be changes in the shape of the unit cell. We shall denote this heat capacity $\tilde{C}_V$. Thus,

$$\tilde{C}_V = C_p - VT\beta^2 K_T, \tag{10.122}$$

for a crystal of any symmetry. The generalisation of (10.121) to anisotropic solids is (Truesdell and Toupin 1960, Barron and Munn 1968)

$$C_\sigma - C_\varepsilon = VT \sum_{i,j=1}^{6} \alpha_i \alpha_j (c_{ij})_T. \tag{10.123}$$

Here the heat capacity C_σ refers to constant stress and C_ε to constant strain. When the specimen is under a hydrostatic pressure p, we have $C_\sigma = C_p$. In theoretical calculations, it is attractive to assume that not only the total volume, but also the shape of the unit cell is kept constant. It is therefore of interest to know how much C_ε differs from $\tilde{C}_V$ calculated by (10.122). In axial lattices, one has (Barron and Munn 1968)

$$\frac{\tilde{C}_V - C_\varepsilon}{C_\varepsilon} = \frac{2(TC_\varepsilon/V)((\gamma_G)_\perp - (\gamma_G)_{\|})^2}{(c_{11}+c_{12})_T + 2(c_{33})_T - 4(c_{13})_T}. \tag{10.124}$$

Finally, we note the following inequality, valid for any lattice symmetry (Barron and Munn 1968)

$$C_p \geqq \tilde{C}_V \geqq C_\varepsilon. \tag{10.125}$$

The last inequality becomes an equality if and only if $\gamma_1 = \gamma_2 = \gamma_3$.

Example: $\tilde{C}_V - C_\varepsilon$ in hexagonal lattices. Barron and Munn (1968) investigated $\tilde{C}_V - C_\varepsilon$ for several non-cubic solids. Table 10.3 is based on some of their results.

Table 10.3
$\tilde{C}_V - C_\varepsilon$ in hexagonal lattices

Material	$(\gamma_G)_\perp$	$(\gamma_G)_\parallel$	$(\tilde{C}_V - C_\varepsilon)_{mol}/3Nk_B$
Magnesium (283 K)	1.54	1.55	0.0000
Zinc (283 K)	2.15	1.98	0.0004
Zinc (600 K)	2.18	1.72	0.0084
Zirconium (300 K)	0.86	1.03	0.0012
Graphite (1000 K)	1.06	0.38	0.0028

11.5 The generalisation of $K_T C_p = K_S C_V$

For an isotropic solid, $K_T C_p = K_S C_V$, but in an anisotropic solid, $K_T C_\sigma \neq K_S C_\varepsilon$ in general. Then the definition of $\gamma_G(T;V)$ is not unique. We may consider

$$\gamma_G(T;\varepsilon) = \beta V K_T / C_\varepsilon = (V/C_\varepsilon)(\partial S/\partial V)_{T,\,\text{isotropic stress}}, \tag{10.126}$$

or

$$\gamma_G(S;\sigma) = \beta V K_S / C_\sigma = -(V/T)(\partial T/\partial V)_{S,\,\text{isotropic stress}}. \tag{10.127}$$

The quantity $\gamma_G(T;\varepsilon)$ has been discussed by Collins and White (1964) and $\gamma_G(S;\sigma)$ by Barron and Munn (1967). For an isotropic solid $\gamma(\boldsymbol{q},\lambda)$ is a constant, but in the expressions above, $\gamma(\boldsymbol{q},\lambda;\varepsilon)$ and $\gamma(\boldsymbol{q},\lambda;\sigma)$ depend on the temperature unless all ratios $s_{ij}(T)/s_{i'j'}(T)$ remain constant. However, the variations in $s_{ij}(T)/s_{i'j'}(T)$ are due to higher-order anharmonic effects and therefore are small. Barron and Munn (1967) estimate that, in zinc, $\partial \ln \omega(\boldsymbol{q},\lambda)/\partial \ln V$ at isotropic stress and constant S varies by a few percent from 0 K to room temperature.

12. Negative thermal expansion

We noted earlier that there are solids which shrink on heating, in particular at low temperatures. Although this phenomenon is not very rare, it is remarkable enough to deserve special attention. For an insulator of cubic lattice symmetry, a negative expansion is equivalent to a negative Grüneisen parameter $\gamma_G(T)$. We recall that $\gamma_G(T)$ is an average over all mode parameters $\gamma(\boldsymbol{q}, \lambda)$, weighted by the heat capacity $C_{har}(\boldsymbol{q}, \lambda)$. Certain solids, e.g. ionic compounds (RbI, RbCl) and covalent crystals (Si, Ge), have $\gamma(\boldsymbol{q}, \lambda) = -(\partial \ln \omega(\boldsymbol{q}, \lambda)/\partial \ln V) < 0$ for some transverse acoustic modes, while $\gamma(\boldsymbol{q}, \lambda) > 0$ for the longitudinal acoustic mode. At low temperatures, γ_G is dominated by the transverse modes (since they have the lowest frequencies and hence the largest C_{har} at a given T). Then γ_G is negative. At high temperatures, $C_{har} \to k_B$ for all modes, and the large and positive values of γ for the longitudinal mode outweigh the negative γ of the transverse branches. That makes $\gamma_G > 0$, and we have a positive thermal expansion. There seems to be no crystalline nonmagnetic solid for which $\gamma_G < 0$ when $T > \Theta_D$. Glassy materials usually behave as crystalline solids. At very low temperatures (< 1 K) some of them have $\beta < 0$, a result which has been interpreted in terms of two-level tunneling states (ch. 7 §2.1), although $\beta < 0$ is not universal (Ackerman and Anderson 1982, Kaspers et al. 1983, Piñango et al. 1983). Magnetic invar alloys have $\beta < 0$ at low temperatures (Steinemann 1978, 1979).

13. Pressure dependence of the expansion coefficient

Macroscopic thermodynamics gives (Wallace 1972)

$$\left(\frac{\partial \beta}{\partial p}\right)_T = \frac{1}{K_T^2}\left(\frac{\partial K_T}{\partial T}\right)_p. \qquad (10.128)$$

Most materials have $(\partial K_T/\partial T)_p < 0$ (ch. 3 §9) and therefore the coefficient of thermal expansion decreases under an external pressure.

14. Factors influencing the expansion coefficient

Usually, the thermal expansion has its major cause in the volume dependence of the phonon frequencies. It has been noted (ch. 4 §12) that the various Debye temperatures are not much affected by lattice defects and they vary smoothly with the composition in concentrated alloys and mixed crystals. The same behaviour is expected for the Grüneisen parameters (ch. 5 §6) and hence also for the expansion coefficient β. An analogous conclusion can be drawn for the electronic contribution to β in metals (§8). The small expansion coefficient in invar alloys depends on the cancellation of a phonon and a magnetic contribution to β. A small change in the alloy composition can give rise to a large relative change in β

CHAPTER 11

THE ELECTRICAL CONDUCTIVITY OF METALS AND ALLOYS

1. Introduction

There are two formulae for the electrical properties of metals, which are found in almost any textbook on solid state physics. One of them is Matthiessen's rule for the resistivity ρ:

$$\rho_{\text{tot}} = \rho_{\text{ep}} + \rho_{\text{def}}. \tag{11.1}$$

The other formula expresses the conductivity σ ($= 1/\rho$) as

$$\sigma = ne^2\tau/m. \tag{11.2}$$

Matthiessen's rule says that the total resistivity, ρ_{tot}, is the sum of contributions from the scattering of conduction electrons by the thermal vibrations (ep stands for electron–phonon) and the scattering by static lattice defects. The relation (11.1) is well obeyed in many cases, but not for highly resistive systems. For instance, it is not even qualitatively correct for zirconium alloys at high temperatures or for stainless steels (§10).

In eq. (11.2), e and m are the electron charge and mass, n is the electron number density and τ is an average relaxation time for the conduction electrons carrying the current. Matthiessen's rule follows from the relation

$$\frac{1}{\tau_{\text{tot}}} = \frac{1}{\tau_{\text{ep}}} + \frac{1}{\tau_{\text{def}}}. \tag{11.3}$$

The relation (11.2) is deceptively simple. It might give the impression that the electron number density n plays a central role. This can be true for doped semiconductors, where n varies by many orders of magnitude, but not for metals. The major problem is hidden in the electron

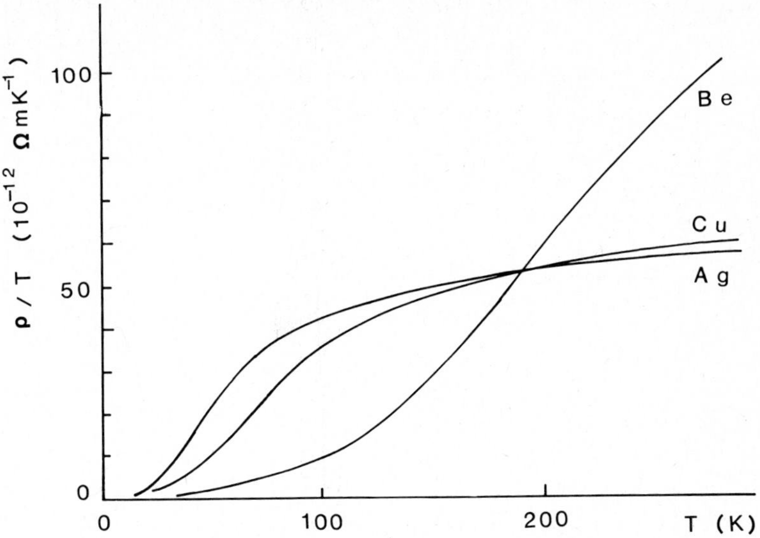

Fig. 11.1. The ratio $\rho(T)/T$ for Cu, Ag and Be. Data from Bass (1982).

relaxation time τ. Let us neglect the effect of impurities on τ and consider the phonon-limited resistivity ρ_{ep} $(= 1/\sigma_{ep})$. Grüneisen (1913) noted that, for pure metals, ρ/T and the heat capacity C_p have approximately the same temperature dependence (see also fig. 15.1). In fact, an Einstein model for the lattice vibrations leads to proportionality between $\rho_{ep}(T)/T$ and C_V; ch. 15, §2.

Figure 11.1 shows $\rho(T)/T$ for Ag, Cu and Be. Silver is the best elemental conductor at room temperature. We see from the figure that this is partly accidental. Beryllium has such a high Debye temperature ($\sim$ 900 K) that 300 K is in the low-temperature region. Figure 11.2 gives the resistivity of some other metals. The "normal" high-temperature behaviour is that ρ increases somewhat faster than linear in T, as for Al and W. Titanium provides an example of a tendency for "resistivity saturation" (§10). In iron, there is a significant contribution to ρ from scattering of electrons by persisting spin disorder above the Curie temperature.

Monographs by Dugdale (1977), Blatt (1968) and Meaden (1966) give a broad, but mostly elementary, survey of the electrical conductivity of metals.

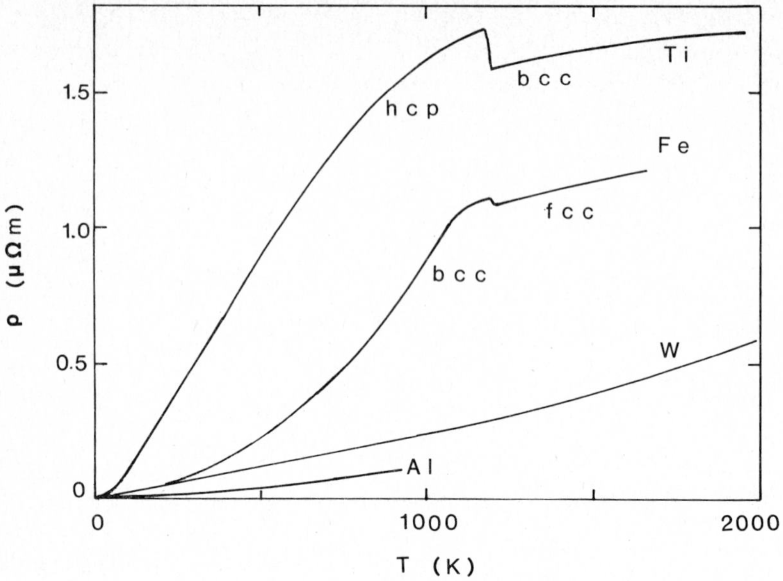

Fig. 11.2. The electrical resistivity $\rho(T)$ of Ti (hcp and bcc), Fe (bcc and fcc), W and Al. Data from Bass (1982).

2. General formulae for the electrical conductivity

2.1. Linear response to an electric field

An electron with the velocity $\boldsymbol{v}_{\boldsymbol{k}}$ carries the electric current $-e\boldsymbol{v}_{\boldsymbol{k}}$. In equilibrium, for each state $\boldsymbol{k}$ there is also a state $-\boldsymbol{k}$, with velocity $-\boldsymbol{v}_{\boldsymbol{k}}$, so that the total electric current vanishes. In the presence of an electric field $\boldsymbol{E}$, the Fermi–Dirac function is changed from its equilibrium value $f_{\boldsymbol{k}}^0$ to the value $f_{\boldsymbol{k}}$. If the field is weak, we can assume that $f_{\boldsymbol{k}}-f_{\boldsymbol{k}}^0$ is linear in $\boldsymbol{E}$ and write

$$f_{\boldsymbol{k}}-f_{\boldsymbol{k}}^0 = e(\boldsymbol{v}_{\boldsymbol{k}}\cdot \boldsymbol{E})\tau(\boldsymbol{k})(\partial f_{\boldsymbol{k}}^0/\partial \varepsilon_{\boldsymbol{k}}), \tag{11.4}$$

where $\varepsilon_{\boldsymbol{k}} = E(\boldsymbol{k})-E_{\mathrm{F}}$. The quantity $\tau(\boldsymbol{k})$ has the dimension of time. We shall call it a relaxation time, although it usually does not fulfill the requirements of such a quantity (viz. that the electron states decay exponentially in t/τ as the field $\boldsymbol{E}$ is turned off at time $t = 0$). In the following we may also write $\tau(\varepsilon_{\boldsymbol{k}})$, $\tau(\varepsilon)$, $\tau(\varepsilon_{\boldsymbol{k}}, \boldsymbol{k})$ or $\tau(\varepsilon, \boldsymbol{k})$, instead of $\tau(\boldsymbol{k})$, to stress that τ is energy dependent or both energy dependent and anisotropic. The actual calculation of $\tau(\boldsymbol{k})$ is difficult, but we can learn much just by assuming that it is a known quantity.

The total electric current density is

$$\boldsymbol{j} = -(2e/V)\sum_{\boldsymbol{k}} \boldsymbol{v}_{\boldsymbol{k}}(f_{\boldsymbol{k}} - f_{\boldsymbol{k}}^{0}). \tag{11.5}$$

The factor of 2 comes from the two spin directions. The current density $\boldsymbol{j}$ is related to the field $\boldsymbol{E}$ through the conductivity tensor $\boldsymbol{\sigma}$, such that

$$\boldsymbol{j} = \boldsymbol{\sigma}\boldsymbol{E}. \tag{11.6}$$

In a system with cubic symmetry, that tensor is diagonal, with all elements $\sigma_{ii} = \sigma$. Then, from (11.4)–(11.6),

$$\sigma = -(2e^2/3V)\sum_{\boldsymbol{k}} (\boldsymbol{v}_{\boldsymbol{k}}\cdot\boldsymbol{v}_{\boldsymbol{k}})\tau(\boldsymbol{k})(\partial f_{\boldsymbol{k}}^{0}/\partial\varepsilon_{\boldsymbol{k}}). \tag{11.7}$$

The usual prescription, eq. (B.1), for turning a sum over $\boldsymbol{k}$ into an integral gives

$$\sigma = -(2e^2/3)(2\pi)^{-3}\int_{S_{\mathrm{F}}} \frac{\boldsymbol{v}_{\boldsymbol{k}}\cdot\boldsymbol{v}_{\boldsymbol{k}}}{|\nabla_{\boldsymbol{k}}E(\boldsymbol{k})|}\,\mathrm{d}S \int_{-\infty}^{\infty} \tau(\varepsilon,\boldsymbol{k})(\partial f_{\boldsymbol{k}}^{0}/\partial\varepsilon_{\boldsymbol{k}})\,\mathrm{d}\varepsilon. \tag{11.8}$$

2.2. Special cases

We now consider several special cases of the main result expressed by eq. (11.7). First, assume that the density of states $N(\varepsilon, \boldsymbol{k})$ can be approximated by the Fermi-level value $N(0, \boldsymbol{k})$. Since n/m is equal to $S^0 v_{\mathrm{F}}^{0}/12\pi^3\hbar$ (S^0 and v_{F}^{0} are free-electron values) and $(-\partial f^0/\partial\varepsilon) = (1/k_{\mathrm{B}}T)f^0(1-f^0)$, (11.8) can be rewritten as

$$\sigma = \frac{ne^2}{mk_{\mathrm{B}}T}\int \frac{\boldsymbol{v}_{\boldsymbol{k}}\cdot\boldsymbol{v}_{\boldsymbol{k}}}{v_{\mathrm{F}}^{0}|\boldsymbol{v}_{\boldsymbol{k}}|}\frac{\mathrm{d}S}{S^0}\int_{-\infty}^{\infty} \tau(\varepsilon,\boldsymbol{k})f^0(\varepsilon)[1-f^0(\varepsilon)]\,\mathrm{d}\varepsilon. \tag{11.9}$$

With the approximation $|\boldsymbol{v}_{\boldsymbol{k}}|/v_{\mathrm{F}}^{0} = m/m_{\mathrm{b}}$, (11.9) is equivalent to

$$\sigma = \frac{ne^2}{m_{\mathrm{b}}}\langle\tau(\varepsilon,\boldsymbol{k})\rangle. \tag{11.10}$$

The average $\langle\tau(\varepsilon, \boldsymbol{k})\rangle$ is defined

$$\langle\tau(\varepsilon, \boldsymbol{k})\rangle = \int_{S_F} \frac{dS}{S^0} \int_{-\infty}^{\infty} \tau(\varepsilon, \boldsymbol{k})(-\partial f^0(\varepsilon)/\partial\varepsilon) d\varepsilon. \tag{11.11}$$

With a constant relaxation time τ, and a free-electron result for (11.11), we recover the familiar form

$$\sigma = ne^2\tau/m_b. \tag{11.12}$$

If $(-\partial f_{\boldsymbol{k}}^0/\partial\varepsilon_{\boldsymbol{k}})$ can be replaced by a delta-function, $\delta(\varepsilon_{\boldsymbol{k}})$, (11.7) becomes

$$\sigma = (2e^2/3V)\sum_{\boldsymbol{k}} v_{\boldsymbol{k}}^2\tau(\boldsymbol{k})\delta(\varepsilon_{\boldsymbol{k}}). \tag{11.13}$$

The Drude plasma frequency ω_{pl} is defined by a Fermi surface average, $\langle \boldsymbol{v}_{\boldsymbol{k}}^2\rangle$, of the electron velocity squared

$$\omega_{pl}^2 = [8\pi e^2 N(E_F)/3V]\langle \boldsymbol{v}_{\boldsymbol{k}}^2\rangle = (8\pi e^2/3V)\sum_{\boldsymbol{k}} \boldsymbol{v}_{\boldsymbol{k}}^2\delta(\varepsilon_{\boldsymbol{k}}). \tag{11.14}$$

If $\tau(\boldsymbol{k})$ in (11.13) can be averaged and separated out as a multiplicative factor τ, we have

$$\sigma = (\omega_{pl}^2/4\pi)\tau. \tag{11.15}$$

For free electrons, $\omega_{pl}^2 = 4\pi ne^2/m$. Then (11.15) is identical to (11.2). Experimental values of ω_{pl}^2 are reviewed by Foiles (1985).

In a nearly-free-electron model, and with $\langle\tau(\boldsymbol{k})\rangle = \tau$, the electron mean free path is

$$\ell = v_F\tau. \tag{11.16}$$

Since $v_F = \hbar k_F/m_b$ we get, by (A.4)

$$\sigma = ne^2\tau/m_b = ne^2\ell/m_b v_F = (3\pi^2)^{-1/3}n^{2/3}(e^2/\hbar)\ell. \tag{11.17}$$

This can be rewritten, with n expressed in terms of the parameter r_s and the Bohr radius a_0 (A.2), in the form

$$\rho = \pi(16\pi/3)^{1/3} r_s^2 \left(\frac{a_0}{\ell}\right)\left(\frac{a_0\hbar}{e^2}\right). \tag{11.18}$$

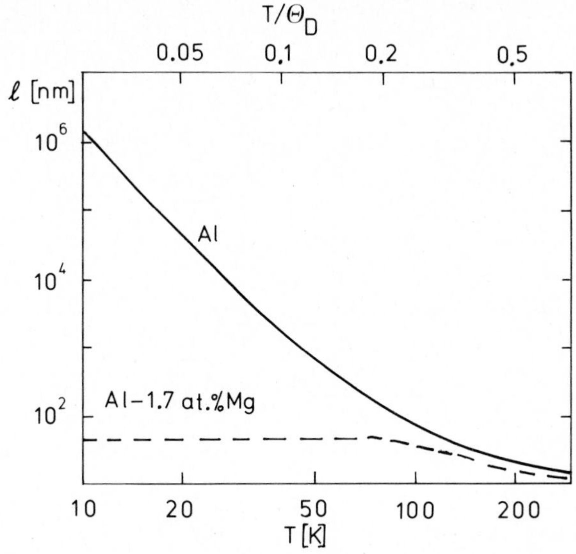

Fig. 11.3 The electronic mean free path in pure aluminium and in Al–1.7at.% Mg. From Grimvall (1981).

Typically, $\pi(16\pi/3)^{1/3} r_s^2 \sim 40$. The quantity $a_0\hbar/e^2 = 0.22\ \mu\Omega\text{m}$ is a fundamental unit of resistivity. From (11.18) we can obtain a rough idea of the mean free path ℓ (in non-transition metals) when the resistivity is known. Figure 11.3 shows ℓ calculated in this way for pure aluminium and an Al alloy.

3. The Boltzmann equation

3.1. Integral equations

The linearised Boltzmann equation gives an expression for $\tau(\boldsymbol{k})$ (Ziman 1960, Grimvall 1981):

$$e(\boldsymbol{v}_{\boldsymbol{k}} \cdot \boldsymbol{E})(\partial f_{\boldsymbol{k}}^0/\partial \varepsilon_{\boldsymbol{k}}) = (1/k_{\mathrm{B}}T) \sum_{\boldsymbol{k}'} (\Phi_{\boldsymbol{k}} - \Phi_{\boldsymbol{k}'}) P(\boldsymbol{k}, \boldsymbol{k}'), \tag{11.19}$$

where

$$\Phi_{\boldsymbol{k}} = -e(\boldsymbol{v}_{\boldsymbol{k}} \cdot \boldsymbol{E})\tau(\boldsymbol{k}), \tag{11.20}$$

and

$$P(\boldsymbol{k}, \boldsymbol{k}') = P^{\mathrm{ep}}(\boldsymbol{k}, \boldsymbol{k}') + P^{\mathrm{def}}(\boldsymbol{k}, \boldsymbol{k}') + P^{\mathrm{r}}(\boldsymbol{k}, \boldsymbol{k}'); \tag{11.21}$$

here P^{ep} refers to the scattering of electrons by phonons, P^{def} to the scattering by various kinds of lattice defects (impurity atoms, vacancies, grain boundaries, etc.) and P^{r} refers to remaining scattering mechanisms (e.g., scattering by other conduction electrons and by magnetic excitations).

The full solution of the Boltzmann equation (11.19) requires the knowledge of the interactions $P(\boldsymbol{k}, \boldsymbol{k}')$, but even without knowing this function we can derive several useful results. The Boltzmann equation can be rewritten in the form of a Fredholm integral equation. Let $\hat{\boldsymbol{E}}$ be a unit vector along the z-axes, so that $v_z(\boldsymbol{k}) = \boldsymbol{v}_{\boldsymbol{k}} \cdot \hat{\boldsymbol{E}}$. Then (11.19) becomes

$$\frac{1}{\tau(\boldsymbol{k})}\left(\frac{\partial f^0}{\partial \varepsilon_{\boldsymbol{k}}}\right) = -\frac{1}{k_{\mathrm{B}}T}\sum_{\boldsymbol{k}'}\left[1 - \frac{v_z(\boldsymbol{k}')}{v_z(\boldsymbol{k})}\frac{\tau(\boldsymbol{k}')}{\tau(\boldsymbol{k})}\right] P(\boldsymbol{k}, \boldsymbol{k}'). \tag{11.22}$$

For a material with cubic lattice symmetry, σ is independent of the direction $\hat{\boldsymbol{E}}$. Hence we can take $\hat{\boldsymbol{E}}$ and $\boldsymbol{k}$ to be parallel. If $\tau(\boldsymbol{k})/\tau(\boldsymbol{k}')$ does not depend strongly on $\boldsymbol{k}$ and $\boldsymbol{k}'$, and if $\boldsymbol{v}(\boldsymbol{k}')$ and $\boldsymbol{k}'$ are parallel, (11.22) yields (θ is the angle between $\boldsymbol{k}$ and $\boldsymbol{k}'$)

$$\frac{1}{\tau(\boldsymbol{k})}\left(\frac{\partial f^0}{\partial \varepsilon_{\boldsymbol{k}}}\right) = -\frac{1}{k_{\mathrm{B}}T}\sum_{\boldsymbol{k}'}(1 - \cos\theta_{\boldsymbol{k}\boldsymbol{k}'}) P(\boldsymbol{k}, \boldsymbol{k}'). \tag{11.23}$$

This formulation allows us to state more precisely the difference between a "scattering time" (or "lifetime"), and the quantity $\tau(\boldsymbol{k})$ to be used in the transport problem. With a factor of 1 instead of $(1-\cos\theta)$ in (11.23), $\tau(\boldsymbol{k})$ would be the lifetime of an electron state labelled $\boldsymbol{k}$. It can be calculated by the Golden Rule of quantum mechanics. In that case all scattering processes out of a state $\boldsymbol{k}$ are of equal importance. In transport problems we have to ask in which direction the electron goes after the scattering, since "backscattering" of electrons gives the largest contribution to the resistivity. This is expressed by the factor in the parenthesis [...] in eq. (11.22), or by the term $1-\cos\theta_{\boldsymbol{k}\boldsymbol{k}'}$ in (11.23).

We next assume the presence of different scattering mechanisms, labelled i. Let $\tau_i(\varepsilon_{\boldsymbol{k}}, \boldsymbol{k})$ be a solution to the Boltzmann equation when only a particular scattering mechanism i is present. If the $\tau_i(\varepsilon_{\boldsymbol{k}}, \boldsymbol{k})$ thus obtained for different i all have the same dependence on the energy $\varepsilon_{\boldsymbol{k}}$ and the wave vector $\boldsymbol{k}$ (i.e. they are just scaled by a factor) they also have the same factor $[1-v_z(\boldsymbol{k}')\tau(\boldsymbol{k}')/v_z(\boldsymbol{k})\tau(\boldsymbol{k})]$ in eq. (11.22). It then immediately follows that

$$\frac{1}{\tau_{\text{tot}}}=\sum_i \frac{1}{\tau_i}. \tag{11.24}$$

This is equivalent to Matthiessen's rule, eq. (11.1).

3.2. A variational solution

The Boltzmann equation is an integral equation, which can be solved by iteration. However, if we only want an approximate solution for the resistivity, we can use a variational procedure. This leaves us with the much simpler (although in practice not trivial) problem of evaluating integrals instead of solving an integral equation. The method was outlined by Kohler (1948, 1949) and Sondheimer (1950) and exploited in detail by Ziman (1960). Like other variational calculations, it gives a bound to the exact result. In our case it can be written

$$\rho < \frac{(V/2k_{\text{B}}T)\int \mathrm{d}^3\boldsymbol{k}\int \mathrm{d}^3\boldsymbol{k}'(\Phi_{\boldsymbol{k}}-\Phi_{\boldsymbol{k}'})^2 P(\boldsymbol{k},\boldsymbol{k}')}{e^2|\int(\boldsymbol{v}_{\boldsymbol{k}}\cdot\Phi_{\boldsymbol{k}})(\partial f^0(\boldsymbol{k})/\partial\varepsilon_{\boldsymbol{k}})\mathrm{d}^3\boldsymbol{k}|^2}, \tag{11.25}$$

where $P(\boldsymbol{k},\boldsymbol{k}')$ is the scattering operator introduced in eq. (11.21) and $\Phi_{\boldsymbol{k}}$ is a trial function to be used in the variational procedure. Since Φ appears squared both in the numerator and the denominator of (11.25), all Φ can be multiplied by the same constant without altering the estimation for ρ. The essence of a variational method is that one can choose Φ to be a function of one or several parameters λ_i. The right-hand side of (11.25), i.e. an upper bound to ρ, is then minimised with respect to all the λ_i. Often, one uses the function

$$\Phi_{\boldsymbol{k}}=\hat{\boldsymbol{E}}\cdot\boldsymbol{k}. \tag{11.26}$$

It has no parameters to vary, but since it gives the exact solution under certain idealised assumptions, it is assumed to give a reasonable value of ρ also in real metals. After some manipulation one finds that (11.25), with Φ as in (11.26), gives

$$\rho < \frac{m_{\text{b}}\langle 1/\tau(\boldsymbol{k})\rangle}{ne^2}. \tag{11.27}$$

This should be compared with the "exact" solution $\rho = 1/\sigma = (m/ne^2)/\langle\tau(\boldsymbol{k})\rangle$. Schwartz' inequality in mathematics implies that $\langle 1/\tau\rangle > 1/\langle\tau\rangle$, which is consistent with the inequality in (11.27). In a real calculation of ρ from the variational formula (11.25), one has to approximate $P(\boldsymbol{k}, \boldsymbol{k}')$, which may introduce errors of unknown sign. Therefore formulae such as (11.25), and relations derived from it, are usually (but not correctly) written with an equality sign.

4. Phonon-limited electrical conductivity

4.1. General considerations

Consider the scattering of an electron from a state $\boldsymbol{k}$ to a state $\boldsymbol{k}'$, under the emission or absorption of a phonon of wave vector $\boldsymbol{q}$ and energy $\hbar\omega(\boldsymbol{q}, \lambda)$. Here λ is a phonon mode index (longitudinal or transverse mode, acoustic or optical branches). Momentum conservation requires that

$$\boldsymbol{k}' = \boldsymbol{k} \pm \boldsymbol{q} + \boldsymbol{G}, \tag{11.28}$$

for phonon absorption and emission, respectively. The reciprocal wave vector $\boldsymbol{G}$ ensures that $\boldsymbol{q}$ lies in the first Brillouin zone. Whenever relevant, the labels $\boldsymbol{k}$ and $\boldsymbol{k}'$ of the electron states should also contain an electron band index (sp-band, d-band, etc.). Energy conservation requires that

$$\varepsilon_{\boldsymbol{k}'} = \varepsilon_{\boldsymbol{k}} \pm \hbar\omega(\boldsymbol{q}, \lambda). \tag{11.29}$$

The scattering probability function $P(\boldsymbol{k}, \boldsymbol{k}')$ introduced in eq. (11.21) can be calculated by the Golden Rule of quantum mechanics. It has the form

$$\begin{aligned} P(\boldsymbol{k}, \boldsymbol{k}') = (2\pi/\hbar) \sum_{\lambda} |g(\boldsymbol{k}, \boldsymbol{k}'; \lambda)|^2 \\ \times \{f^0(\varepsilon_{\boldsymbol{k}})[1 - f^0(\varepsilon_{\boldsymbol{k}'})]n(\boldsymbol{q}, \lambda)\delta(\varepsilon_{\boldsymbol{k}} - \varepsilon_{\boldsymbol{k}'} + \hbar\omega(\boldsymbol{q}, \lambda)) \\ + f^0(\varepsilon_{\boldsymbol{k}})[1 - f^0(\varepsilon_{\boldsymbol{k}'})][1 + n(\boldsymbol{q}, \lambda)]\delta(\varepsilon_{\boldsymbol{k}} - \varepsilon_{\boldsymbol{k}'} - \hbar\omega(\boldsymbol{q}, \lambda))\}. \end{aligned} \tag{11.30}$$

Here $g(\boldsymbol{k}, \boldsymbol{k}'; \lambda)$ is a matrix element for the scattering of an electron by a phonon. It does not depend explicitly on the temperature but may do so indirectly, for example through the thermal expansion. The explicit temperature dependence of ρ arises entirely from the Fermi–Dirac and Bose–Einstein factors, f^0 and n respectively. See also Ziman (1960) or Grimvall (1981) for details.

4.2. The Éliashberg transport coupling function $\alpha_{\mathrm{tr}}^2F(\omega)$

Consider the scattering of an electron from a state $\boldsymbol{k}$ to all states $\boldsymbol{k}'$ which have an energy $\varepsilon_{\boldsymbol{k}'} = \varepsilon_{\boldsymbol{k}} \pm \hbar\omega$. It is convenient to introduce a coupling function $\alpha_{\mathrm{tr}}^2F(\omega, \boldsymbol{k})$

$$\alpha_{\mathrm{tr}}^2F(\omega, \boldsymbol{k}) = \frac{V}{\hbar(2\pi)^3} \sum_{\lambda} \int_{S_{\mathrm{F}}} \frac{\mathrm{d}^2\boldsymbol{k}'}{|\boldsymbol{v}_{\boldsymbol{k}'}|}$$

$$\times \left(1 - \frac{\boldsymbol{v}_{\boldsymbol{k}} \cdot \boldsymbol{v}_{\boldsymbol{k}'}}{|\boldsymbol{v}_{\boldsymbol{k}}|^2}\right) |g(\boldsymbol{k}, \boldsymbol{k}'; \lambda)|^2 \delta(\hbar\omega - \hbar\omega(\boldsymbol{q}, \lambda)). \tag{11.31}$$

The fact that the integral is carried out on the Fermi surface S_{F} means that $\boldsymbol{k}$ and $\boldsymbol{k}'$ are assumed to be so close to S_{F} that we can take the Fermi surface values of $g(\boldsymbol{k}, \boldsymbol{k}'; \lambda)$, $\boldsymbol{v}_{\boldsymbol{k}}$ and $\boldsymbol{v}_{\boldsymbol{k}'}$. Sometimes, it is a good approximation to take

$$1 - \frac{\boldsymbol{v}_{\boldsymbol{k}} \cdot \boldsymbol{v}_{\boldsymbol{k}'}}{|\boldsymbol{v}_{\boldsymbol{k}}|^2} = 1 - \cos\theta_{\boldsymbol{v}\boldsymbol{v}'}, \tag{11.32}$$

where $\theta_{\boldsymbol{v}\boldsymbol{v}'}$ is the angle between $\boldsymbol{v}$ and $\boldsymbol{v}'$. In free-electron-like metals, it is common to take $1 - \boldsymbol{v}_{\boldsymbol{k}} \cdot \boldsymbol{v}_{\boldsymbol{k}'}/|\boldsymbol{v}_{\boldsymbol{k}}|^2 = 1 - \cos\theta_{\boldsymbol{k}\boldsymbol{k}'}$, where $\theta_{\boldsymbol{k}\boldsymbol{k}'}$, is the angle between $\boldsymbol{k}$ and $\boldsymbol{k}'$.

We also define a function $\alpha_{\mathrm{tr}}^2F(\omega)$. It gives the coupling between all electron states $\boldsymbol{k}$ and $\boldsymbol{k}'$ near the Fermi level, with an energy difference $|\varepsilon_{\boldsymbol{k}} - \varepsilon_{\boldsymbol{k}'}| = \hbar\omega$, but irrespective of their wave vectors:

$$\alpha_{\mathrm{tr}}^2F(\omega) = \int_{S_{\mathrm{F}}} \alpha_{\mathrm{tr}}^2 F(\omega, \boldsymbol{k}) \frac{\mathrm{d}^2\boldsymbol{k}}{|\boldsymbol{v}_{\boldsymbol{k}}|} \Big/ \int_{S_{\mathrm{F}}} \frac{\mathrm{d}^2\boldsymbol{k}}{|\boldsymbol{v}_{\boldsymbol{k}}|}. \tag{11.33}$$

The notation $\alpha_{\mathrm{tr}}^2F(\omega)$ is chosen because if $g(\boldsymbol{k}, \boldsymbol{k}'; \lambda)$ and $|\boldsymbol{v}_{\boldsymbol{k}}|$ are independent of $\boldsymbol{k}$ and $\boldsymbol{k}'$, $\alpha_{\mathrm{tr}}^2F(\omega)$ can be regarded as a coupling function which measures the importance of the phonons of energy ω in limiting the electrical conductivity. The label "tr" refers to "transport processes". A similar expression, $\alpha^2F(\omega)$, enters in Éliashberg's (1962) treatment of electron–phonon many-body effects; see Grimvall (1981). We obtain $\alpha^2F(\omega)$ by dropping the term $\boldsymbol{v}_k \cdot \boldsymbol{v}_{k'}/|\boldsymbol{v}_k|^2$ in (11.31). The Éliashberg coupling function allows us to write the total resistivity in a very simple, although approximate, form. In the variational expression for ρ, (11.25), we take $\Phi_k = \hat{\boldsymbol{E}} \cdot \boldsymbol{v}_k$. Using (11.14) for the Drude plasma frequency we then get

$$\rho = \frac{(4\pi)^2}{\omega_{\mathrm{pl}}^2} \int_0^{\omega_{\max}} \frac{(\hbar\omega/k_{\mathrm{B}}T)\alpha_{\mathrm{tr}}^2F(\omega)}{[\exp(\hbar\omega/k_{\mathrm{B}}T)-1][1-\exp(-\hbar\omega/k_{\mathrm{B}}T)]}\,\mathrm{d}\omega. \tag{11.34}$$

It is worth commenting on the accuracy of the expression (11.34). First, it requires that the Boltzmann equation is valid. This is true for many metals and alloys, but there are notable exceptions among highly resistive systems (§10). Secondly, (11.34) is based on the variational method. However, we do not vary anything, but simply take the estimation of ρ that results from the trial function $\hat{\boldsymbol{E}} \cdot \boldsymbol{v}_k$. The accuracy of this procedure is not very well known for transition metals, but it may be acceptable in free-electron-like metals. Since a true variational approach is very complicated, and still of somewhat uncertain value because its success depends on the choice of trial functions, there is usually no practical alternative to the use of $\Phi_k = \hat{\boldsymbol{E}} \cdot \boldsymbol{v}_k$ or $\Phi_k = \hat{\boldsymbol{E}} \cdot \boldsymbol{k}$. Further, we have assumed in (11.34) that $\alpha_{\mathrm{tr}}^2F(\omega)$ and ω_{pl}^2 are well approximated by their values exactly at the Fermi level. We therefore do not allow for the effect of a rapidly varying electron density of states $N(E_{\mathrm{F}})$ at the Fermi level. (However, there is still an energy dependence of $\tau(\varepsilon)$ arising from the functions f^0 and n.) In an actual calculation of the resistivity, it of course remains to find $g(\boldsymbol{k}, \boldsymbol{k}'; \lambda)$. This is not too difficult for free-electron-like metals but presents serious problems for transition metals.

4.3. Bloch–Grüneisen and related resistivity formulae

For about half a century, the Bloch–Grüneisen formula has been used to describe the temperature dependence of the phonon-limited electrical resistivity in metals. It is often given in the form

$$\rho_{\mathrm{BG}}(T) = \frac{c_1}{T} \int_0^{q_{\mathrm{D}}} \frac{q^5\,\mathrm{d}q}{[\exp(\hbar Cq/k_{\mathrm{B}}T)-1][1-\exp(-\hbar Cq/k_{\mathrm{B}}T)]}. \tag{11.35}$$

Here c_1 is a constant, specific for the metal under consideration, q is a phonon wave number lying between 0 and the Debye value q_D, C is the sound velocity and $\hbar Cq$ is a long-wavelength phonon energy. The subscript BG stands for "Bloch–Grüneisen". Putting $z = \hbar Cq/k_B T$ in (11.35) we have

$$\rho_{BG}(T) = \frac{c_2}{\theta_D}\left(\frac{T}{\theta_D}\right)^5 \int_0^{\theta_D/T} \frac{z^5\,dz}{(e^z-1)(1-e^{-z})} = \frac{c_2}{\theta_D}\left(\frac{T}{\theta_D}\right)^5 J_5(\theta_D/T), \tag{11.36}$$

where c_2 is another constant and J_5 is the Debye integral of order 5. At low temperatures, we need the value $J_5(\infty) = 124.4$, and at high temperatures ($T \gg \theta_D$) we need $J_5(T/\theta_D) \approx (1/4)(\theta_D/T)^4$. Then we can write

$$\rho_{BG}(T) = (124.4c_2/\theta_D)(T/\theta_D)^5, \qquad T \ll \theta_D, \tag{11.37}$$

and

$$\rho_{BG}(T) = (c_2/4\theta_D)(T/\theta_D), \qquad T \gg \theta_D. \tag{11.38}$$

In two famous papers, Bloch (1928, 1930) derived first the high-temperature and then the low-temperature results, (11.38) and (11.37). Grüneisen (1933) noted, empirically, that Bloch's low-temperature formula could be matched to the high-temperature version and thus provide a good account of the electrical resistivity at all temperatures.

The original derivation of eq. (11.36) rests on a number of assumptions which are usually not very well fulfilled, even for free-electron-like metals. However, it is well known that the Bloch–Grüneisen formula describes $\rho(T)$ for many real systems with a remarkable accuracy. This is easy to explain if we start from the variational estimation of ρ, expressed through the transport electron–phonon coupling $\alpha_{tr}^2F(\omega)$. As a model approximation, we represent this function by a power law in ω,

$$\alpha_{tr}^2F(\omega) = C_n\omega^n \tag{11.39}$$

for $0 < \omega < \omega_{max}$. If now $n = 4$ and $\hbar\omega_{max} = k_B\theta_D$, the expression (11.34), based on (11.25), yields exactly the Bloch–Grüneisen formula (11.36). The true function $\alpha_{tr}^2F(\omega)$ contains all complications regarding the phonon spectrum, the electron–phonon matrix elements, coupling to transverse phonons, Umklapp processes etc. Except at very low energies

ω, $\alpha_{tr}^2F(\omega)$ is reminiscent of the phonon density of states $F(\omega)$. Because ρ results from an integration over ω, the resistivity is determined by the gross features of $\alpha_{tr}^2F(\omega)$ but is insensitive to its finer details. This is analogous to the fact that a Debye model can give a good account of the phonon heat capacity, although $F(\omega)$ may deviate substantially from an ω^2 shape. The model (11.39) allows us to define generalised Bloch–Grüneisen formulae $\rho_{BG}(T;n)$. They all have $\rho \sim T$ at high temperatures. In the low temperature limit, $\rho_{BG}(T;n) \sim T^{n+1}$. Contrary to a widespread belief, the experimental resistivity usually does not vary as T^5 at low temperatures – not even for the free-electron-like metals Na and K. In fact, the full variational formula (11.25) is not quite adequate at low temperatures since there may be complications due to "phonon drag" and other effects (cf. Grimvall 1981). At intermediate and high temperatures (say $T > \theta_D/4$) any reasonable representation of $\alpha_{tr}^2F(\omega)$ gives a good account of the resistivity and ρ_{BG} is just one possible choice, which has no deeper justification. In the extreme case of an Einstein model, i.e.

$$\alpha_{tr}^2F(\omega) = C_0\delta(\omega-\omega_E), \tag{11.40}$$

where C_0 is a constant, ρ/T becomes proportional to the Einstein model heat capacity. Even this crude approximation gives a reasonable description of $\rho(T)$ in many metals, at not too low temperatures (fig. 15.1).

The high temperature limit of (11.34) can be cast in an interesting form. We write

$$\rho = \frac{(4\pi)^2(k_BT/\hbar)}{\omega_{pl}^2}\int_0^{\omega_{max}}\frac{\alpha_{tr}^2F(\omega)}{\omega}\,d\omega = \frac{8\pi^2k_BT}{\hbar\omega_{pl}^2}\lambda_{tr}, \tag{11.41}$$

where

$$\lambda_{tr} = 2\int_0^{\omega_{max}}\frac{\alpha_{tr}^2F(\omega)}{\omega}\,d\omega. \tag{11.42}$$

We noted above that the essential difference between $\alpha_{tr}^2F(\omega)$ and $\alpha^2F(\omega)$ is a factor which is approximately $\cos\theta_{\boldsymbol{kk}'}$. Therefore, it is not unreasonable to approximate $\alpha_{tr}^2F(\omega)$ by

$$\alpha_{tr}^2F(\omega) = \langle 1-\cos\theta\rangle\,\alpha^2F(\omega), \tag{11.43}$$

where $\langle 1-\cos\theta \rangle$ is some properly defined average. This leads to the high-temperature form

$$\rho = \frac{8\pi^2 k_B T \langle 1-\cos\theta \rangle}{\hbar\omega_{pl}^2}\lambda = \frac{2\pi m k_B T \langle 1-\cos\theta \rangle}{\hbar n e^2}\lambda. \tag{11.44}$$

In the last equality we assumed a free-electron expression for ω_{pl}^2. The parameter $\lambda (= 2\int F(\omega)(d\omega/\omega))$ is the same as that appearing in the electron–phonon enhancement factor $1+\lambda$ of the electronic heat capacity; ch. 8. Obviously, we can identify τ with $\hbar/2\pi\lambda k_B T$; cf. ch. 8 §3.3.

5. Non-cubic lattices

Quite generally, the conductivity $\boldsymbol{\sigma}$ (or resistivity $\boldsymbol{\rho}$) is a 3×3 tensor of rank two, with elements σ_{ij} (ρ_{ij}). If a tensor of rank two has cubic symmetry, it reduces to a scalar, i.e. $\sigma_{ii} = \sigma$ while $\sigma_{ij} = 0$ when $i \neq j$ ($\rho_{ii} = \rho$ and $\rho_{ij} = 0$). In the general case,

$$\boldsymbol{j} = \boldsymbol{\sigma}\boldsymbol{E}, \qquad \boldsymbol{E} = \boldsymbol{\rho}\boldsymbol{j}, \tag{11.45}$$

and the current $\boldsymbol{j}$ is not parallel to the field $\boldsymbol{E}$ except in certain directions of high symmetry.

We shall exemplify a non-scalar conductivity by a material of hexagonal symmetry (the same relations hold for trigonal and tetragonal symmetry). Let θ be the angle between the current $\boldsymbol{j}$ and the c-axis. The resistivity $\rho(\theta) = E(\theta)/|\boldsymbol{j}|$ is

$$\rho(\theta) = \rho_{||}\cos^2\theta + \rho_{\perp}\sin^2\theta. \tag{11.46}$$

This describes current flow in a wire, for example. Next we let the field $\boldsymbol{E}$ have a specified angle θ to the c-axis (e.g., $\boldsymbol{E}$ is applied across a slab). The conductivity $\sigma(\theta) = j(\theta)/|\boldsymbol{E}|$ is

$$\sigma(\theta) = \sigma_{||}\cos^2\theta + \sigma_{\perp}\sin^2\theta. \tag{11.47}$$

Note that although $\sigma_{\perp} = 1/\rho_{\perp}$ and $\sigma_{||} = 1/\rho_{||}$, $\sigma(\theta) \neq 1/\rho(\theta)$. A theoretical account of the anisotropy of ρ_{ep} is difficult, because it requires a detailed consideration of the scattering process over the Fermi surface (e.g., Chan

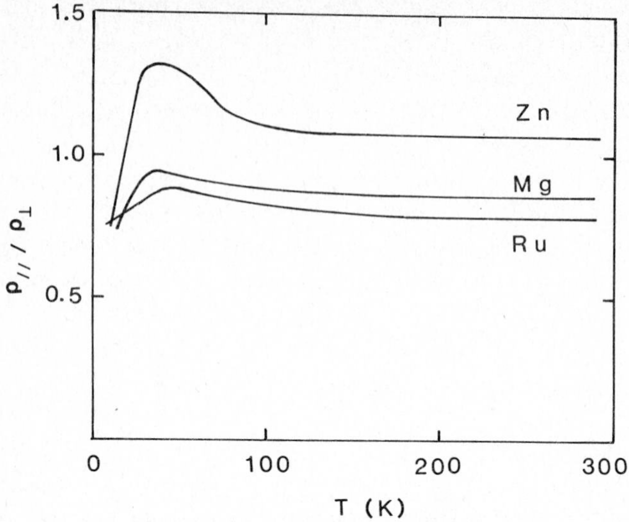

Fig. 11.4. The anisotropy of the electrical resistivity, given as $\rho_{\|}(T)/\rho_{\perp}(T)$, for hcp Zn, Mg and Ru. Data from Bass (1982).

1978, Lawson and Guénault 1982). Figure 11.4 shows how $\rho_{\|}/\rho_{\perp}$ may vary with the temperature in pure metals.

6. Defect scattering

6.1. Dilute alloys

Matthiessen and Vogt (1864) noted that the temperature dependence of the resistivity was not affected by small amounts of alloying elements:

$$\mathrm{d}\rho_{\mathrm{tot}}(T, c)/\mathrm{d}T = \mathrm{d}\rho_{\mathrm{ep}}(T)/\mathrm{d}T. \tag{11.48}$$

The rule is usually quoted in the integrated form

$$\rho_{\mathrm{tot}}(T, c) = \rho_{\mathrm{ep}}(T) + \rho_{\mathrm{def}}(c). \tag{11.49}$$

Corrections to this result are known as "deviations from Matthiessen's rule" (DMR). Mathematically, we can trace their origin to the relation

$$\frac{1}{\tau_{\mathrm{tot}}(\varepsilon, \boldsymbol{q})} = \frac{1}{\tau_{\mathrm{ep}}(\varepsilon, \boldsymbol{q})} + \frac{1}{\tau_{\mathrm{def}}(\boldsymbol{q})}. \tag{11.50}$$

If we consider $\rho \sim \langle 1/\tau_{\text{tot}} \rangle$, as in the variational expression (11.27), Matthiessen's rule is obtained from (11.50). However, we should rather consider $\rho \sim 1/\langle \tau_{\text{tot}} \rangle = 1/\langle \tau_{\text{ep}}\tau_{\text{def}}/(\tau_{\text{ep}} + \tau_{\text{def}}) \rangle$, which becomes $\langle 1/\tau_{\text{tot}} \rangle$ only if $\tau_{\text{ep}}(\varepsilon, \boldsymbol{k})$ and $\tau_{\text{def}}(\varepsilon, \boldsymbol{k})$ differ by a (temperature dependent) factor. In this way we see how (11.50) gives rise to a DMR. The fact that $\tau_{\text{ep}}(\varepsilon, \boldsymbol{q})$ is energy dependent, through ε, while $\tau_{\text{def}}(\boldsymbol{q})$ refers to elastic scattering and is independent of ε, causes a DMR which is peaked at $T \sim 0.1\theta_{\text{D}}$ and is of the order of $0.1\rho_{\text{def}}$ or (much) less (Engquist and Grimvall 1980). Differences in the anisotropy of τ_{ep} and τ_{def} also lead to a DMR. A magnetic field may reduce the DMR (Mitchel et al. 1980). Bass (1972) and Cimberle et al. (1974) have reviewed work on DMR. Bass (1982) gives numerical results for the resistivity in dilute alloys, also covering other aspects than DMR.

Impurities not only give a static scattering but they also change the vibrational properties and hence alter $\alpha_{\text{tr}}^2 F(\omega)$. Usually there is no dramatic effect caused by heavy (Kus and Taylor 1980) or light (Kus and Taylor 1982) impurities. An interesting case is the ordered compound $Al_{10}V$. There is a localised mode of low energy, with a resistivity contribution which is well described by an Einstein model (Caplin and Nicholson 1978).

6.2. *Concentrated alloys*

For low impurity concentrations, c, ρ_{def} varies linearly with c:

$$\rho_{\text{def}}(c) = c\rho^*_{\text{def}}. \tag{11.51}$$

Some values of ρ^*_{def} are given in table 11.1 (Bass 1982).

Table 11.1
Impurity resistance in dilute alloys of Al and Ti, in units of $\mu\Omega$ cm/at. %

Al–Si	Al–Fe	Al–Mg	Al–Cu	Al–Ti	Ti–Al	Ti–Fe	Ti–Zr
0.76	6.1	0.44	0.80	5.9	13	6	1.8

In a system $A_{1-c}B_c$, the linear dependence on the impurity content must hold for both $c \approx 0$ and $1 - c \approx 0$. The simplest relation which interpolates between these limits is Nordheim's (1931) rule,

$$\rho_{\text{def}}(c) = c(1-c)\rho^*_{\text{def}}. \tag{11.52}$$

Qualitatively, this rule is observed in concentrated alloys, but the transport process is too complicated to allow a more precise, and yet accurate, relation.

6.3. Lattice defects

The effect of vacancies and other intrinsic defects is similar to that of impurity atoms. In thermal equilibrium the magnitudes of these effects are very small. For instance, the resistivity per 1 at.% vacancies in elemental metals is a few $\mu\Omega$ cm (Wollenberger 1982). The resistivity increase in heavily cold-worked copper is $\sim 2\times10^{-4}$ $\mu\Omega$m (Powell et al. 1959). Still, such small changes in the resistivity are of interest in studies of the recovery of a specimen on annealing, after it has been cold-worked, irradiated etc. Complications in the interpretation arise when point defects migrate to dislocation cores and grain boundaries (e.g., Kasen 1972). Grain boundaries are considered in §7. Scattering by precipitates, in particular Guinier–Preston zones, has attracted much interest, and the field is controversial (Hillel et al. 1975, Hillel 1983, Rossiter 1976, Osamura et al. 1982).

7. Size effects

When the electron mean free path ℓ becomes comparable with a characteristic dimension of the specimen, surface scattering gives corrections to the resistivity of large specimens, ρ_{bulk}. (This requires pure materials and low temperatures.) For thin films, Fuchs (1938) and Sondheimer (1952) have developed a simple model, which gives the first-order correction in the small parameter ℓ_{bulk}/d, where d is the film thickness,

$$\rho_{\text{film}} = \rho_{\text{bulk}}[1+(3/8)(\ell_{\text{bulk}}/d)(1-p)]. \tag{11.53}$$

The parameter p $(0<p<1)$ measures the amount of specular character in the surface scattering. Very similar formulae hold for thin wires. This early theory has been improved (Soffer 1967) to give good agreement with experiments (Stesmans 1982). Thin films may be very fine-grained. Then, grain boundary scattering is not negligible (van Attekum et al. 1984, Mayadas et al. 1969). Bass (1982) gives a brief theoretical introduction and an extensive survey of experimental results on size effects in electron scattering.

8. Pressure dependence

We start from the expression for the phonon-limited resistivity, (11.41):

$$\rho_{ep} = (8\pi^2 k_B T/\hbar\omega_{pl}^2)\lambda_{tr}. \qquad (11.54)$$

In our treatment of the electronic part of the expansion coefficient (ch. 10 §8) we discussed the volume dependence of the electron–phonon interaction parameter λ. Since λ_{tr} is closely related to λ we can take the same approach here, giving

$$\mathrm{d}\ln\rho/\mathrm{d}\ln V = -2(\mathrm{d}\ln\omega_{pl}/\mathrm{d}\ln V)+2\gamma. \qquad (11.55)$$

where γ may be (crudely) approximated by the Grüneisen parameter γ_G. In a free-electron model, $\omega_{pl}^2 = 4\pi ne^2/m$, i.e. $\mathrm{d}\ln\omega_{pl}/\mathrm{d}\ln V = -1/2$. Then the right-hand side of (11.55) becomes $1+2\gamma_G$. Similar results have been obtained for alkali metals, in pioneering work by Dugdale and Gugan (1962). In transition metals, there are other important terms in $\mathrm{d}\ln\lambda_{tr}/\mathrm{d}\ln V$ than just $2\gamma_G$, and a detailed consideration of the various contributions to $\mathrm{d}\rho_{ep}/\mathrm{d}V$ is necessary (Sundqvist et al. 1985). Dugdale and Myers (1985) have briefly reviewed the theory of the pressure (i.e. volume) dependence of the electrical resistivity and they give an extensive survey of experiments.

9. Relation to the electron density of states

In the simplest model, the band structure of the electrons enters the expression of ρ in the form of the Fermi velocity v_F. More generally, ρ is proportional to the Drude plasma frequency ω_{pl}, but there is also a dependence on the electron band structure in the quantum mechanical matrix element for the electron–phonon interaction, i.e. in our parameter λ_{tr}. Without a very detailed analysis, one cannot say how the resistivity is affected by certain features in the band structure, but generally one expects that ρ increases with $N(E_F)$ (Mott 1972, Inoue and Shimizu

1976). In this context, one may note that a strong dependence of $N(E)$ on E gives a less pronounced variation in ω_{pl} (Mattheiss et al. 1978).

10. Saturation effects

According to the simple theory, (11.38) or (11.44), ρ increases linearly in T when $T > \theta_{\mathrm{D}}$. Thermal expansion, combined with the fact that ρ varies with the volume mainly through phonon frequencies, gives an additional slight increase of ρ with the temperature. This is the "normal" behaviour, shown, for example, by Al and W (fig. 11.2). Titanium (fig. 11.2) has an anomalous temperature dependence of ρ which cannot be explained by a rapidly varying $N(E)$ near the Fermi level. Instead, we have an example of what has been termed "resistivity saturation" (Fisk and Webb 1976). Empirically, the shape of $\rho(T)$ is very well approximated by the "shunt resistor model" (Wiesmann et al. 1977):

$$\frac{1}{\rho(T)} = \frac{1}{\rho_{\mathrm{ideal}}(T)} + \frac{1}{\rho_{\mathrm{sat}}}. \tag{11.56}$$

Here ρ_{ideal} is the "normal" resistivity, and ρ_{sat} is a value which the actual resistivity would take when ρ_{ideal} becomes very large (either because T is high or because ρ_{def} is large). A consequence of (11.56) is that $\mathrm{d}\rho/\mathrm{d}T < \mathrm{d}\rho_{\mathrm{ideal}}/\mathrm{d}T$. In fact, $\mathrm{d}\rho/\mathrm{d}T$ is experimentally found to decrease with increasing ρ and becomes negative for highly resistive materials ($\rho \sim 1.5\,\mu\Omega\,\mathrm{m}$), such as certain steels and amorphous metals. This is known as Mooij's rule (Mooij 1973). Qualitatively one may understand the phenomenon of saturation as a result of the short electron mean free path ℓ. Naturally, it cannot be shorter than the distance between two scattering centers, i.e. ℓ must be larger than the diameter of an atom. This sets a universal (but crude) upper limit to ρ. In spite of much effort (see reviews by Allen 1980a, b) there is no generally accepted theory which explains saturation, and its relation to phenomena like Anderson localisation (e.g., Kaveh and Mott 1982, Howson 1984).

11. Scattering by magnetic disorder

The resistivity of iron (fig. 11.2) shows a tendency to saturation. Although the mechanism discussed in the previous section may be applicable, the main cause of a large, and apparently temperature independent, resistivity is scattering by spin disorder above the Curie temperature T_c. At first sight, this may seem to be at variance with theories of magnetism. In the usual band model of magnetism in metals, as presented in elementary textbooks, the bands for "spin up" and "spin down" states are displaced (on an energy scale) in the magnetically ordered state, but when $T > T_c$, the two spin-states are equally populated. Then there should be no difference between, say, Fe and W as regards the nature of the paramagnetic state. However, there is clear evidence of short-range spin disorder also well above T_c. The effect is particularly pronounced in Fe, where it seems to be present even at the melting temperature T_m. Other ferromagnets may have a much more rapid decay of the disorder as one goes beyond T_c. These problems are controversial, and it is not possible to give a simple theoretical account of the resistivity.

12. Magnetoresistance

A magnetic field H usually increases the electrical resistance, by an amount $\Delta\rho = \rho(H) - \rho(H = 0)$. There are many intricate problems related to magnetoresistance, and we just quote the Kohler (1938) rule ($\rho_0 = \rho(H = 0)$):

$$\Delta\rho/\rho_0 = F(H/\rho_0). \tag{11.57}$$

Here $F(H/\rho)$ is a universal function, which only depends on the metal and the geometrical configuration of field and specimen. The condition of a weak field may be written on the dimensionless form $(eH/mc)\tau \ll 1$. Here we recognise the well-known cyclotron frequency eH/mc.

13. Theoretical calculations

A theoretical calculation of ρ_{ep} in non-transition metals is in principle straight-forward, but in practice there are many details to take into consideration. Among many papers we may note those of Shukla and Taylor (1976) on Na and K and Tomlinson (1979) on (anisotropic) Zn. There are very few papers which give a realistic treatment of transition metals, for instance Yamashita and Asano (1974) on Mo and Nb, Yamashita et al. (1975) on ferromagnetic Ni and Pinski et al. (1978) on Pd. The work by Mertig et al. (1982) exemplifies a calculation of the resistivity due to impurity atoms (transition metals in Cu).

CHAPTER 12

THERMAL CONDUCTIVITY

1. Introduction

Metals are usually characterised by a high thermal conductivity κ. The heat is carried by the conduction electrons, and an account of the thermal conductivity closely parallels that of the electrical conductivity σ. The Wiedemann–Franz law, $\kappa = L_e \sigma T$, connects the two phenomena through the Lorenz number L_e. But it is an insulator – diamond – which has the highest known thermal conductivity of any material at room temperature. In that case, the heat conduction is by phonons.

Figure 12.1 exemplifies the characteristic temperature dependence of the thermal conductivity in a pure metallic element (Al), a concentrated alloy (stainless steel), a pure insulator (MgO) and a strongly disordered insulator (glass). Pure metals and insulators have a maximum in the thermal conductivity, much below room temperature. With increasing lattice disorder of various kinds, this maximum decreases in height and shifts to somewhat higher temperatures. In very impure or disordered materials, the maximum is absent.

In this book, the emphasis is on what affects the thermal conductivity of real materials. More detailed theoretical treatments are found in reviews by Klemens (1958, 1969) and Beck et al. (1974). Parrot and Stuckes (1975) and Berman (1976) give general reviews of the thermal conductivity, partly along the lines presented here, while Slack (1979) has reviewed the thermal conductivity of insulators at high temperatures. Ziman (1960) gives a detailed account, with ample references to work before 1958.

It will be characteristic of our theoretical models, in particular those referring to the lattice part of the conductivity, that they only give qualitative descriptions. Thus, they provide a scheme for semiempirical analysis and for the establishment of trends, but we cannot yet hope for accurate numerical predictions.

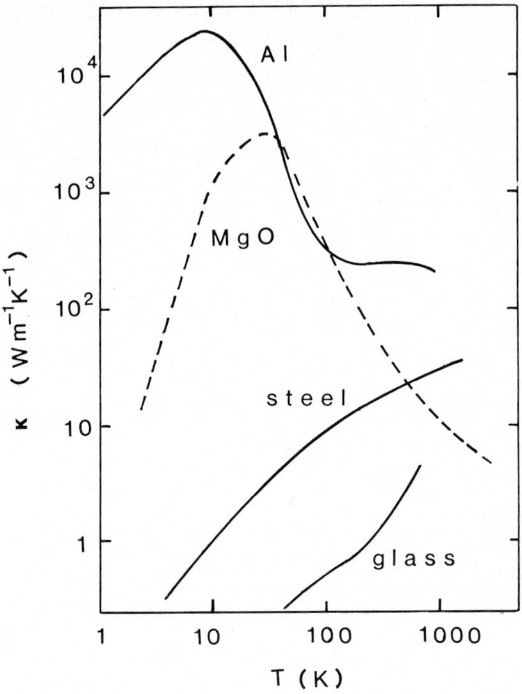

Fig. 12.1. The thermal conductivity $\kappa(T)$ of pure aluminium, pure MgO, a glass and a stainless steel. Data from Touloukian et al. (1970).

2. Macroscopic relations

2.1. Thermal conductivity

The heat flow $\boldsymbol{Q}$ in a temperature gradient ∇T is, according to Fourier's law,

$$\boldsymbol{Q} = -\kappa(\nabla T). \tag{12.1}$$

To be more precise, κ is a tensor with components κ_{ij} $(i, j = x, y, z)$ and

$$Q_i = -\sum_{j=1}^{3} \kappa_{ij}(\nabla T)_j. \tag{12.2}$$

For a material in which κ_{ij} has cubic symmetry, $\kappa_{ii} = \kappa$ when $i = 1, 2, 3$

and $\kappa_{ij} = 0$ when $i \neq j$. Then, the heat flow is parallel to the temperature gradient and (12.1) holds with a scalar κ. Often we shall refer to the thermal resistivity W. It is the inverse of the conductivity:

$$W = 1/\kappa. \tag{12.3}$$

In non-cubic lattices, the tensors $\boldsymbol{\kappa}$ and $\boldsymbol{W}$ have the same symmetry properties as the corresponding electrical quantities, ch. 11 §5. Table 12.1 gives $\kappa_{\perp}$ and $\kappa_{\parallel}$ in some axial crystals (data from Touloukian et al. 1970).

Table 12.1
Components of the thermal conductivity in some axial crystals (at 300 K)

	Sn (white)	Cd	Bi	TiO_2	Graphite (pyrolytic)
$\kappa_{\perp}$[W/m K]	74	104	9	7	2000
$\kappa_{\parallel}$[W/m K]	52	83	5	10	10

In metals, the heat transport by electrons and phonons can be considered as independent. Then the total thermal conductivity is

$$\kappa_{\text{tot}} = \kappa_{\text{e}} + \kappa_{\text{ph}}. \tag{12.4}$$

Equation (12.4) does not mean that the electron–phonon interaction has been neglected, but only that we may consider separate Boltzmann equations for the electrons and the phonons. Note that the thermal resistivities of different heat carriers are not additive,

$$W_{\text{tot}} \neq W_{\text{e}} + W_{\text{ph}}, \tag{12.5}$$

although that may hold for the contributions from various scattering mechanisms for electrons or phonons, considered separately (Matthiessen's rule).

2.2. Thermal diffusivity

The physicist is primarily interested in the thermal conductivity κ, while measurements often give the thermal diffusivity a. The two quantities are related by

$$a = \frac{\kappa}{c_p \rho}. \tag{12.6}$$

Here c_p is the specific (i.e. per unit mass) heat capacity at constant pressure and ρ is the mass density. Fourier's equation (the heat equation) contains the thermal diffusivity:

$$\frac{\partial T}{\partial t} - a(\nabla^2 T) = 0. \tag{12.7}$$

When the thermal conductivity is not uniform throughout the sample, (12.7) is replaced by

$$c_p\rho\frac{\partial T}{\partial t} - \nabla(\kappa\nabla T) = 0. \tag{12.8}$$

Note that κ may be spatially varying not only because of an inhomogeneous material but also because κ is temperature dependent and the sample is in a temperature gradient. The heat equation coupled to the equation of state (i.e. with allowance for thermal expansion) was considered in ch. 10 §3.

Example: Relative magnitude of thermal conductivity and diffusivity. When $T > \theta_D$, C_p is approximately $3k_B$ per atom. If M is the (average) mass of an atom in the specimen, N the total number of atoms, V the specimen volume and $\Omega_a = V/N$ the volume per atom, we have from (12.6) ·

$$\frac{\kappa}{a} = c_p\rho \approx \left(\frac{3k_B}{M}\right)\left(\frac{M}{\Omega_a}\right) = \frac{3k_B}{\Omega_a}. \tag{12.9}$$

Since Ω_a does not vary nearly as much as the thermal conductivity between different materials, the material with the higher thermal conductivity usually also has the higher thermal diffusivity, provided that $T > \theta_D$. Figure 12.2 exemplifies how $a(T)$ and $\kappa(T)$ co-vary at intermediate and high temperatures.

3. Lattice thermal conductivity

3.1. General considerations

The heat flow by phonons will be referred to as the lattice thermal conductivity. In textbooks, its mathematical description usually starts

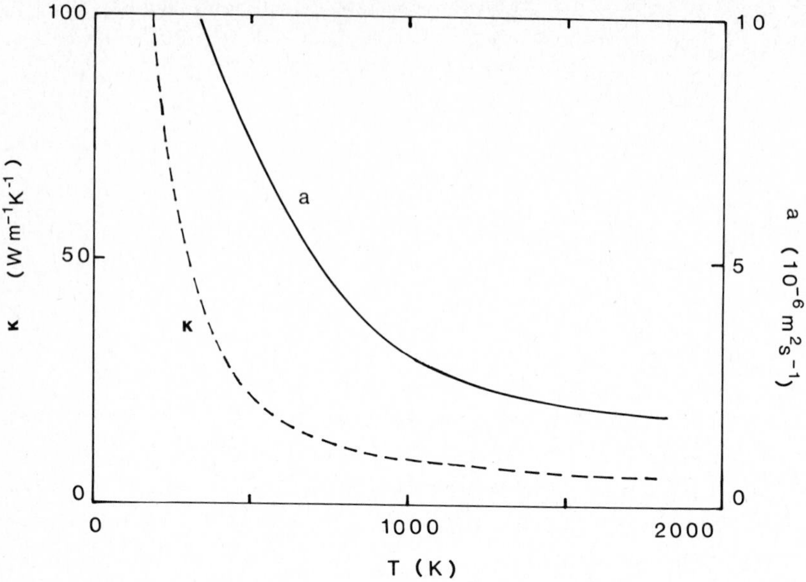

Fig. 12.2. The thermal diffusivity $a(T)$, right scale, and the thermal conductivity $\kappa(T)$, left scale, of MgO. Data from Touloukian et al. (1970, 1973).

from the theory of thermal conduction in a classical gas. This does not mean that one has resorted to a classical description, since the scattering rate is calculated using quantum mechanics. We may think of phonons as a gas of particles, characterised by a momentum $\hbar \boldsymbol{q}$ ($\boldsymbol{q}$ is the wave vector) and an energy $\hbar\omega(\boldsymbol{q}, \lambda)$. In an ideal gas, there are no collisions between the particles. That leads, in the case of a classical gas, to an infinite thermal conductivity. Likewise, the lattice thermal conductivity is infinite if the phonon mean free path is not limited by collisions – between phonons themselves or between phonons and lattice imperfections. In a collision event, energy and momentum are conserved, but since the phonons are massless particles they may be annihilated or created. The simple picture just outlined allows us to reach a qualitative understanding of characteristic features in the temperature dependence of κ. Figure 12.3 shows a typical temperature dependence of κ in a non-metallic crystal. There are four regions, denoted A, B, C and D, which we shall explain.

In a classical gas one has the well-known formula

$$\kappa = (1/3) n c_V \bar{C} \ell. \tag{12.10}$$

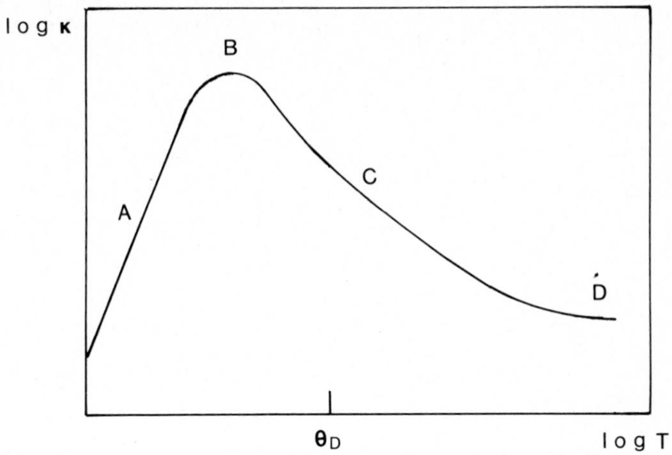

Fig. 12.3. A schematic plot of log(κ) versus T. The four regions A, B, C and D are discussed in the text.

Here n is the number of particles per volume, c_V is the heat capacity per particle (i.e. $C_V = nc_V$ is the heat capacity per volume), $\bar{C}$ is the (average) velocity of a gas particle in the thermal equilibrium, and ℓ is its mean free path. In a solid, the lattice thermal conductivity is given by an expression of the form (12.10) but we must sum over all phonon modes $(\boldsymbol{q}, \lambda)$ and let C_V, $\bar{C}$ and ℓ depend on $(\boldsymbol{q}, \lambda)$. The phonons are not true particles but waves, characterised by a phase velocity and a group velocity. In this case, $\bar{C}$ in (12.10) should be replaced by the group velocity $C_g(\boldsymbol{q}, \lambda) = |\nabla_{\boldsymbol{q}}\omega(\boldsymbol{q}, \lambda)|$. Then we have for the lattice thermal conductivity $\kappa = \kappa_{ph}$:

$$\kappa_{ph} = (1/3)\sum_{\boldsymbol{q},\lambda} C_V(\boldsymbol{q}, \lambda)C_g(\boldsymbol{q}, \lambda)\ell(\boldsymbol{q}, \lambda). \tag{12.11}$$

An accurate evaluation of this expression is a formidable task, but it is easy to draw some qualitative conclusions. The phonon mean free path ℓ is limited by collisions with other phonons and with lattice defects of various kinds. We assume that these mechanisms are additive so that

$$\frac{1}{\ell_{tot}(\boldsymbol{q}, \lambda)} = \frac{1}{\ell_{ph-ph}(\boldsymbol{q}, \lambda)} + \sum \frac{1}{\ell_{ph-def}(\boldsymbol{q}, \lambda)}. \tag{12.12}$$

The first term on the right-hand side refers to phonon–phonon collisions and the next terms to phonon scattering by faults such as point defects, dislocations, grain and phase boundaries and the finite size of the sample. The discussion is simplified if only one of the terms in (12.12) dominates at a time and we make that approximation. A calculation of κ_{ph} from (12.11) requires so many simplifying approximations that there is no point in using the precise phonon spectrum. We shall later be content with a Debye model, but we first discuss, qualitatively, the four regions in fig. 12.3.

Region C. The number of phonons $(\boldsymbol{q}, \lambda)$ present in thermal equilibrium is given by the Bose–Einstein statistical factor

$$n(\boldsymbol{q}, \lambda) = \frac{1}{\exp[\hbar\omega(\boldsymbol{q}, \lambda)/k_B T] - 1} \approx \frac{k_B T}{\hbar\omega(\boldsymbol{q}, \lambda)}. \tag{12.13}$$

When $T > \theta_D$, we take the leading high-temperature term. Phonon collisions which limit the thermal conductivity must involve three or more phonons. The simplest case arises when two phonons combine, in a "collision" event, and form one new phonon (fig. 12.4a) or when one phonon annihilates to form two phonons (fig. 12.4b). It is intuitively clear that the probability for the process depicted in fig. 12.4a varies as

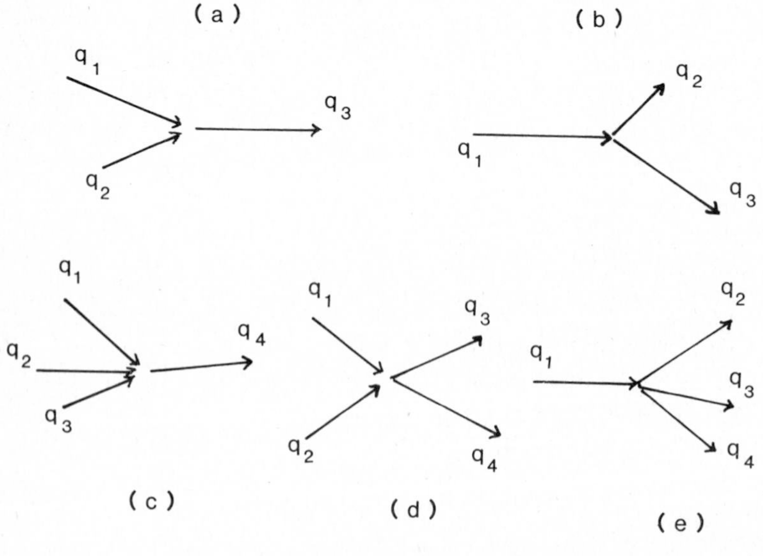

Fig. 12.4. Three-phonon (a, b) and four-phonon (c, d, e) scattering processes.

the number of phonons present to collide with, i.e. as some average n of all $n(\boldsymbol{q},\lambda)$. Since $n \sim T$ at high temperatures, we then expect that scattering such as in fig. 12.4a (and, it can be argued, such as in fig. 12.4b) leads to $\ell \sim 1/T$. There may also be four-phonon processes – figs. 12.4c d, e. For them, $\ell \sim 1/T^2$. Usually, the phonon–phonon scattering at high temperatures is dominated by three-phonon processes (figs. 12.4a, b). The heat capacity at high temperatures is well approximated by its classical value, $3k_B$ per atom. Hence κ_{ph} varies with T as $1/T$. This is close to the observed temperature dependence in region C of fig. 12.3.

Region A. We now turn to region A in fig. 12.3, where $\log(\kappa_{ph})$ varies linearly with $\log(T)$, i.e. $\kappa_{ph} \sim T^n$. Typically, $n \sim 3$. Since there are few phonons excited at low temperatures, collision processes between them are rare. The mean free path is limited by phonon scattering against defects. Let us take ℓ to be independent of $(\boldsymbol{q},\lambda)$ and T. This would be the case if ℓ is limited by grain boundary scattering or the finite sample size. The heat capacity at low T varies as T^3 for all crystalline solids, which yields $\kappa \sim T^3$.

Region B. The maximum of $\kappa_{ph}(T)$, region B, occurs roughly where $\ell_{ph-ph} \sim \ell_{ph-def}$. Since ℓ_{def} depends on the nature and number of defects in the crystal, the temperatures covered by region B are characteristic of the specimen but usually not of the material. As a crude rule of thumb, the maximum lies at temperatures just below $0.1\theta_D$.

Region D. At very high temperatures, $\kappa_{ph}(T)$ often decreases less rapidly than as $1/T$. There seems to be a "saturation" so that $\kappa_{ph}(T)$ does not fall below a certain value. This is analogous to the electrical conductivity in metals and alloys and it has a similar explanation. The mean free path decreases with increasing T, but ℓ cannot be shorter than the distance between two neighbouring atoms. (A phonon is a wave packet defined by the displacements of discrete atoms. It would be meaningless to consider ℓ shorter than a wavelength.) Hence, from (12.11), κ_{ph} cannot be smaller than of the order of $(1/3)(3Nk_B/V)C_Da \sim k_BC_D/a^2$, where N is the number of atoms in the volume V and $a^3 \sim V/N$. Our argument is only qualitative because the usual Boltzmann equation breaks down when ℓ approaches a. However, it gives a physical explanation for the observed weak temperature dependence of $\kappa_{ph}(T)$ at very high T or in very defect lattices.

In the following sections we shall deal separately, and in more detail, with the scattering mechanisms which dominate $\kappa_{ph}(T)$ in the regions A, C and D of fig. 12.3. Before that, we formulate the Boltzmann equation for phonon transport and derive a general expression for κ_{ph}.

3.2. The Boltzmann equation

Each excited phonon mode is associated with an energy $\hbar\omega(\boldsymbol{q},\lambda)$ and a group velocity $\boldsymbol{C}_g(\boldsymbol{q},\lambda) = \nabla_q\omega(\boldsymbol{q},\lambda)$. The total heat current carried by these excitations is

$$\boldsymbol{Q} = (1/V)\sum_{\boldsymbol{q},\lambda} n(\boldsymbol{q},\lambda)\hbar\omega(\boldsymbol{q},\lambda)\boldsymbol{C}_g(\boldsymbol{q},\lambda), \tag{12.14}$$

where n is the Bose-Einstein statistical factor (12.13). For each $(\boldsymbol{q},\lambda)$, there is a mode $(-\boldsymbol{q},\lambda)$ with $\omega(\boldsymbol{q},\lambda) = \omega(-\boldsymbol{q},\lambda)$, $n(\boldsymbol{q},\lambda) = n(-\boldsymbol{q},\lambda)$ and $\boldsymbol{C}_g(\boldsymbol{q},\lambda) = -\boldsymbol{C}_g(-\boldsymbol{q},\lambda)$. Hence, positive and negative terms cancel in (12.14) and there is no net heat flow in a system of uniform temperature.

To have a non-vanishing heat flow $n(\boldsymbol{q},\lambda)$ must vary in space. Consider $n(\boldsymbol{q},\lambda;T)$ at a certain point R and at time t. At a later time, $t+\Delta t$, the phonons have moved the distance $\boldsymbol{C}_g\Delta t$ away from R and are replaced by phonons coming from a region where the Bose–Einstein factor is

$$n(\boldsymbol{q},\lambda)-[\partial n(\boldsymbol{q},\lambda)/\partial T](\boldsymbol{C}_g\cdot\nabla T)\Delta t. \tag{12.15}$$

The rate of change of n at R, due to this drift of phonons, is

$$\left(\frac{\partial n}{\partial t}\right)_{\text{drift}} = -(\partial n/\partial T)(\boldsymbol{C}_g\cdot\nabla T). \tag{12.16}$$

In a steady state, $(\partial n/\partial t)_{\text{drift}}$ is balanced by scattering processes. This gives the Boltzmann equation

$$\left(\frac{\partial n(\boldsymbol{q},\lambda)}{\partial t}\right)_{\text{drift}} + \left(\frac{\partial n(\boldsymbol{q},\lambda)}{\partial t}\right)_{\text{scattering}} = 0. \tag{12.17}$$

There is no simple method to solve the Boltzmann equation exactly and we have to resort to approximations. The most common approach is the relaxation time approximation. Let now n refer to the thermal average (12.13) and $\tilde{n}$ to a non-equilibrium situation. If $\tilde{n}-n$ is small, we make the usual assumption that the return to equilibrium is governed by a relaxation time τ,

$$\left(\frac{\partial \tilde{n}(\boldsymbol{q},\lambda)}{\partial t}\right)_{\text{scattering}} = -\frac{[\tilde{n}(\boldsymbol{q},\lambda)-n(\boldsymbol{q},\lambda)]}{\tau(\boldsymbol{q},\lambda)}. \tag{12.18}$$

If we let $\partial n/\partial T$ in (12.16) be approximated by the thermal equilibrium value, if we combine (12.16)–(12.18) to get $\tilde{n}$ to be used in (12.14) and recall that the equilibrium n does not contribute to $\boldsymbol{Q}$ in (12.14), the total heat current becomes

$$\boldsymbol{Q} = -(1/V)\sum_{\boldsymbol{q},\lambda}[\boldsymbol{C}_{\mathrm{g}}(\boldsymbol{q},\lambda)\cdot\nabla T]\boldsymbol{C}_{\mathrm{g}}(\boldsymbol{q},\lambda)\tau(\boldsymbol{q},\lambda)C_{\mathrm{E}}(\boldsymbol{q},\lambda). \tag{12.19}$$

Here

$$C_{\mathrm{E}} = k_{\mathrm{B}}\frac{x^2\mathrm{e}^x}{(\mathrm{e}^x-1)^2}, \tag{12.20}$$

with $x = \hbar\omega(\boldsymbol{q},\lambda)/k_{\mathrm{B}}T$, i.e. an Einstein heat capacity. Comparison of eqs. (12.2) and (12.19) gives the tensor components of the thermal conductivity as

$$\kappa_{ij} = (1/V)\sum_{\boldsymbol{q},\lambda}[C_{\mathrm{g}}(\boldsymbol{q},\lambda)]_i[C_{\mathrm{g}}(\boldsymbol{q},\lambda)]_j\tau(\boldsymbol{q},\lambda)C_{\mathrm{E}}(\boldsymbol{q},\lambda). \tag{12.21}$$

In an isotropic system, $\omega(\boldsymbol{q},\lambda)$ and $\tau(\boldsymbol{q},\lambda)$ depend only on $|\boldsymbol{q}|$ and λ. Then $\boldsymbol{C}_{\mathrm{g}}(\boldsymbol{q},\lambda) = (\boldsymbol{q}/|\boldsymbol{q}|)C_{\mathrm{g}}(\lambda)$. When $i \neq j$ the sum over $\boldsymbol{q}$ in (12.21) yields $\kappa_{ij} = 0$, as expected. The mean free path $\ell(\boldsymbol{q},\lambda)$ is

$$\ell(\boldsymbol{q},\lambda) = C_{\mathrm{g}}(\boldsymbol{q},\lambda)\tau(\boldsymbol{q},\lambda). \tag{12.22}$$

With (12.22) in (12.19) we recover the formulation (12.11) of κ_{ph} in terms of ℓ. The prefactor (1/3) in (12.11) arises from (12.19) because, in an isotropic system, $[C_{\mathrm{g}}]_i[C_{\mathrm{g}}]_i = (1/3)[C_{\mathrm{g}}]^2$ on the average.

It is often more natural to express $\tau(\boldsymbol{q},\lambda)$ and $\ell(\boldsymbol{q},\lambda)$ as a function of the phonon frequencies ω of the modes $(\boldsymbol{q},\lambda)$. Then sums over $(\boldsymbol{q},\lambda)$ can be replaced by an integral over ω containing the phonon density of states $F(\omega)$. For instance, from (12.21)

$$\begin{aligned}\kappa_{\mathrm{ph}} &= (n/3)\bar{C}_{\mathrm{g}}\int_0^{\omega_{\max}} C_{\mathrm{E}}(\omega)\ell(\omega)F(\omega)\mathrm{d}\omega\\ &= (n/3)\bar{C}_{\mathrm{g}}^2\int_0^{\omega_{\max}} C_{\mathrm{E}}(\omega)\tau(\omega)F(\omega)\mathrm{d}\omega.\end{aligned} \tag{12.23}$$

Here $\bar{C}_g$ is some average phonon group velocity, $C_E(\omega)$ is the heat capacity of a single phonon mode, (12.20), and n is the number of atoms per. volume. We recall our normalisation convention $\int F(\omega)\mathrm{d}\omega = 3$.

4. Phonon-limited lattice conductivity. Anharmonic effects

4.1. General results

If the lattice vibrations are strictly harmonic, the phonons are represented by non-interacting waves. Then there is no mechanism for phonon scattering and the lifetime, $\tau_{\mathrm{ph-ph}}$, is infinite (in a perfect crystal of infinite size). In order to have a finite $\tau_{\mathrm{ph-ph}}$ one must retain the anharmonic terms when the potential energy Φ is expanded in the atomic displacements u (cf. (C.1)):

$$\Phi = \Phi_0 + (1/2) \sum_{\alpha\beta,ij} \Phi_{ij} u_\alpha u_\beta + (1/3!) \sum_{\alpha\beta\gamma,ijk} \Phi_{ijk} u_\alpha u_\beta u_\gamma$$

$$+ (1/4!) \sum_{\alpha\beta\gamma\delta,ijkl} \Phi_{ijkl} u_\alpha u_\beta u_\gamma u_\delta + \cdots. \tag{12.24}$$

Here i, j, k, l label atoms and $\alpha, \beta, \gamma, \delta$ label Cartesian coordinates. It can be shown that the terms in (12.24) which are cubic in u correspond to three-phonon scattering, figs. 12.4a, b. The terms quartic in u correspond to four-phonon processes, figs. 12.4c, d, e. To calculate the phonon-limited lattice thermal conductivity from (12.21) we first must find the lifetimes $\tau(\boldsymbol{q}, \lambda)$. Consider scattering into and out of a state 1, as shown schematically in fig. 12.4a, b. Let the statistical distribution function of the state 1 be perturbed from its equilibrium value $n(1)$ to $\tilde{n}(1) = n(1) + \Delta n(1)$ but let the states 2 and 3 be described by their equilibrium functions $n(2)$ and $n(3)$. The probability for a scattering event $1 \to 2+3$ contains the temperature explicitly through the factors $\tilde{n}[n(2)+1][n(3)+1]$. (The term $n+1$ arises because of the stimulated emission of bosons, i.e. phonons.) Similar expressions hold for events $2+3 \to 1$ etc. One ends up with an expression

$$\frac{1}{\tau(1)} = \frac{2\pi}{\hbar} \sum_{2,3} |H(1,2,3)|^2 [n(2)+n(3)+1], \tag{12.25}$$

where $|\langle H(1,2,3)\rangle|^2$ is short for $|\langle 1|\Delta H|2,3\rangle|^2 \delta(\boldsymbol{q}_1 - \boldsymbol{q}_2 - \boldsymbol{q}_3 - \boldsymbol{G})\delta(\hbar\Delta_\omega)$,

$\Delta_\omega = \omega_1 - \omega_2 - \omega_3$ or similar terms, $|\ldots \Delta H \ldots|$ is a quantum mechanical matrix element for the transition, and delta functions account for wave-number and energy conservation. We now rewrite the last factor in (12.25). By the mathematical identity $[(e^x - 1)^{-1} + (e^y - 1)^{-1} + 1] = (e^{x+y} - 1)(e^x - 1)^{-1}(e^y - 1)^{-1}$, and the conditions $\omega_1 = \omega_2 + \omega_3$ etc. from the delta functions, (12.25) becomes

$$\frac{1}{\tau(1)} = \frac{2\pi}{\hbar} \sum_{2,3} |H(1,2,3)|^2 \frac{n(2)n(3)}{n(1)}. \qquad (12.26)$$

It is very difficult to evaluate $|H(1,2,3)|^2$ accurately, but simple assumptions (e.g., Leibfried and Schlömann 1954, Klemens 1958, 1969), and even dimensional arguments, lead to a form

$$|H(1,2,3)|^2 = A \frac{\hbar^2 a \gamma^2}{3MN} \frac{\omega_1 \omega_2 \omega_3}{C_g^3}. \qquad (12.27)$$

Here A is a dimensionless constant, roughly of the order of unity. Further, we have assumed a monatomic lattice with lattice parameter a and atomic mass M. The parameter γ in (12.27) is related to the Grüneisen parameters, but is not equal to any particular $\gamma(n)$. N is the number of atoms and $3N$ is the number of phonon modes. The evaluation of $\tau(1)$ requires a summation over modes 2 and 3. This cannot be done analytically, so it is not possible to give a closed-form expression $\kappa_{\mathrm{ph-ph}}(T)$ valid at all temperatures. Instead we discuss separately the cases of high and low temperatures.

The four-phonon processes, figs. 12.4c, d, e, are much less important than the three-phonon processes (Ecsedy and Klemens 1977). This is contrary to the case of anharmonic shifts in the vibrational entropy (eq. (5.49)), where the three- and four-phonon interactions are of roughly equal importance.

4.2. Umklapp scattering

Peierls (1929) derived the very important result that only scattering with $\boldsymbol{G} \neq 0$ in (12.25) contributes to the thermal resistance. A simplified argument is as follows: Let $\omega_i = C_g q_i$ for $i = 1, 2, 3$, with an isotropic $|\boldsymbol{C}_g|$. Consider the case when one phonon, denoted by 1, is annihilated and two phonons, 2 and 3, are created in a scattering event for which $\boldsymbol{G} = 0$. The heat flow in mode 1 before the scattering is $\boldsymbol{Q}(1) = \hbar\omega(1)\boldsymbol{C}_g(1)$

$= \hbar C_g^2 \boldsymbol{q}_1$ and the heat flow after scattering is $\boldsymbol{Q}(2)+\boldsymbol{Q}(3) = \hbar\omega(2)\boldsymbol{C}_g(2)+\hbar\omega(3)\boldsymbol{C}_g(3) = \hbar C_g^2(\boldsymbol{q}_2+\boldsymbol{q}_3)$. But momentum conservation requires that $\boldsymbol{q}_1 = \boldsymbol{q}_2+\boldsymbol{q}_3 \pm \boldsymbol{G}$. Hence, the heat flow is unchanged unless $\boldsymbol{G} \neq 0$. This argument can be extended to an arbitrary phonon dispersion relation (e.g., Berman 1976). It follows that not all $\boldsymbol{q}_i$ ($i = 1, 2, 3$) can be small compared to the Debye wave vector. When $\boldsymbol{G} \neq 0$, one speaks of Umklapp scattering. This is the reason why κ_{ph-ph} is sometimes called the Umklapp term. The common labelling κ_g, instead of our κ_{ph}, stems from the German word *Gitter* (lattice).

4.3. *Low temperatures*

The combination $n(2)n(3)/n(1)$ of Bose–Einstein factors in (10.26) must be considered in some detail. At a given (low) temperature, only phonon states with $\hbar\omega_1 < k_B T$ are sufficiently populated to be of importance in the scattering process $1 \rightarrow 2+3$. By Peierls' argument, only Umklapp scattering matters, so at least one of $|\boldsymbol{q}_2|$ and $|\boldsymbol{q}_3|$ must be of the order of $|\boldsymbol{G}|$, i.e. of the order of q_D. This also implies that at least one of ω_2 and ω_3 is of the order of the Debye frequency ω_D. Hence, when $T \ll \theta_D$, the factor $n(2)n(3)/n(1)$ contains a term of the order

$$\exp(-\hbar\omega_D/k_B T) = \exp(-\theta_D/T), \tag{12.28}$$

Such an exponential term dominates the temperature dependence of $\kappa_{ph-ph}(T)$ and, in a qualitative description, we can neglect any prefactors containing powers of T. Therefore, approximately,

$$\kappa_{ph-ph} = \kappa_0 \exp(-\theta_D/T). \tag{12.29}$$

The prefactor κ_0 is, in principle, to be calculated from $\tau(1)$ in (12.26). In practice this is not feasible, and we are left with κ_0 as a free parameter. Its order of magnitude may be estimated by joining $\kappa_{ph-ph}(T)$ at high temperatures (eq. 12.32) to the low temperature form (12.29), with $T = \Theta_D$. One should note that the Debye temperature Θ_D in (12.28) only measures, in a rough way, the magnitude of typical phonon energies; Θ_D is not equal to any particular $\Theta_D(n)$. Sometimes a reasonable account of $\kappa_{ph-ph}(T)$ in the transition region between B and C in fig. 12.3 is obtained with $\Theta_D = \Theta_D(-3)/b$, where b is of the order of two.

4.4. High temperatures

Let the temperature be much higher than the Debye temperature Θ_D, i.e. $k_B T \gg \hbar\omega_i$ for $i = 1, 2, 3$. Then $n(2)n(3)/n(1) \approx (k_B T/\hbar)(\omega_1/\omega_2\omega_3)$. When this is inserted in (12.26), the explicit dependence on ω_2 and ω_3 vanishes, by (12.27). There is still a dependence on 2 and 3 in $H(1, 2, 3)$ but that has been allowed for in an average manner in the constant A in (12.27). The summation over the mode 2 simply gives a multiplicative factor $3N$, and the summation over the mode 3 reduces to a single term because of the delta functions in (12.25). One then obtains

$$\frac{1}{\tau(1)} = 2\pi A \frac{a\gamma^2 k_B T}{M} \frac{[\omega(1)]^2}{C_g^3}. \tag{12.30}$$

Finally, it remains to integrate over all modes 1, to obtain κ_{ph-ph} from the expression (12.23). With a Debye phonon spectrum, and using the result that $C_g/a = \omega_D/(6\pi^2)^{1/3}$ and $(4\pi/3)a^3 = \Omega_a$, one obtains

$$\kappa_{ph-ph} = \frac{B}{(2\pi)^3} \frac{Mak_B^3\Theta_D^3}{\hbar^3\gamma^2 T}, \tag{12.31}$$

where B is a dimensionless constant. This is a key result. Formally, it is valid when $T \gg \Theta_D$, but in practice the $1/T$ dependence is often well obeyed for $T > \Theta_D/3$. A relation of the form (12.31) has been derived by Leibfried and Schlömann (1954) (corrected for an error by a factor of two by Julian (1965)), by Klemens (1958, 1969) and others. These theoretical approaches give essentially the same combination of characteristic parameters, but they differ in the numerical prefactor B. This is natural, since an evaluation of B requires a large number of simplifying, and to some extent arbitrary, assumptions. Klemens (1969) suggests that $B = 1.61$, a value which usually gives an adequate description of the measured conductivity of non-metals at high temperatures (Slack 1979). In view of the theoretical uncertainties about the absolute magnitude of κ_{ph-ph}, one may have to analyse high-temperature experiments using a semiempirical formula

$$\kappa_{ph-ph} = \kappa^*(T_m/T)^\eta. \tag{12.32}$$

Here κ^* is a fitted prefactor, T_m is the melting temperature and the exponent η is near unity. One can derive η from a plot of $\ln[\kappa_{ph-ph}(T)]$

vs. ln (T). Neglecting any temperature dependence other than the explicit factor $1/T$ and the indirect temperature dependences of θ_D due to the thermal expansion, we have from (12.31) near T_m

$$\eta = 1-(\mathrm{d}\ln\kappa_{\mathrm{ph-ph}}/\mathrm{d}\ln T) \approx 1-3(\partial\ln\theta_D/\partial\ln V)(\mathrm{d}\ln V/\mathrm{d}\ln T)$$

$$\approx 1+3\beta\gamma T_m. \tag{12.33}$$

Since Θ_D in our model is not related to any particular moment of the phonon frequencies, we use an unspecified Grüneisen parameter γ. For many solids, $\beta T_m \sim 0.06$, eq. (15.8), and $\gamma = 1.5$. Then $\eta \approx 1.3$. Thus, thermal expansion causes η to be somewhat larger than unity. The four-phonon processes in fig. 12.4c, d, e give a scattering rate for which $\tau \sim 1/T^2$. That would also make η larger than 1, but the effect appears to be small (Escedy and Klemens 1977).

Example: Temperature dependence of the phonon–phonon scattering. Figure 12.5 shows $\log \kappa_{\mathrm{tot}}$ vs. $\log T$, from experiments (Touloukian et al. 1970). Since κ_{tot} for these solids is mainly limited by phonon-phonon scattering, the slope of the curves give an estimation of the parameter η in eq. (12.32).

4.5. *Several atoms per primitive cell*

In the derivation above, we have assumed a Debye spectrum for all phonon modes. This is questionable when there are several atoms per primitive cell. Consider the extreme case that the dispersion curves for the optical branches are "flat", as in fig. 12.6. The group velocity $\boldsymbol{C}_g = \nabla_{\boldsymbol{q}}\omega(\boldsymbol{q},\lambda) = 0$ for these branches, and therefore they do not contribute to the heat flow. Still the optical modes affect the thermal conductivity since they interact with the heat-carrying acoustic modes and thus limit the relaxation time. Slack and Oliver (1971) assumed that one should sum over only the acoustic modes in (12.19) and they neglected the influence of the optical modes on τ for the acoustic branches, while Roufosse and Klemens (1973) included the effect of optical modes on τ. When the high-temperature formula (12.31) is applied to lattices with several atoms, p, per primitive cell, one has to be careful in the definition of Θ_D. Typical acoustic phonon energies are $k_B\Theta_D/p^{1/3}$, and not $k_B\Theta_D$. Slack (1979) has considered the thermal conduction in lattices with $p > 1$ in some detail. He finds that a result calculated as if there were

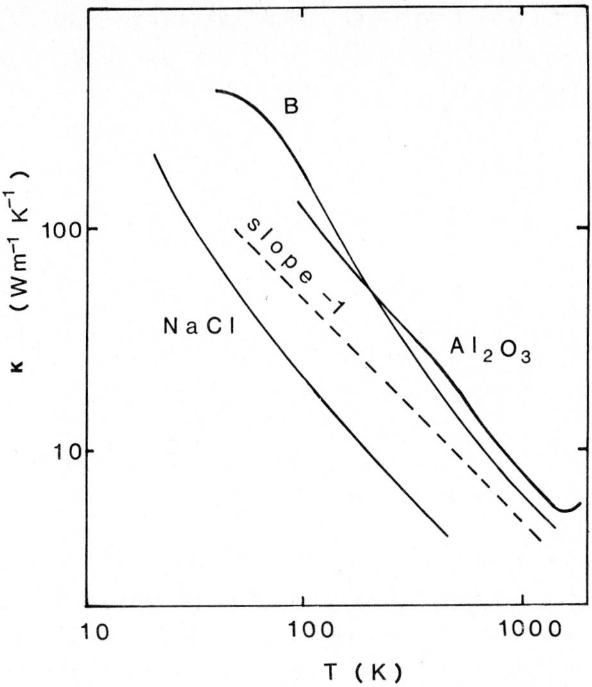

Fig. 12.5. Experimental values (Touloukian et al. 1970) of $\log\kappa$ versus $\log T$, in the high-temperature region. The dashed line has the slope -1, i.e. the slope corresponding to $\kappa \sim 1/T$.

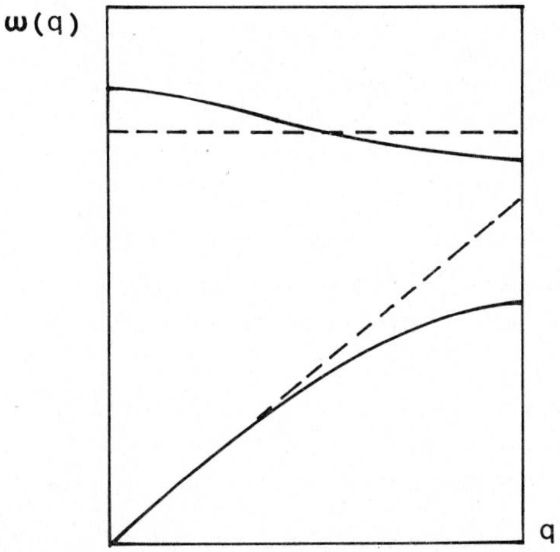

Fig. 12.6. A schematic phonon dispersion curve for a solid with two atoms per primitive cell. The dashed line is a model dispersion curve, corresponding to a Debye model for the acoustic branch and an Einstein model for the optical branch.

one atom per cell should be reduced by a factor $p^{-2/3}$. This is no strict rule, but it gives the trend. In an analogous way, Θ_D must be reduced in the factor $\exp(-\Theta_D/T)$ that accounts for the freezing-out of Umklapp scattering as the temperature is decreased.

5. Defect-limited lattice conductivity

5.1. General considerations

We shall consider the scattering of phonons by:
(i) Point defects; substitutional atoms, interstitials, vacancies, isotopic defects etc.
(ii) Line defects; thin cylindrical inclusions.
(iii) Planar defects; thin platelets.
(iv) Volume defects; the strain field from dislocations.
(v) Interface boundaries and sample boundaries.
The scattering cross section has a different dependence on the phonon wavelength in each of the five cases (i)–(v). It is well known, for example from the Rayleigh scattering in acoustics and optics, that the scattering cross section for a point defect varies as $1/\lambda^4$, where λ is the wavelength. (Note that in this chapter, and elsewhere, λ is also a mode index for the phonon state $(\boldsymbol{q}, \lambda)$.) For line defects, (ii), the cross section varies as $1/\lambda^3$, for planar defects, (iii), as $1/\lambda^2$, and for the strain field due to dislocations, (iv), as $1/\lambda$. In the case (v), it is practically independent of λ.

The scattering by defects is of particular importance at low temperatures. Since only phonons with $\hbar\omega(\boldsymbol{q}, \lambda) \lesssim k_B T$ are excited, it is a good approximation to take $\omega(\boldsymbol{q}, \lambda)$ linear in $|\boldsymbol{q}|$. Then, $\omega \sim 1/\lambda$. It follows that the scattering cross sections for the cases (i)–(v) vary as ω^n where $n = 4$, 3, 2, 1 and 0, respectively. The corresponding relaxation times, τ, vary as ω^{-n}. We now assume an isotropic Debye spectrum for $F(\omega)$, with $k_B\Theta_D = \hbar\omega_{\max}$, and make the substitution $\hbar\omega/k_B T = x$. Then from (12.23),

$$\kappa_{ph}(T) = \frac{1}{2\pi^2} \frac{k_B^4 T^3}{\hbar^3 C_g} \int_0^{\Theta_D/T} \frac{x^4 e^x \tau(x)}{(e^x - 1)^2} dx. \tag{12.34}$$

With $\tau \sim \omega^{-n} \sim x^{-n}$, and at low temperatures, we have

$$\kappa_{ph}(T) \sim T^{3-n} \int_0^{\infty} \frac{x^4 x^{-n} e^x}{(e^x - 1)^2} dx. \tag{12.35}$$

The integral in (12.35) diverges at $x = 0$ when $n = 3$ or 4, i.e. for point and line defects. One therefore must introduce a cut-off x_c as the lower integration limit. Physically this cut-off means that the finite size of the specimen, or other defects, dominates the scattering of very low frequency phonons. When $n = 0$, 1 or 2, the integrand in (12.35) has its maximum at, very roughly, $x \sim 1$. This means that the phonons which carry most of the heat have $\hbar\omega \sim k_B T$, i.e. a wavelength $\lambda \sim (\Theta_D/T)\lambda_D \sim (\Theta_D/T)a$ where a is an atomic diameter.

After this general introduction to defect scattering of phonons, we discuss each defect separately.

5.2. Point defect scattering

It is instructive first to recall Rayleigh's (1894) classical theory of sound wave scattering against small objects. Some complications arise in solids because there are longitudinal as well as transverse modes (Ziman 1960). We can write, for macroscopic spherical inclusions,

$$\frac{1}{\tau} = n_{\text{def}}\left(\frac{AG}{\lambda^4}\right)V^2[(\Delta\rho/\rho)^2 + 6(\Delta G/G)^2]. \tag{12.36}$$

Here n_{def} is the number of defects per volume, V is the volume of a single defect, $(\Delta\rho/\rho)$ is the change in the mass density of the inclusion, $(\Delta G/G)$ refers to the shear modulus G (which relates to the sound velocity) and A is a dimensionless constant of the order of $4\pi^3$. Note that $(\Delta\rho/\rho)$ and $(\Delta G/G)$ need not be small.

The archetype of a point defect is a substitutional atom in a lattice. It creates several kinds of disturbances for a propagating phonon. The atomic mass and the interatomic forces are changed, and atoms are more or less displaced. We first discuss the mass change (isotope scattering), because that is the simplest case to treat theoretically.

Klemens (1955) derived an expression for the scattering time $\tau_{\Delta M}$, as limited by a change of an atomic mass from M to $M + \Delta M$. In a monatomic lattice, and with an isotropic Debye model for the phonons, he obtained

$$\frac{1}{\tau_{\Delta M}(\omega)} = c(\text{def})\frac{V\omega^4}{4\pi N C_g^3}\left(\frac{\Delta M}{M}\right)^2. \tag{12.37}$$

This is equivalent to (12.36) when $\Delta G/G = 0$; V/N is the volume per atom and $c(\text{def})$ (with $0 < c(\text{def}) < 1$) is the defect concentration in the lattice (i.e. volume or site fraction occupied by defects). As in (12.36), eq. (12.37) does not require $\Delta M/M$ to be small. A more detailed formulation of the scattering problem (Takeno 1963, Callaway 1963, Krumhansl 1965, Elliott and Taylor 1964, McCombie and Slater 1964, Klein 1966) leads to essentially the same result.

The combined effect of mass and force constant changes have been studied theoretically by Krumhansl and Matthew (1965) and Yussouff and Mahanty (1966, 1967). In one dimension (a linear chain), Krumhansl and Matthew found that the factor $(\Delta M/M)^2$ in the scattering rate should be replaced by

$$\left[\frac{\Delta M}{M} + \frac{2(\Delta k/k)}{1+(\Delta k/k)}\right]^2, \tag{12.38}$$

if the force constants between nearest neighbours at the defect site are changed from k to $k+\Delta k$. Thus there is in this model a cancellation of the two effects when $(\Delta k/k) = -(\Delta M/M)/[2+(\Delta M/M)]^{-1}$. No such simple result seems to be known for a three-dimensional system, but it is obvious that the effects of mass and force constant changes on the thermal conductivity are not additive.

An impurity atom may cause a dilatation of the lattice. Its effect on the thermal conductivity can be estimated from the classical theory of Rayleigh scattering, if one relies on some Grüneisen parameter to estimate changes in C_g. Also in this case there is the possibility of a cancellation between the two effects. In spite of much work (e.g. Carruthers 1961), it is still uncertain how to handle the dilatation term (Klemens 1983a).

In our treatment of point defects, it has been required that $\lambda \gg L$, where L is a characteristic size of the scattering object. Nothing has been said about the distance D between these objects. If the phonon mean free path $\ell \gg D$, a phonon will "see" very many point-like scattering centres and there may be interference effects. The total scattering amplitude is the sum of amplitudes from each scattering center. However, the scattered intensity depends on the total scattering amplitude squared. If the scattering centres are randomly distributed, the interference effect averages to zero and the total scattered intensity is the sum of the intensity calculated from each scattering centre separately. (A better known example of this phenomenon is the blue colour of the sky. It is caused by Rayleigh scattering of light against air molecules, but it is essential that the number-density of molecules is spatially fluctuating.)

Example: Vacancy scattering. Vacancies can be regarded as point defects with the mass change $\Delta M/M = -1$ and the force constant change $\Delta k/k \approx -1$. The equilibrium number of thermally generated vacancies is much too small ($< 10^{-3}$) to have any measurable effect on the lattice thermal conductivity. A larger concentration is found in doped ionic solids, e.g. KCl doped with Ca (Slack 1957), where charge neutrality requires that vacancies are formed. The vacancies are often adjacent to the impurity atoms. Their combined effect on κ_{ph} is small, but may be detected in accurate experiments at low temperatures (Schwartz and Walker 1967a). The theory of vacancy scattering is in fair agreement with experiments (Klemens 1983a).

Example: Clustering of small defects. Assume that clusters are formed, each containing p of the original small defects. Then, n_{def} in (12.36) decreases by a factor p while V^2 increases by a factor p^2. Thus, within Rayleigh's classical model, τ (and hence κ_{ph}) decreases by a factor of p. However, the Rayleigh formula is not valid when the characteristic size L of the scattering objects becomes comparable to the wavelength. We expect a breakdown of (12.36) when $qL > 1$, i.e., when $\omega > \bar{C}_g/L$. The theory describing the transition from Rayleigh scattering to the "geometrical" scattering against large objects is complicated (Schwartz and Walker 1967b).

5.3. Dislocation scattering

The influence of dislocations on the lattice thermal conductivity has received continuous interest for a long time (e.g., Klemens 1955, Bross et al. 1963, Eckhardt and Wasserbäch 1978, Anderson 1983). Because of the strain field surrounding a dislocation, they do not scatter like line defects. The effect of dislocations on the thermal conductivity is controversial and theoretical calculations are in poor agreement with experiments. It is only at very low temperatures ($T < \theta_D/10$) that dislocation scattering in heavily deformed materials may be of importance.

5.4. Boundary scattering

At very low temperatures, the phonon mean free path ℓ may be comparable to a characteristic sample dimension d (the diameter of a wire or the diameter of small crystals). Then ℓ is independent of λ (or ω) and κ_{ph} varies as T^3.

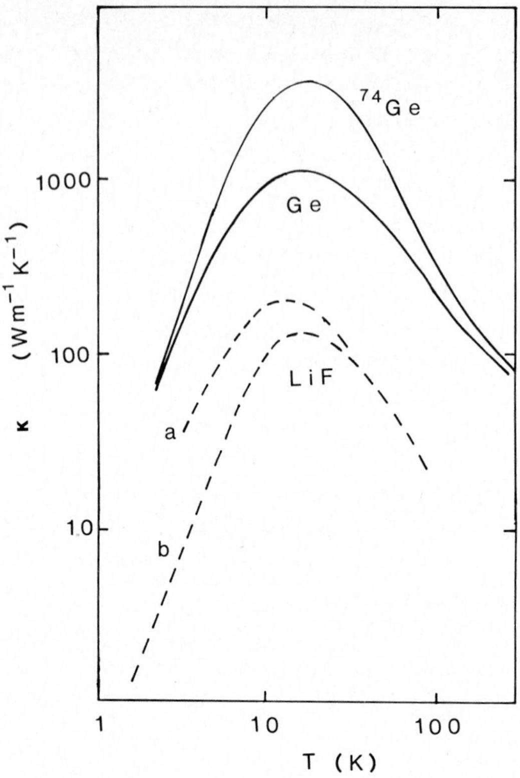

Fig. 12.7. The thermal conductivity κ of a crystal of the pure isotope ^{74}Ge and a crystal of natural Ge (Geballe and Hull 1958). Also shown is κ of irradiated LiF in two crystal sizes—a is larger than b (Pohl 1960).

Example: Scattering by defects in LiF and Ge. Figure 12.7 shows the measured thermal conductivity of Ge (Geballe and Hull 1958) and LiF (Pohl 1960). For Ge, $\kappa(T)$ is shown for an isotopically almost pure sample of ^{74}Ge and for natural Ge. The difference in κ can be explained by the isotope scattering model in §5.2. At very low temperatures, boundary scattering gives κ a T^3-dependence. The two curves for LiF refer to samples of different sizes ($6.7 \times 0.8 \times 40$ mm and $0.7 \times 0.8 \times 40$ mm). F-centers, introduced by irradiation, contribute to the scattering at T near the maximum in κ.

5.5. Several scattering processes acting simultaneously

In (12.12) we wrote the inverse phonon mean free path as a sum of inverse mean free paths for each scattering mechanism. This is equivalent to a corresponding relation for the lifetimes $\tau(\omega)$:

$$\frac{1}{\tau_{\text{tot}}(\omega)} = \frac{1}{\tau_{\text{ph-ph}}(\omega)} + \sum \frac{1}{\tau_{\text{ph-def}}(\omega)}. \tag{12.39}$$

Let us assume that there are only two scattering mechanisms, each associated with a constant relaxation time. Then, with $\tau_{\text{tot}} = (1/\tau_1 + 1/\tau_2)^{-1}$ inserted in (12.23),

$$\begin{aligned} W_{\text{tot}} = 1/\kappa_{\text{tot}} &= \left[(n/3)\bar{C}_{\text{g}}^2 \int_0^{\omega_{\max}} C_{\text{E}}(\omega)F(\omega)\mathrm{d}\omega\right]^{-1}(1/\tau_1 + 1/\tau_2) \\ &= 1/\kappa_1 + 1/\kappa_2 = W_1 + W_2. \end{aligned} \tag{12.40}$$

If τ depends on ω, the arguments above can be repeated if $\tau_1(\omega)/\tau_2(\omega)$ is independent of ω. Only then does Matthiessen's rule hold for the phonon parts of the thermal conductivity. However, (12.39) is the natural relation to start from, even when Matthiessen's rule is not obeyed, since it expresses the fact that the total scattering rate is the sum over the various scattering processes that are present.

5.6. *Concentrated alloys*

The electrical resistivity of a concentrated alloy roughly follows the rule $\rho_{\text{tot}} = \rho_{\text{ep}}(T) + c(1-c)\rho^*$, where the defect resistance $c(1-c)\rho^*$ varies parabolically with the composition $A_{1-c}B_c$ of the alloy and is independent of the temperature. (This behaviour excludes cases with "saturation".) In phonon transport such a simple result does not hold. The reason is to be sought in the point defect scattering, which gives a divergence in κ_{ph} from low frequency phonons. The divergence is taken care of by the anharmonic (phonon–phonon) scattering. Because of the very different behaviour of $\tau(\omega)$ for the two scattering mechanisms, Matthiessen's rule is strongly violated. The qualitative result, at high T and not too small c, is (Klemens 1960)

$$W_{\text{tot}} \sim [c(1-c)]^{1/2} T^{1/2}. \tag{12.41}$$

Such a behaviour has been observed, e.g., in experiments on Ge–Si (Abeles 1963).

6. The phonon mean free path and saturation effects

A (temperature dependent) characteristic phonon mean free path $\bar{\ell}$ is sometimes calculated from the measured thermal conductivity and heat capacity by

$$\bar{\ell} = 3\kappa V / C_g C_V. \tag{12.42}$$

Here, κ and C_V should only contain the phonon contributions. A similar expression, $\bar{\ell} = 3a/C_g$, involves the thermal diffusivity a. However, (12.42) can be quite misleading. Since ℓ and C_g may vary strongly with $(\boldsymbol{q}, \lambda)$, the weighting of modes $(\boldsymbol{q}, \lambda)$ that enter $\bar{\ell}$ is not clear. Further, $\bar{\ell}$ has both an explicit temperature dependence and an indirect variation with T through C_V, which makes an interpretation of $\bar{\ell}$ difficult (MacDonald and Anderson 1983).

We noted in §3.1, in the discussion of region D, that the mean free path $\ell(\boldsymbol{q}, \lambda)$ cannot be shorter than the wavelength of the phonon, and the wavelength cannot be shorter than a typical distance between two neighbouring atoms. Hence, $\kappa_{ph} > k_B C_D / a^2$, where C_D is the Debye velocity. We rewrite the right-hand side of this inequality as

$$(\kappa_{ph})_{sat} \sim \frac{k_B(C_D q_D)}{a^2 q_D} \sim \frac{k_B \omega_D}{4a} = \frac{k_B^2 \theta_D}{4a\hbar}, \tag{12.43}$$

where we have used the result that $q_D = (6\pi^2)^{1/3}/a$. Taking the typical values $\theta_D = 300$ K and $a = 3 \times 10^{-10}$ m, we have for the "saturation" value of the thermal conductivity $(\kappa_{ph})_{sat} \sim 0.5$ W/m K. This is an underestimation of $(\kappa_{ph})_{sat}$, since we have gone to the extreme limit $\ell \sim a$ for all phonon modes. However, it explains why the conductivity of many strongly disordered materials of quite different kinds (alloys, ceramics, ionic solids, polymers), and also the high-temperature conductivity of many pure solids, seems to saturate at a rather universal value. Kittel (1949) used these ideas to explain the conductivity of glasses and Slack (1979) has reviewed the field in some detail.

7. Electronic contribution to the thermal conductivity

7.1. Introduction

The expression for the conductivity of a classical gas was the starting point for our discussion of the phonon part of the thermal conductivity in solids. The same fundamental expression can be used in a treatment of the electronic contribution, κ_e. One can write

$$\kappa_e = \frac{1}{3} n C_e v_e \ell_e, \tag{12.44}$$

where C_e is the electronic heat capacity (per electron), n the number of conduction electrons per volume, v_e the electron velocity and ℓ_e the electron mean free path. Now let the electrons be described by a free-electron model, i.e. $v_e = v_F$ and $C_e = \pi^2 k_B^2 T/mv_F^2$. The mean free path can be expressed in the electron lifetime τ as $\ell = v_F \tau$. The thermal conductivity then becomes

$$\kappa_e = \frac{\pi^2 n k_B^2 T \tau}{3m}. \tag{12.45}$$

Compare this with the result (11.2) for the electrical conductivity, $\sigma = ne^2\tau/m$. We get

$$\frac{\kappa_e}{\sigma} = \frac{\pi^2 n k_B^2 T \tau/3m}{ne^2\tau/m} = \frac{\pi^2}{3}\left(\frac{k_B}{e}\right)^2 T = L_0 T. \tag{12.46}$$

L_0 ($= 2.44 \times 10^{-8}$ $W\Omega K^{-2}$, or V^2K^{-2}) is the Lorenz number for a free-electron model in which the lifetime τ is taken to be the same for electrical and thermal transport. The proportionality between κ_e and σ was discovered by Wiedemann and Franz (1853). Lorenz (1881) noted that the proportionality constant is linear in T. Therefore (12.46) is known as the Wiedemann–Franz or the Wiedemann–Franz–Lorenz law. In a real metal, we *define* the Lorenz number by

$$L_e = \kappa_e/\sigma T; \tag{12.47}$$

L_e is temperature dependent, but when $T > \theta_D/2$, L_e often does not deviate more than 20% from L_0 (Laubitz and Matsumura 1972, Laubitz et al. 1976). An experimental determination of the Lorenz number requires that the phonon part κ_{ph} is subtracted from the total thermal conductivity $\kappa = \kappa_{ph} + \kappa_e$. The magnitude of κ_{ph} is very difficult to calculate accurately. In fig. 12.8 we therefore plot the quantity $L(T) = \kappa/\sigma T$, from experimental data on σ and the total thermal conductivity κ. Obviously, the Wiedemann–Franz law, with $L_e = L_0$, is

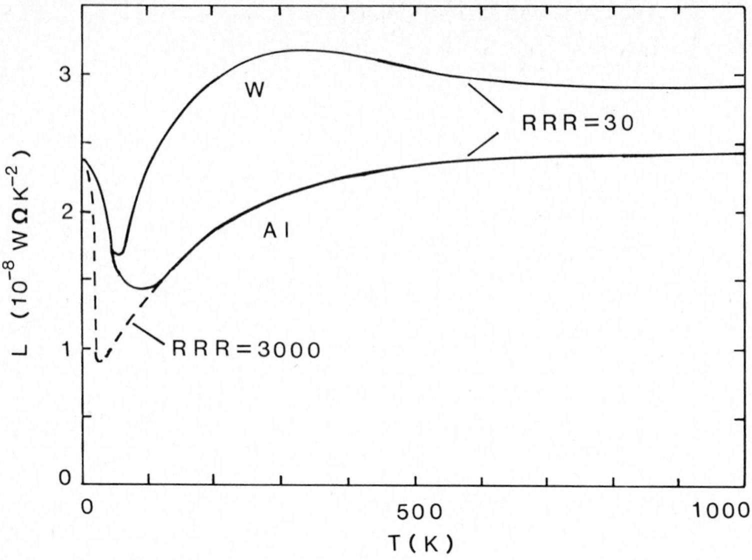

Fig. 12.8. The Lorenz number $L = \kappa/\sigma T$, from measured κ and σ (Touloukian et al. 1970, Bass 1982). RRR is the residual resistance ratio.

approximately obeyed also for these real materials. The minimum at low T is discussed in §7.4.

7.2. *Fundamental expressions for* κ_e

In ch. 11, dealing with the electrical conductivity σ, we obtained the expression

$$\sigma = \frac{ne^2}{m_b}\langle\tau(\varepsilon, \boldsymbol{k})\rangle. \tag{12.48}$$

Here m_b is an effective electron band mass and $\tau(\varepsilon, \boldsymbol{k})$ is an electron lifetime which depends both on the direction of the wave vector $\boldsymbol{k}$ and on the energy distance ε to E_F. The brackets $\langle\,\rangle$ denote an average over all electron states, with a weight factor $-(\partial f(\varepsilon_{\boldsymbol{k}})/\partial\varepsilon_{\boldsymbol{k}})$ which implies that we only sample electron states $\boldsymbol{k}$ near the Fermi level. The analogue of (12.48) for the electronic part of the thermal conductivity is

$$\kappa_e = \frac{nk_B^2T}{m_b}\left\langle\left(\frac{\varepsilon_k - E_F}{k_BT}\right)^2 \tau(\varepsilon, \boldsymbol{k})\right\rangle. \tag{12.49}$$

The lifetime $\tau(\varepsilon, \boldsymbol{k})$ solves the Boltzmann equation in a thermal gradient.

We shall now see that (12.49) contains, as a special case, the Wiedemann–Franz law. Let $\tau(\varepsilon, \boldsymbol{k})$ be isotropic and energy independent, i.e. $\tau(\varepsilon, \boldsymbol{k}) = \tau$. (This still allows τ to be temperature dependent.) Then, from (12.49),

$$\kappa_e = \frac{nk_B^2 T\tau}{m_b} \int_{-\infty}^{\infty} \left(\frac{\varepsilon - E_F}{k_B T}\right)^2 \left(-\frac{\partial f(\varepsilon)}{\partial \varepsilon}\right) \frac{d\Omega}{4\pi} d\varepsilon. \tag{12.50}$$

The integral over $d\Omega$ gives 4π and the integral over ε can be solved exactly. One obtains the desired result

$$\kappa_e = \frac{\pi^2 nk_B^2 T\tau}{3m_b} = \sigma L_0 T. \tag{12.51}$$

7.3. κ_e *expressed in electron–phonon coupling functions*

In ch. 11 §4.2, we expressed the electron lifetime, and the resulting electrical resistivity, in integrals involving the transport Éliashberg coupling function $\alpha_{tr}^2 F(\omega)$. The corresponding expression for κ_e contains both $\alpha_{tr}^2 F(\omega)$ and the ordinary Éliashberg coupling function $\alpha^2 F(\omega)$. One has (Leung et al. 1977, Tomlinson 1979, Grimvall 1981)

$$\frac{1}{\kappa_e} = \frac{1}{L_0 T} \frac{(4\pi)^2}{\omega_{pl}^2} \int_0^{\omega_{max}} \frac{\hbar\omega/k_B T}{[\exp(\hbar\omega/k_B T) - 1][1 - \exp(-\hbar\omega/k_B T)]}$$
$$\times \left\{\left[1 - \frac{1}{2\pi^2}\left(\frac{\hbar\omega}{k_B T}\right)^2\right]\alpha_{tr}^2 F(\omega) + \frac{3}{2\pi^2}\left(\frac{\hbar\omega}{k_B T}\right)^2 \alpha^2 F(\omega)\right\} d\omega. \tag{12.52}$$

We consider this formula in some special cases. It is then convenient to use the thermal resistivity $W_e = 1/\kappa_e$.

High-temperature limit: Let $\hbar\omega_{max}/k_B T \gg 1$ (in practice, even $T/\theta_D > 1$ may be enough) and take the free-electron result $\omega_{pl}^2 = 4\pi ne^2/m$. Then

$$W_e = \frac{2\pi mk_B}{L_0 ne^2\hbar} \int_0^{\omega_{max}} \frac{2\alpha_{tr}^2 F(\omega)}{\omega} d\omega = \frac{2\pi mk_B \lambda_{tr}}{L_0 ne^2 \hbar}. \tag{12.53}$$

This is now compared with the corresponding expression for the electrical conductivity, $1/\sigma = 2\pi k_B m T \lambda_{tr}/\hbar ne^2$. Again, we recover the Wiedemann–Franz law.

Einstein phonon spectrum: Let the phonons be described by an Einstein model. Then

$$\alpha_{\mathrm{tr}}^2 F(\omega) = A\delta(\omega - \omega_{\mathrm{E}}), \qquad \alpha^2 F(\omega) = B\delta(\omega - \omega_{\mathrm{E}}), \tag{12.54}$$

where A and B are constants. The integral in (12.52) just picks up the delta-function contributions and one obtains

$$W_{\mathrm{e}} = k_{\mathrm{E}} C(T/\theta_{\mathrm{E}}) \left[\frac{A}{B} + \left(\frac{\theta_{\mathrm{E}}}{T}\right)^2 \frac{1}{2\pi^2} \left(3 - \frac{A}{B}\right) \right]. \tag{12.55}$$

Here k_{E} is a constant and $C_{\mathrm{E}}(T/\theta_{\mathrm{E}})$ is the lattice heat capacity in an Einstein model. In ch. 11 §4.3, we made the approximation $\alpha_{\mathrm{tr}}^2 F(\omega) = \langle 1 - \cos\theta \rangle \alpha^2 F(\omega)$, where $\langle 1 - \cos\theta \rangle$ is a geometrical factor and θ is an electron scattering angle. Typically, $\langle 1 - \cos\theta \rangle \approx 1$, i.e. $A/B \approx \langle 1 - \cos\theta \rangle \approx 1$. Motakabbir and Grimvall (1981) discussed W_{e} on the basis of an Einstein model, with A/B as a free parameter. In particular, they were able to account for the minimum in the thermal conductivity of pure metals which is observed, e.g., for Al, Na and Zn at $T \sim 0.4\theta_{\mathrm{D}}$.

7.4. The Wiedemann–Franz law

Comparing $1/\kappa_{\mathrm{e}}$ in (12.52) with $\rho = 1/\sigma$ in (11.34), it is obvious that the Wiedemann–Franz law is valid if we can make the approximation

$$\left[1 - \frac{1}{2\pi^2}\left(\frac{\hbar\omega}{k_{\mathrm{B}}T}\right)^2\right] \alpha_{\mathrm{tr}}^2 F(\omega) + \frac{3}{2\pi^2}\left(\frac{\hbar\omega}{k_{\mathrm{B}}T}\right)^2 \alpha^2 F(\omega) \approx \alpha_{\mathrm{tr}}^2 F(\omega), \tag{12.56}$$

i.e. if we can neglect a term

$$\frac{1}{2\pi^2}\left(\frac{\hbar\omega}{k_{\mathrm{B}}T}\right)^2 [3\alpha^2 F(\omega) - \alpha_{\mathrm{tr}}^2 F(\omega)] \tag{12.57}$$

in the integrand. When $T > \theta_{\mathrm{D}}$, the factor preceding $[\cdots]$ in (12.57) is < 0.05. At low temperatures (in pure metals), there are large deviations from the Wiedemann–Franz law (fig. 12.8). The prime reason is the inelastic scattering of electrons by phonons, i.e. the scattering with energy loss or gain. It is often stated that the Wiedemann–Franz law holds in the limit of low and high temperatures. At low T, this is true if T

is so low that the (elastic) impurity scattering dominates. At high T, the (dominating) electron–phonon scattering is basically inelastic, but the energy changes, $\sim \hbar\omega_D$, are small compared to the energy $k_B T$ and the scattering therefore appears to be elastic.

For completeness, we remark that we have left out a correction of a few percent or less, related to the thermoelectric power (e.g. Laubitz and Matsumura 1972).

7.5. Thermal conductivity in impure metals

We first recall the basic results for the electrical conductivity of impure metals. Let ρ_{tot} obey Matthiessen's rule, $\rho_{tot} = \rho_{ep} + \rho_{def}$. We may also write this as $\sigma_{tot} = ne^2\tau_{tot}/m$, with $1/\tau_{tot} = 1/\tau_{ep} + 1/\tau_{def}$. The Wiedemann–Franz law gives

$$W_{tot} = \frac{m}{ne^2 L_0 T\tau_{tot}} = \frac{m}{ne^2 L_0 T}\left(\frac{1}{\tau_{ep}} + \frac{1}{\tau_{def}}\right). \tag{12.58}$$

Thus, Matthiessen's rule holds also for the thermal conductivity

$$W_{tot} = W_{ep} + W_{def}. \tag{12.59}$$

This argument can be repeated with a more fundamental expression than $ne^2\tau_{tot}/m$ for σ_{tot}.

Assuming that the Wiedemann–Franz law is valid, the effect of various static lattice defects on the electron part of the thermal conductivity follows from our discussion of the electrical conductivity, ch. 11. See also Motakabbir and Grimvall (1981) for a simple model calculation of $W_{tot}(T)$ in an alloy, with allowance for the energy dependence of $\tau_{ep}(\varepsilon)$.

7.6. Saturation effects

In ch. 11 §10, we disscussed saturation effects in the electrical resistivity. Since they suggest a breakdown of the usual Boltzmann equation, one may wonder if the Wiedemann–Franz law still holds. Figure 12.9 shows $L_0 T/\kappa$ and ρ for metals which do (V, U) and do not (W) show saturation. The result is consistent with the idea that the thermal conductivity is almost entirely due to the electrons and that the Wiedemann–Franz law still holds.

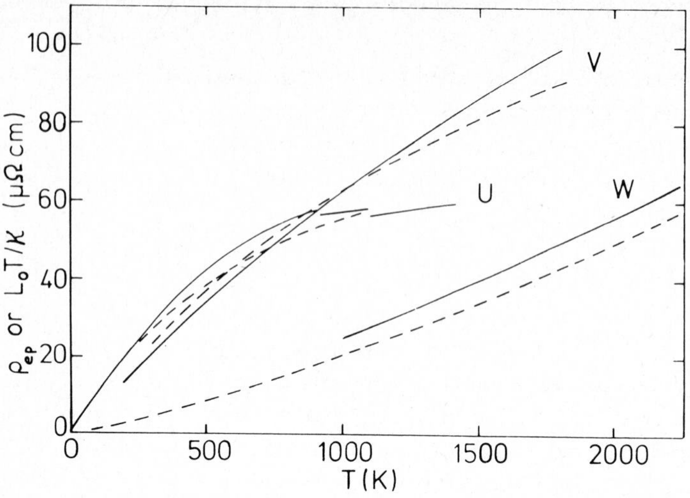

Fig. 12.9. The electrical resistivity ρ (solid lines) and the quantity L_0T/κ (dashed lines) as obtained from measured ρ and κ. From Grimvall (1984a). (Reproduced with permission from Physica.)

8. Miscellaneous transport mechanisms

8.1. Both electron and phonon transport

In alloys, heat is conducted by both the electrons and the phonons. It follows from fig. 12.9 that the high-temperature conductivity of pure metals is dominated by the electrons, also when the elemental metal is not a very good electrical conductor. Even in a concentrated alloy like stainless steel the electrons carry most of the heat, except in the temperature region of the maximum in the phonon part, i.e. roughly at $T \sim 0.1\theta_D$ (e.g. White 1969). In particular, this means that metals and alloys have a low-temperature limit $\kappa \sim T$. (Use Matthiessen's rule and the fact that impurity resistance always dominates ρ at low enough temperatures.)

8.2. Magnons

Magnons carry heat much like phonons. The effect is significant only at very low temperatures. See, for example, Berman (1976).

8.3. *Photons*

In materials which are semitransparent to infrared radiation, there is a high temperature contribution to κ which increases with T as T^3. This term is the likely cause of the upturn in κ at high T in Al_2O_3; fig. 12.5. The problem has been discussed by, inter alia, Men' and Sergev (1973). It appears that the mechanism may be significant even at room temperature, e.g., for In_2Te_3 (Petrusevich et al. 1960).

8.4. *Porous materials*

Conduction in inhomogeneous materials is considered in chs. 13 and 14. Here we only mention that in porous materials both contact resistance, gas conduction and radiation may be important. See Larkin and Churchill (1959) for a discussion of insulating fiber materials and Klemens (1983b) for a treatment of metal powders.

9. Pressure dependence

The volume (pressure) dependence of the thermal conductivity of metals follows from our discussion of the electrical conductivity and with reference to the Wiedemann–Franz law. The volume dependence of κ_{ph} was, indirectly, discussed in §4.4 where we noted that thermal expansion affected the high temperature κ_{ph} mainly through θ_D. Many materials, both metals and insulators, have $-d\ln\kappa/d\ln V$ in the range 4 to 8. For a more detailed discussion, see Ross et al. (1984), Bäckström (1985) and Mooney and Steg (1969).

10. Theoretical calculations

10.1. *Phonon transport*

It is exceedingly difficult to calculate the phonon part of the thermal conductivity with high accuracy, without the use of fitted parameters. Among the few ambitious attempts to obtain κ_{ph-ph}, we note work on solid noble gases (Julian 1965) and alkali halides (Logachev and Yur'ev 1972, Petterson 1986).

10.2. Electron transport

A calculation of the electron contribution to the thermal conductivity is not much different from the corresponding calculation of the electrical conductivity. See, for instance, Leung et al. (1977) and Tomlinson (1979), who considered Al and Zn, respectively, Leavens (1977) calculated κ_e in potassium, with full consideration of the Fredholm integral equation for $\tau(\varepsilon)$.

CHAPTER 13

TRANSPORT, ELASTIC AND THERMAL-EXPANSION PARAMETERS OF COMPOSITE MATERIALS

1. Introduction

Aluminium–silicon alloys are typical two-phase materials, as there is almost no mutual solubility of silicon and aluminium. A specimen of an Al–Si alloy therefore is a mixture of Al and Si grains. Suppose that the grains are small compared with the size of the specimen and that they are in an isotropic statistical distribution on a large scale. Such a material has an isotropic electrical and thermal conductivity, bulk modulus, shear modulus, thermal expansion coefficient, etc. Handbook data for these properties are often sparse and not very accurate. For instance, the American Institute of Physics Handbook (1972) gives the same shear modulus ($G = 26.5$ GPa) for pure aluminium and cast Al alloys with 5 and 12% Si. One obviously needs methods to estimate the properties of the composite system when the corresponding properties of the constituent phases are known and one has some information about their geometrical distribution. This is the theme of this chapter.

Consider the following well-known relations:

electrical conduction $\qquad \boldsymbol{j} = \sigma \boldsymbol{E}, \qquad (13.1)$

thermal conduction $\qquad \boldsymbol{Q} = -\kappa \nabla T, \qquad (13.2)$

dielectric displacement $\qquad \boldsymbol{D} = \varepsilon \boldsymbol{E}, \qquad (13.3)$

magnetic induction $\qquad \boldsymbol{B} = \mu \boldsymbol{H}, \qquad (13.4)$

diffusion $\qquad \boldsymbol{Q} = -D \nabla c. \qquad (13.5)$

These equations are all of the same mathematical structure. The left-hand side gives the response to a disturbance, and the proportionality constant (σ, κ etc.) is a tensor of rank two. For a system of cubic symmetry the tensor reduces to a single constant which is the case assumed above. Suppose that we have solved the problem of finding the average conductivity σ of a composite material, in terms of the conductivities of the constituent phases and their geometrical distribution. The same mathematical expression can then be used to find, for instance, the average dielectric constant ε, if we know ε of the pure phases. It therefore suffices to consider one of the properties in eqs. (13.1)–(13.5), and we shall often choose the electrical conductivity.

The elastic properties are more complicated. The counterpart of the transport equations above is

$$\text{Hooke's law} \qquad \boldsymbol{\sigma} = \boldsymbol{c}\boldsymbol{\varepsilon}, \tag{13.6}$$

where now $\boldsymbol{\sigma}$ is the stress and $\boldsymbol{\varepsilon}$ the strain tensor. The elasticity tensor $\boldsymbol{c}$ never reduces to a single constant. For a single crystal of a material with a cubic lattice structure, $\boldsymbol{c}$ has three independent components, c_{11}, c_{12} and c_{44}. For an isotropic system, two elastic parameters suffice. In solid mechanics one often chooses them to be the bulk modulus K and the shear modulus G. They are related to Young's modulus E and Poisson's ratio ν through equations given in ch. 3 §2. In principle, one may proceed as for the electrical conductivity, and obtain an estimation of the elastic properties of a composite material in terms of the properties of the constituent phases.

The strain $\boldsymbol{\varepsilon}$ associated with a temperature increase ΔT is given by the equation for

$$\text{thermal expansion} \qquad \boldsymbol{\varepsilon} = \boldsymbol{\alpha}\Delta T. \tag{13.7}$$

The second-rank tensor $\boldsymbol{\alpha}$ reduces to a constant in a material of cubic symmetry. In a composite material, an estimation of α has to include not only the expansion coefficient of the constituent phases but also their elastic properties, expressed, for example, by K and G.

One of the first extensive treatments of eqs. (13.1)–(13.5) for two-phase materials is that of Bruggeman (1935), who was mainly interested in dielectric properties. Part of the material in this chapter is covered in a review by Hale (1976). Reviews by van Beek (1967), Landauer (1978) and Bergman (1978) concentrate on dielectric properties while Christensen (1979), Watt et al. (1976) and Laws (1980) consider the elastic properties; Watt et al. with special reference to geophysical applications. Hashin (1983), Walpole (1981) and Willis (1981) discuss

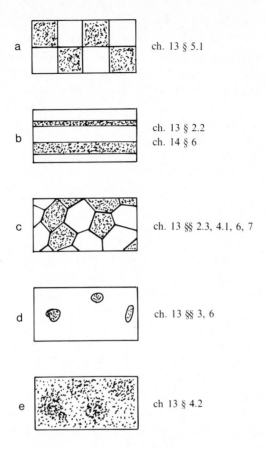

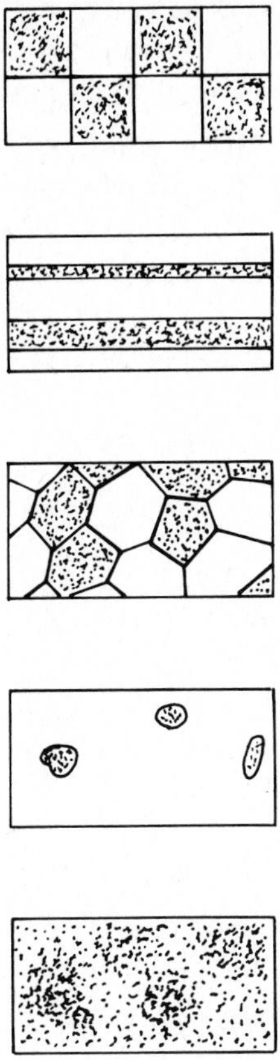

Fig. 13.1. Some schematic structures, and the section where these geometries are treated.

mainly the elastic properties of composites, but also thermal expansion and transport. An overview of the geometries treated in this and the following chapter is given in fig. 13.1.

2. Rigorous bounds

2.1. General aspects

Figures 13.1a, c and d show, schematically, the microstructure of statistically isotropic and homogeneous two-phase materials. A sample of such a material, if large compared with the characteristic dimension of the phase geometry, behaves as a homogeneous and isotropic one-phase material. The volume fractions of the phases 1 and 2 are f_1 and f_2 ($=1-f_1$). Lamellar or fibrous materials (fig. 13.1b) represent, as we shall see, extreme cases, with properties that cannot be exceeded by materials with other phase geometries. Throughout this chapter it is assumed that there are no surface effects caused by the phase boundaries. In transport properties this means that we neglect boundary scattering, contact potentials, etc. In the case of elastic properties and thermal expansion, we neglect grain boundary sliding and other anelastic behaviour.

2.2. Absolute bounds

In this section we consider bounds which are always valid, whatever is the phase geometry of the composite material. Let σ_1 and σ_2 be the conductivities of phases 1 and 2, and σ_e the effective conductivity of the composite material. Then (Wiener 1912, Jackson and Coriell 1968)

$$(f_1/\sigma_1+f_2/\sigma_2)^{-1} \leqq \sigma_e \leqq f_1\sigma_1+f_2\sigma_2. \tag{13.8}$$

The lower bound in (13.8) follows from an assumption of a spatially uniform current through the sample and the upper bound from an assumption of a uniform electric field. The bounds in eq. (13.8), which may be referred to as the series and parallel model, are the best possible since they are attained for conduction in fibrous or lamellar geometries. The generalisation to N phases is

$$\left(\sum_{i=1}^{N} f_i/\sigma_i\right)^{-1} \leqq \sigma_e \leqq \sum_{i=1}^{N} f_i\sigma_i. \tag{13.9}$$

Analogous results hold for the bulk modulus and the shear modulus (Paul 1960, Hill 1963). Thus,

$$K_R \leqq K_e \leqq K_V, \tag{13.10}$$

$$G_R \leqq G_e \leqq G_V, \tag{13.11}$$

with

$$K_V = f_1K_1+f_2K_2, \tag{13.12}$$

$$\frac{1}{K_R} = \frac{f_1}{K_1} + \frac{f_2}{K_2}, \tag{13.13}$$

and analogous relations for G_V and G_R. The upper bounds are obtained if one assumes a uniform strain in the composite, and the lower bounds if the stress is uniform. These assumptions, in the case of one-phase polycrystalline materials, lead to the Voigt and Reuss bounds (ch. 14 §4), motivating our use of the labels V and R above. We can obtain bounds to Young's modulus by the general relation (3.10) between E, K and G:

$$\frac{1}{E_R} = \frac{1}{3G_R} + \frac{1}{9K_R}, \qquad \frac{1}{E_V} = \frac{1}{3G_V} + \frac{1}{9K_V}. \tag{13.14}$$

From the bounds to K and G, one finds

$$E_R \leqq E_e \leqq E_V. \tag{13.15}$$

Note, however, that $E_V \neq f_1E_1 + f_2E_2$. After a little algebra, eqs. (13.12)–(13.14) give

$$E_V = f_1E_1 + f_2E_2 + \frac{27f_1f_2(G_1K_2 - G_2K_1)^2}{(3K_V + G_V)(3K_1 + G_1)(3K_2 + G_2)}. \tag{13.16}$$

The special case $\nu_1 = \nu_2$ implies, (eq. 3.8), $G_1K_2 = G_2K_1$. Then one recovers the familiar form $E_V = f_1E_1 + f_2E_2$ that is often used for fibrous materials. The relation $1/E_R = f_1/E_1 + f_2/E_2$ (which gives E perpendicular to lamellae) holds for any ν_1 and ν_2. There is no inequality like (13.10) for Poisson's ratio ν.

2.3. Hashin–Shtrikman bounds

Often one is interested in the properties of a composite material which is, in a statistical sense, isotropic and homogeneous. This excludes the fibrous and lamellar geometries for which the bounds of the previous section are relevant. Instead there are more narrow bounds. They are the best possible for σ and K, in the sense that the bounds are attained in certain geometries (§5.2.). It appears not to be known if the bounds to G are the best possible (Hashin 1983). Many of the relations given below can be cast in a variety of algebraic forms. Although it is often not quite transparent, they are all equivalent. However, these results should not be confused with similar, but not stringent, formulae that are abundant in the literature.

Transport properties: Hashin and Shtrikman (1962a) derived bounds to the conductivity σ. We choose the labelling such that $\sigma_2 > \sigma_1$. The upper bound is

$$\sigma_u = \sigma_2 + \frac{f_1}{1/(\sigma_1 - \sigma_2) + f_2/3\sigma_2}, \tag{13.17}$$

and the lower bound is

$$\sigma_\ell = \sigma_1 + \frac{f_2}{1/(\sigma_2 - \sigma_1) + f_1/3\sigma_1}. \tag{13.18}$$

Note that σ_ℓ is obtained from σ_u if indices 1 and 2 are interchanged. Analogous relations hold for the thermal conductivity, the dielectric constant, the magnetic susceptibility and the diffusion constant. The number 3 in the denominator of the right-hand side of (13.17) and (13.18) will be replaced by 2 for a two-dimensional and by 1 for a one-dimensional system (e.g. Milton 1980, Bergman 1982).

Elastic properties: Hashin and Shtrikman (1963a), Walpole (1966) and others have used various mathematical methods to derive bounds to the elastic parameters. One has, for $K_2 > K_1$,

$$K_u = K_2 + \frac{f_1}{1/(K_1 - K_2) + 3f_2/(3K_2 + 4G_2)}. \tag{13.19}$$

and

$$G_u = G_2 + \frac{f_1}{1/(G_1 - G_2) + 6f_2(K_2 + 2G_2)/5G_2(3K_2 + 4G_2)}, \tag{13.20}$$

The lower bounds, K_ℓ and G_ℓ, are obtained from (13.19) and (13.20 with indices 1 and 2 interchanged. If $(K_2 - K_1)(G_2 - G_1) < 0$, the bounds to K are reversed (Hill 1963). The bounds to G, when $(K_2 - K_1)(G_2 - G_1) < 0$, are similar to (13.20) (Walpole 1966). Bounds to

Young's modulus are obtained through the relation (3.10) between E, K and G. There are no analogous bounds to Poisson's ratio ν.

Thermal expansion: Levin (1967) and, independently, Rosen and Hashin (1970) and Schapery (1968) have shown that the linear expansion coefficient of an isotropic two-phase composite can be written in the exact form

$$\alpha = f_1\alpha_1 + f_2\alpha_2 + \frac{\alpha_1 - \alpha_2}{1/K_1 - 1/K_2}[1/K_e - f_1/K_1 - f_2/K_2]. \tag{13.21}$$

If we now use that $K_\ell < K_e < K_u$, eqs. (13.19) and (13.21) give an upper bound to α:

$$\alpha_u = \alpha_2 - f_1(\alpha_2 - \alpha_1)\frac{K_1(3K_2 + 4G_2)}{K_2(3K_1 + 4G_2) + 4f_1G_2(K_1 - K_2)}. \tag{13.22}$$

Again, the lower bound is obtained if indices 1 and 2 are interchanged. Equation (13.22) holds if $(\alpha_2 - \alpha_1)(K_2 - K_1) > 0$ and $G_2 > G_1$. If this is not the case, the bounds are reversed so that (13.22) gives α_ℓ. When $\alpha_2 = \alpha_1$ or $G_2 = G_1$, the upper and lower bounds to α coincide. $G_2 = G_1 (= G)$ leads to

$$\alpha = f_1\alpha_1 + f_2\alpha_2 - f_1(1 - f_1)(\alpha_1 - \alpha_2) \times \frac{4G(K_2 - K_1)}{3K_1K_2 + 4G(f_1K_1 + f_2K_2)}. \tag{13.23}$$

Properties of multi-phase systems: The derivation by Hashin and Shtrikman (1962a, 1963a) allowed for an arbitrary number of phases. To illustrate the mathematical structure of the bounds for this general case we quote two relations. The upper bound to the conductivity is

$$\sigma_u = \sigma_N + A_N\left[1 - \frac{A_N}{3\sigma_N}\right]^{-1}, \tag{13.24}$$

where

$$A_N = \sum_{i=1}^{N-1} f_i \left[\frac{1}{\sigma_i - \sigma_N} + \frac{1}{3\sigma_N}\right]^{-1}, \tag{13.25}$$

The upper bound to the bulk modulus is

$$K_u = K_N + A_N \left[1 - \frac{3A_N}{3K_N + 4G_N}\right]^{-1}, \tag{13.26}$$

where

$$A_N = \sum_{i=1}^{N-1} f_i \left[\frac{1}{K_i - K_N} + \frac{3}{3K_N + 4G_N}\right]^{-1}. \tag{13.27}$$

Here $i = 1-N$ denotes the phases, numbered so that σ_N, K_N and G_N are largest.

Bounds have also been derived for non-isotropic systems. They are considered in ch. 14. Milton (1980, 1981a, b, c) and Bergman (1980, 1982) have extended the conductivity bounds to complex σ and shown how the Wiener and Hashin–Shtrikman bounds are the lowest-order results in a general description which includes information about phase geometries.

Example: Comparison of bounds to σ. In fig. 13.2 are shown the Wiener

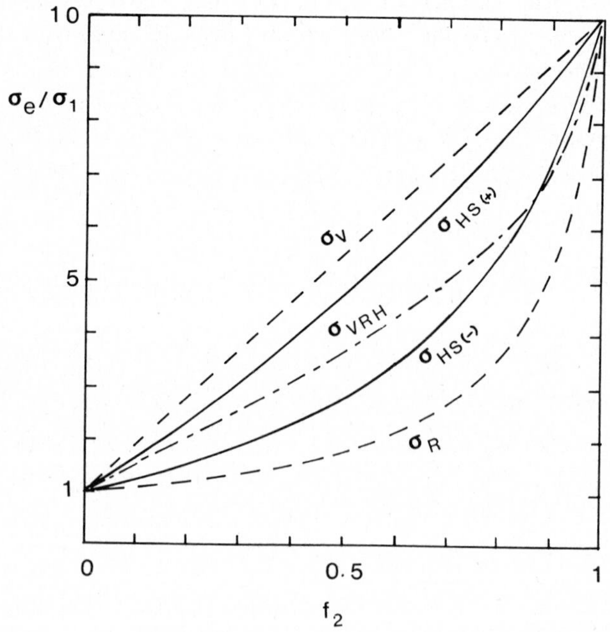

Fig. 13.2. The Wiener bounds, σ_V and σ_R, the average $\sigma_{VRH} = (\sigma_V + \sigma_R)/2$, and the upper and lower Hashin–Shtrikman bounds, σ_{HS}, to the effective conductivity σ_e, when $\sigma_2/\sigma_1 = 10$.

bounds to the conductivity σ, the average of these bounds (which is sometimes taken as an estimation of σ) and the Hashin–Shtrikman bounds, all for the case $\sigma_2/\sigma_1 = 10$. If the bounds are applied to a real system, with phases A and B, one should note that the conductivity of A and B may be significantly influenced, e.g. by the diffusion of (small amounts of) atoms from A into B and vice versa. This complication does not arise in the case of elastic or thermal expansion properties.

Example: The aggregate method. One may determine approximately a property of a material A from measurements on a mixture A and B, where B is a material with known properties. Brace et al. (1969) formed aggregates of several minerals with copper or lead as a matrix, and determined the effective bulk modulus K_e of the aggregate. Assuming that $K_e = (K_V + K_R)/2$ and with known elastic properties of Cu and Pb, they calculated the bulk modulus of the minerals. Since K_e can have any value between K_V and K_R, the estimation has obvious uncertainties. It is improved if one uses the Hashin–Shtrikman bounds. See Watt et al. (1976) for further discussions of the aggregate method.

Example: Bounds to the bulk modulus of aluminium–silicon alloys. Aluminium and silicon have a very low mutual solid solubility, and the Al–Si system consists of almost pure Al and Si. Figure 13.3 shows the Voigt, Reuss and Hashin–Shtrikman bounds to the bulk modulus K. Hill (1952) suggested that the arithmetic mean of the Voigt and Reuss limits (the so-called Voigt–Reuss–Hill approximation, K_{VRH}) could make a good estimate

$$K_{VRH} = (1/2)(K_V + K_R), \tag{13.28}$$

with analogous relations for G and E.

Example: Bounds to the expansion coefficient in aluminium–silicon alloys. The Hashin–Shtrikman bounds to the linear expansion coefficient α in Al–Si are barely resolved in the main part of fig. 13.4. The inset shows the upper and lower bounds, α_u and α_ℓ, as well as the average $\alpha = f_1\alpha_1 + f_2\alpha_2$. The circles are experimental results (Touloukian et al. 1975).

Example: Thermal expansion of a material with voids. Consider a one-phase material with voids. The voids can be regarded as a second phase, with

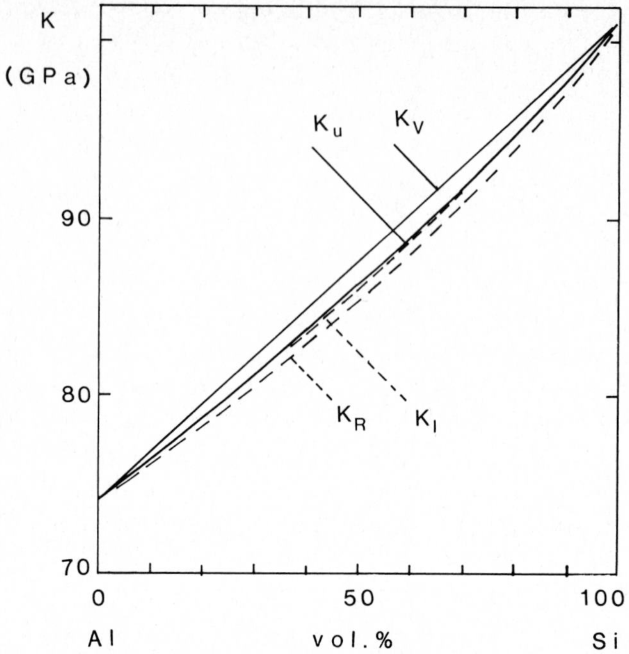

Fig. 13.3. The Voigt (K_V) and Reuss (K_R) and the upper and lower Hashin–Shtrikman bounds (K_u, K_l) to the bulk modulus K in Al–Si alloys.

$K_2 = 0$. Then, the expression (13.21) for α_e reduces to

$$\alpha_e = f_1\alpha_1 + f_2\alpha_1 = \alpha_1, \tag{13.29}$$

i.e. the linear expansion coefficient is the same as for the material without voids.

3. Dilute suspensions

We shall consider the limiting case when the concentration $f_1 \ll 1$, and phase 1 is in the form of a dilute suspension. The shape of the suspended particles can be spherical, ellipsoidal, rod- or disc-like. We exclude systems for which phase 1 is continuous through the material, even though $f_1 \ll 1$.

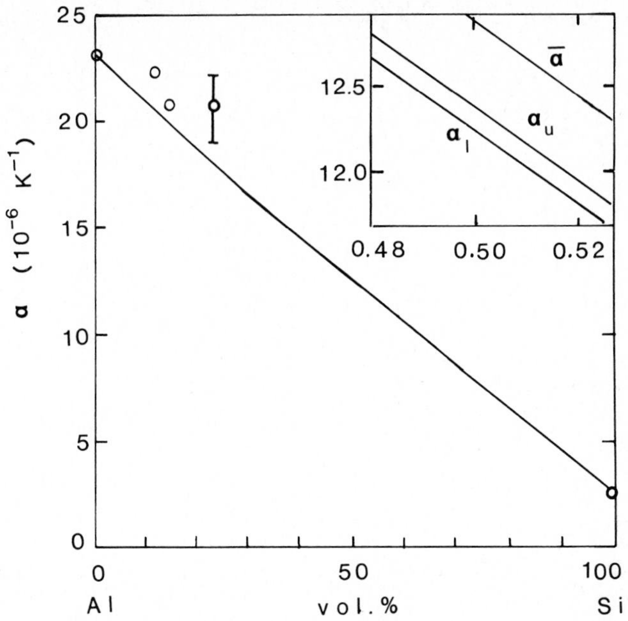

Fig. 13.4. The Hashin–Shtrikman bounds to the linear expansion coefficient α in Al–Si alloys almost coincide (solid line). The inset shows the upper and lower bounds, α_u and α_l, and also the arithmetic average $\bar{\alpha} = f_1\alpha_1 + f_2\alpha_2$.

3.1. Spherical inclusions

Transport properties: Rayleigh (1892) and others (see the review by Grosse and Greffe (1979)) have considered spherical inclusions in a matrix. In the dilute limit ($f_1 \ll 1$), but for any ratio σ_2/σ_1, one has

$$\sigma_e = \sigma_2\left[1 - f_1\frac{3(\sigma_2-\sigma_1)}{2\sigma_2+\sigma_1}\right]. \tag{13.30}$$

This result coincides with the lower Hashin–Shtrikman bound (for $f_1 \ll 1$) when $\sigma_2/\sigma_1 < 1$ (a dilute suspension of conducting spheres) and with the upper Hashin–Shtrikman bound when $\sigma_2/\sigma_1 > 1$ (a dilute suspension of insulating spheres).

In this context, it can be noted that the electric field $\boldsymbol{E}$ is uniform inside a sphere, or more generally inside an ellipsoid, embedded in a large uniform matrix (e.g. Stratton 1941). If an external field is along the i-axis of the ellipsoid, the average field strength E_i inside the ellipsoid is

$$E_i = E\sigma_2/[\sigma_2 + A_i(\sigma_1 - \sigma_2)]. \tag{13.31}$$

In the case of heat conduction, the thermal gradient ∇T is constant inside an ellipsoidal inclusion in a large uniform matrix. Similar results hold in the elastic case (Eshelby 1957). The dimensionless constants A_i are given in table 13.1.

Table 13.1
Depolarising factors

	A_1	A_2	A_3
Sphere $a_1 = a_2 = a_3$	1/3	1/3	1/3
Very prolate ellipsoid $a_1 \gg a_2 = a_3$	0	1/2	1/2
Very oblate ellipsoid $a_1 \ll a_2 = a_3$	1	0	0

Note: $\sum_{i=1}^{3} A_i = 1$.

Elastic properties: The bulk modulus K_e and the shear modulus G_e have been derived, independently, by Dewey (1947), Eshelby (1957) and Hashin (1959). The result is

$$K_e = K_2 - f_1(K_2 - K_1)\frac{3K_2 + 4G_2}{3K_1 + 4G_2}, \tag{13.32}$$

$$G_e = G_2 - f_1(G_2 - G_1)\frac{5(3K_2 + 4G_2)}{9K_2 + 8G_2 + 6(K_2 + 2G_2)G_1/G_2}. \tag{13.33}$$

In analogy to the transport case, K_e and G_e coincide with a Hashin–Shtrikman bound for $f_1 \ll 1$.

Thermal expansion: From eqs. (13.21) and (13.32) we obtain the linear expansion coefficient for a suspension of spheres in a matrix. Neglecting $(K_1 - K_2)f_1$ compared with K_2, one obtains

$$\alpha_e = \alpha_2 - f_1(\alpha_2 - \alpha_1)\frac{K_1(3K_2 + 4G_2)}{K_2(3K_1 + 4G_2)}. \tag{13.34}$$

3.2. Ellipsoidal inclusions

For a dilute distribution of ellipsoids,

$$\sigma_e = \sigma_2 - f_1 \frac{\sigma_2 - \sigma_1}{3} \sum_{i=1}^{3} \frac{\sigma_2}{\sigma_2 + A_i(\sigma_1 - \sigma_2)}, \tag{13.35}$$

where A_i are depolarising factors along the ellipsoid axes $i = x$, y and z. Tables of A_i have been published by Stoner (1945), Osborn (1945) and Fricke (1953). Some special cases are given in table 13.1. For a sphere, which has all $A_i = 1/3$, we recover eq. (13.30). Corresponding expressions for the elastic properties of a dilute suspension of ellipsoids are given by Berryman (1980) and Wu (1966).

3.3. Rods and discs

Rods, needles and fibres can be viewed as limiting cases of elongated ellipsoids of revolution. Similarly, circular discs and plates can be viewed as oblate ellipsoids.

Transport properties: From eq. (13.35), and with A_i from table 13.1, we have for the effective conductivity of a dilute suspension of randomly oriented rods

$$\sigma_e = \sigma_2 - f_1 \frac{(\sigma_2 - \sigma_1)(5\sigma_2 + \sigma_1)}{3(\sigma_2 + \sigma_1)}. \tag{13.36}$$

The corresponding relation for a dilute suspension of randomly oriented circular thin discs is

$$\sigma_e = \sigma_2 - f_1 \frac{(\sigma_2 - \sigma_1)(\sigma_2 + 2\sigma_1)}{3\sigma_1}. \tag{13.37}$$

Elastic properties: The elastic properties of a matrix with a dilute suspension of randomly oriented rods and circular discs have been treated by Walpole (1969), Watt et al. (1976) and Berryman (1980). In the low concentration limit of phase 1 one has, for stiff discs,

$$K_e = K_2 - f_1(K_2 - K_1)\frac{3K_2 + 4G_1}{3K_1 + 4G_1}, \tag{13.38}$$

$$G_e = G_2 - f_1(G_2 - G_1)\frac{G_2 + F_1}{G_1 + F_1}, \tag{13.39}$$

where $F_1 = (G_1/6)(9K_1 + 8G_1)/(K_1 + 2G_1)$.

In tne case of rods,

$$K_e = K_2 - f_1(K_2 - K_1)\frac{K_2 + G_2 + G_1/3}{K_1 + G_2 + G_1/3}. \tag{13.40}$$

The corresponding expression for G_e is algebraically complicated.

3.4. Inclusions with extreme properties

The inclusions may have properties which are very different from those of the matrix; voids is one extreme case. Liquid inclusions, with zero shear modulus G_1 but a finite compressibility K_1, give another example. In this section we shall take the expressions in §3 for transport properties to the limits $\sigma_1 \gg \sigma_2$ and $\sigma_1 \ll \sigma_2$, with an analogous discussion of K and α.

The conductivity, the bulk modulus and the linear expansion coefficient can be written, when $f_1 \ll 1$,

$$\sigma_e = \sigma_2(1 - f_1 H_\sigma), \tag{13.41}$$

$$K_e = K_2(1 - f_1 H_K). \tag{13.42}$$

$$\alpha_e = \langle\alpha\rangle - f_1(\alpha_2 - \alpha_1)H_\alpha, \tag{13.43}$$

The functions H_σ, H_K and H_α, given in tables 13.2–13.4, are obtained as limiting cases of the formulae in §3.1 and §3.3.

Table 13.2
The function H_σ

Type of inclusions	$\sigma_2 \gg \sigma_1$	$\sigma_2 \ll \sigma_1$
Spheres	3/2	-3
Rods	5/3	$-\sigma_1/3\sigma_2$
Discs	$\sigma_2/3\sigma_1$	$-2\sigma_1/3\sigma_2$

We can summarise the implications of table 13.2 in some general statements. When highly conducting inclusions are added to a matrix, the overall conductivity of course increases. The smallest increase (σ_e coinciding with the lower H-S bound σ_ℓ) occurs for spherical inclusions.

Table 13.3
The function H_K

Type of inclusions	Soft inclusions $K_2 \gg K_1,\ G_2 \gg G_1$	Stiff inclusions $K_2 \ll K_1,\ G_2 \ll G_1$
Spheres	$\dfrac{3K_2+4G_2}{3K_1+4G_2}$	$-\dfrac{3K_2+4G_2}{3K_2}$
Rods	$\dfrac{K_2+G_2}{K_1+G_2}$	$-\dfrac{K_1(3K_2+G_1)}{K_2(3K_1+G_1)}$
Discs	$\dfrac{3K_2+4G_1}{3K_1+4G_1}$	$-\dfrac{K_1(3K_2+4G_1)}{K_2(3K_1+4G_1)}$

If, instead, a poorly conducting phase is added, the overall conductivity decreases, and the effect is smallest (σ_e coinciding with the upper H-S bound σ_u) when the dilute phase has the shape of spheres. A dispersion of highly conducting discs in a matrix provides easy paths for the current and the conductivity has its largest possible value (the upper H-S bound) for the given volume fraction of inclusion. Also rods provide easy paths for the current but their effect (for a given volume fraction) is only half that of the discs (cf. table 13.2). Poorly conducting discs are effective in cutting off the current and this shape of the inclusions leads to the largest decrease in σ (i.e. the lower H-S bound). The same volume

Table 13.4
The function H_α

Type of inclusions	Soft inclusions $K_2 \gg K_1,\ G_2 \gg G_1$	Stiff inclusions $K_2 \ll K_1,\ G_2 \ll G_1$
Spheres	$-\dfrac{4G^2}{3K_1+4G_2}$	$\dfrac{4G_2}{3K_2}$
Rods	$-\dfrac{G_2}{K_1+G_2}$	$\dfrac{G_1K_1}{K_2(3K_1+G_1)}$
Discs	$-\dfrac{4G_1}{3K_1+4G_1}$ [a]	$\dfrac{4K_1G_1}{K_2(3K_1+4G_1)}$

[a] Let $f_1 \to 0$ if $G_1, K_1 \to 0$.

of rod-like inclusions does not force the current to make large "detours" in the composite and therefore does not strongly affect σ_e.

Analogous results hold for the elastic properties. The smallest change (K_e equals a H-S bound) is obtained for spherical inclusions and the largest change (K_e equals the other H-S bound) is obtained for disc-like inclusions. Very soft inclusions ($K_1 \ll G_2$) correspond to $H_\alpha = 1$ for all shapes (cf. the example on p. 265). Stiff discs and rods have a large influence on the expansion coefficient, while stiff spheres give $H_\alpha \approx 1$. When, as for voids, $\sigma_1 = K_1 = G_1 = 0$, the Hashin–Shtrikman lower bounds are all zero since the voids may form a continuous phase even though their volume fraction is low.

Example: Steel shot in epoxy resin. Lee and Taylor (1978) measured the thermal diffusivity of steel shot dispersed in epoxy resin, as a function of the volume fraction f_1 of the steel shot and of their radius r. The effective thermal diffusivity is

$$a_e = \kappa_e / C_e \rho_e, \tag{13.44}$$

where κ_e is the thermal conductivity of the composite, C_e its specific heat (heat capacity per mass unit) and ρ_e its mass density; C and ρ are additive, i.e.

$$C_e \rho_e = (f_1 C_1 + f_2 C_2)(f_1 \rho_1 + f_2 \rho_2). \tag{13.45}$$

Figure 13.5 shows a_e from eqs. (13.30) and (13.45) and the experimental results. The Hashin–Shtrikman bounds form such a wide interval that they are not of much use (shaded area in the inset). Since the inclusions are spherical, κ_e (and a_e) is well approximated by the bound σ_ℓ(HS) – the solid line in the main figure.

Expressions like (13.30) depend on the shape and volume concentration of the inclusions but not on their absolute size. This is born out in fig. 13.5, in which the measured a_e is seen to be independent of the radius of the shot.

Example: Elastic properties of cast iron. Cast iron has graphite embedded in a matrix, which is itself a composite of ferrite and cementite. Here we shall regard the matrix as a uniform material. When the graphite is mainly in the form of sheets, the material is referred to as

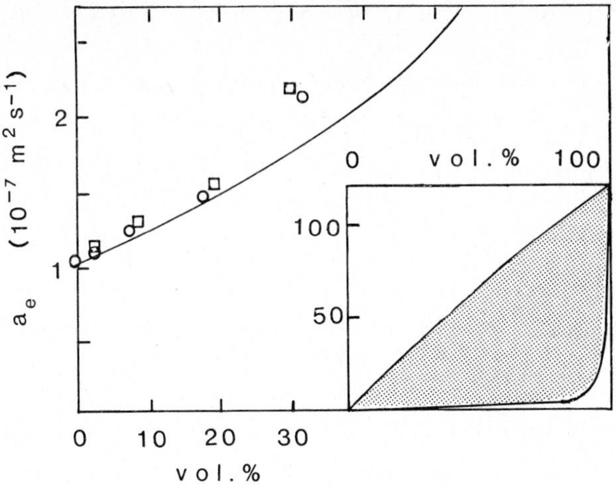

Fig. 13.5. The thermal diffusivity, a_e, of steel shot in epoxy, plotted versus the volume fraction of the steel inclusions. The shaded area of the inset is based on the Hashin–Shtrikman bounds. The full curve in the main figure is the lower Hashin–Shtrikman bound. Circles and squares correspond to different diameters of the steel spheres; 584 μm and 279 μm. After Lee and Taylor (1978); with permission.

gray cast iron and when it is essentially spherical one refers to ductile cast iron. Graphite itself is elastically highly anisotropic, but we follow Speich et al. (1980) and take for the graphite $K_1 = 6.7$ GPa and $G_1 = 3.3$ GPa. For the matrix, $K_2 = 162$ GPa and $G_2 = 81$ GPa. Figure 13.6 shows experimental data for cast iron of varying graphite concentration. The solid and dashed lines are the upper and lower Hashin–Shtrikman bounds. The circles refer to ductile iron, which has spherical graphite inclusions. The squares refer to gray cast iron, which has flake-like graphite inclusions. In the dilute limit of inclusions, spheres and plates give properties coinciding with the upper and lower Hashin–Shtrikman bounds (after Anand 1982).

Example: Elastic properties of porous NbC. Speck and Miccioli (quoted by Toth (1971)) measured the shear modulus $G(p)$ and Young's modulus $E(p)$ for sintered $NbC_{0.97}$, as a function of the porosity p. Figure 13.7 shows their data and the expressions for G and E that result from (13.32) and (13.33) if we take $f_1 = p$ and $K_1 = G_1 = 0$ for the pores.

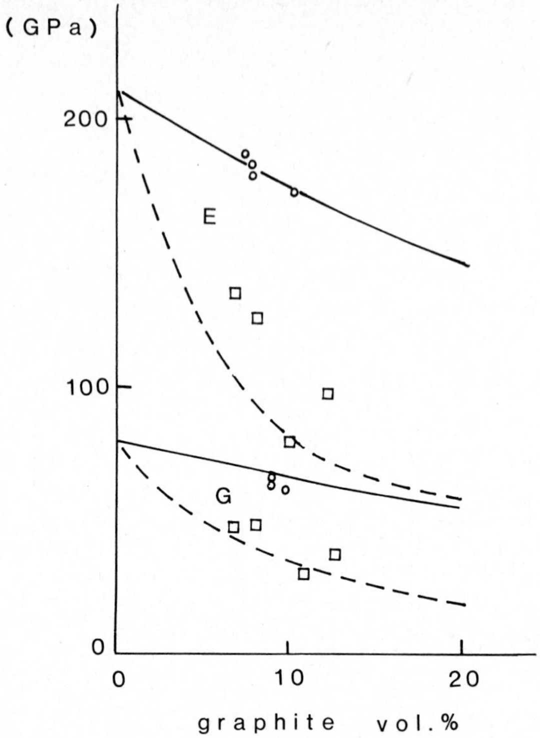

Fig. 13.6. Young's modulus E and the shear modulus G of cast iron. Solid and dashed lines are the upper and lower Hashin–Shtrikman bounds. Circles and squares refer to spherical and plate-like graphite inclusions (after Anand 1982).

4. Weakly inhomogeneous materials

4.1. Two-phase materials

When the conductivities of the phases are almost equal, we can expand σ_e in powers of the small difference $\delta\sigma = \sigma_2 - \sigma_1$ and use perturbation theory to estimate σ_e. To lowest non-vanishing order in $\delta\sigma$, the upper and lower Hashin–Shtrikman bounds coincide and we get

$$\sigma_e \approx \langle\sigma\rangle - f_1 f_2 \frac{(\delta\sigma)^2}{3\langle\sigma\rangle}. \tag{13.46}$$

The corresponding relations for the bulk modulus K_e, the shear modulus

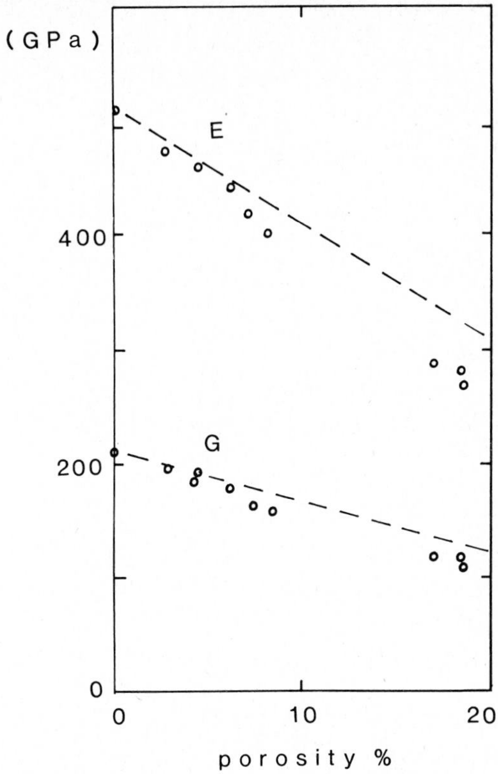

Fig. 13.7. Young's modulus E and the shear modulus G plotted versus the porosity p of porous NbC; after data quoted by Toth (1971). The straight lines are extrapolations of the result for a dilute suspension of spherical voids.

G_e and the linear expansion coefficient α_e are

$$K_e \approx \langle K\rangle - f_1 f_2 \frac{3(\delta K)^2}{\langle 3K+4G\rangle}, \tag{13.47}$$

$$G_e \approx \langle G\rangle - f_1 f_2 \frac{6\langle K+2G\rangle(\delta G)^2}{5\langle G\rangle\langle 3K+4G\rangle}, \tag{13.48}$$

$$\alpha_e \approx \langle\alpha\rangle - f_1 f_2 \frac{4(\alpha_1-\alpha_2)(\delta K)}{\langle K\rangle\langle 3K+4G\rangle}. \tag{13.49}$$

Note that σ_e, K_e and G_e lie below the arithmetic averages $\langle\sigma\rangle$, $\langle K\rangle$ and

$\langle G \rangle$, respectively. This is necessary since $\langle \sigma \rangle$ etc. were shown in §2.2 to be upper bounds. The difference $\alpha_e - \langle \alpha \rangle$, on the other hand, may lie either above or below $\langle \alpha \rangle$. To lowest order it is independent of δG, in accordance with the fact that α_u and α_ℓ coincide when $G_1 = G_2$.

Series expansions of σ_e to the third and fourth order in $\delta\sigma$ have been worked out by Phan-Thien and Milton (1982), including a generalisation to an arbitrary number of phases. Corresponding results for the elastic parameters are given by Milton and Phan-Thien (1982).

To lowest order in $\delta\sigma$, eq. (13.46) is equivalent with

$$(\sigma_e)^{1/3} = f_1(\sigma_1)^{1/3} + f_2(\sigma_2)^{1/3}. \tag{13.50}$$

Occasionally, this relation may give a good account of σ_e for real materials even when $\delta\sigma/\sigma_e$ is not small, but it has no deeper theoretical justification.

4.2. One-phase materials

Consider a one-phase material in which the composition, and therefore also the electrical conductivity, is (isotropically) weakly inhomogeneous. The variations are assumed so small that a local conductivity σ can be defined everywhere in the material. Let $\langle \cdots \rangle$ denote a spatial average over the entire specimen. The effective conductivity σ_e is (see Brown (1955), Nedoluha (1957), Herring (1960), Landau and Lifshitz (1960), Beran and Molyneux (1963), Beran (1965), Hori (1973a, b) for proofs and generalisations),

$$\sigma_e = \langle \sigma \rangle \left\{ 1 - \frac{1}{3} \frac{\langle (\sigma - \langle \sigma \rangle)^2 \rangle}{\langle \sigma \rangle^2} \right\}. \tag{13.51}$$

The corresponding relation for the electrical resistivity ρ is

$$\rho_e = \langle \rho \rangle \left\{ 1 - \frac{2}{3} \frac{\langle (\rho - \langle \rho \rangle)^2 \rangle}{\langle \rho \rangle^2} \right\}. \tag{13.52}$$

Equation (13.46) for a two-phase material is closely related to eq. (13.51), since we can write $\langle \sigma^2 \rangle = f_1(\sigma_1)^2 + f_2(\sigma_2)^2$ and $(\langle \sigma \rangle)^2 = (f_1\sigma_1 + f_2\sigma_2)^2$. Note that although $\rho_e = 1/\sigma_e$, one has $\langle \rho \rangle = \langle 1/\sigma \rangle \neq 1/\langle \sigma \rangle$.

The approach in this chapter relies on a macroscopic description of the conduction. Therefore, in averages such as $\langle(\sigma-\langle\sigma\rangle)^2\rangle$ one should not consider fluctuations on a length scale shorter than the (electron) mean free path.

The bulk modulus in weakly inhomogeneous materials is (Molyneux and Beran 1965)

$$K_e = \langle K\rangle - \frac{3\langle(K-\langle K\rangle)^2\rangle}{\langle 3K+4G\rangle}. \tag{13.53}$$

This is obviously another version of eq. (13.47).

5. Exact results

In a few cases, it is possible to give mathematically exact formulae for σ, K, G or α of a composite material. Although they hold under very special assumptions, the results are of interest, since they provide test cases in the assessment of approximate methods.

5.1. Symmetric cell materials

A two-phase cell material (Miller 1969) is defined as follows: The composite material is subdivided by closed surfaces into closed regions, or cells. Each cell is randomly assigned physical properties with probabilities f_1 and f_2 referring to one or the other of the two phases. The cells are distributed in such a way that the material is statistically isotropic and homogeneous.

A fibrous material when considered transverse to the fibres may fulfill the requirements of a two-dimensional symmetric cell material. Let its conductivity be $\sigma_e(1,2)$. Now, interchange the phases 1 and 2, without altering the phase boundaries. The new conductivity is denoted $\sigma_e(2,1)$. Then (Dykhne 1970)

$$\sigma_e(1,2)\sigma_e(2,1) = \sigma_1\sigma_2. \tag{13.54}$$

Relation (13.54) holds also for σ_e given by the effective medium theory, i.e. eq. (13.60) (with 2 in the denominator replaced by 1 in two dimensions). The Hashin–Shtrikman bounds in two dimensions obey an analogous relation:

$$\sigma_u(f_1)\sigma_\ell(1-f_1) = \sigma_1\sigma_2. \tag{13.55}$$

It follows that materials with $f_1 = f_2 = 0.5$, and such that the same material is obtained if the conductivities σ_1 and σ_2 are interchanged (without altering the grain boundaries), have $\sigma_e = \sqrt{(\sigma_1\sigma_2)}$. Special cases of this are the ordinary chequer board geometry and the same geometry with "black" and "white" colours randomly distributed over the squares. A regular alternate stacking of cubes of phases 1 and 2 has $\sigma_e = 2\sqrt{(\sigma_1\sigma_2)}$ when $\sigma_1 \gg \sigma_2$ (Söderberg and Grimvall 1983). This is consistent with the following relation for a three-dimensional symmetric cell material (Schulgasser 1976a):

$$\sigma_e(1,2)\sigma_e(2,1) \geqq \sigma_1\sigma_2. \tag{13.56}$$

5.2. *Attained Hashin–Shtrikman bounds*

The Hashin–Shtrikman conductivity bounds are attained in the "coated sphere" geometry ("composite sphere assemblage"). Spheres of phase 1 and radius R_1 are coated with shells of phase 2 and radius R_2, such that R_1 and R_2 enclose the relative amounts f_1 and f_2 of the two phases. All space is now filled with such coated spheres, which requires a distribution of radii, including infinitesimally small values. This geometry has an effective conductivity equal to the upper Hashin–Shtrikman bound. If spheres of phase 2 are coated by phase 1, the effective conductivity is equal to the lower Hashin–Shtrikman bound (Hashin and Shtrikman 1962a). It has been shown (Hashin 1962) that also K_ℓ and K_u are attained in the coated sphere geometry, but it appears not yet certain if this is true for the shear modulus (Hashin 1983). Further, when K_e is known α_e is also known exactly. Grimvall (1984b) has generalised the coated sphere results to eutectic-like systems.

5.3. *The case of equal shear moduli*

Hill (1963) derived an exact result for the bulk modulus when $G_1 = G_2 = G$:

$$K_e = \langle K \rangle - f_1 f_2 \frac{3(K_2 - K_1)^2}{3f_1K_2 + 3f_2K_1 + 4G}. \tag{13.57}$$

We have also noted (in §2.3) that the lower and upper Hashin–Shtrikman-type bounds to the coefficient of thermal expansion coincide when $G_1 = G_2$ $(= G)$. This yields the exact result, (13.23).

6. Effective medium theories

6.1. Introduction

In the first part of this chapter we established bounds to σ_e, K_e, G_e and α_e for two-phase materials. These bounds may lie far apart and be of little practical value. It is of interest to have a model which gives a single estimated value. In the preceding section we noted that there are very few geometries which allow an exact solution. This fact has lead to a rich literature aiming at approximate but closed-form expressions for σ_e, K_e, etc. Much of that work relies on an empirical fitting to certain functions or on dubious theoretical asumptions. However, there is one approach, the effective medium theory, which is algebraically simple and yet physically well founded and we shall concentrate on this. (One may remark that effective-medium theories can be defined in slightly different ways. We shall not dwell on that point.)

6.2. Transport properties

In §3.1 we considered a dilute suspension of spheres or ellipsoids in a matrix. If the volume fraction f_1 is not small, one can still use a similar approach but let a sphere of phase 1 be surrounded by a medium with the effective conductivity σ_e instead of σ_2. Then

$$\sigma_e = \sigma_2 - f_1 \frac{3(\sigma_2 - \sigma_1)\sigma_e}{2\sigma_e + \sigma_1}. \tag{13.58}$$

This formula, which can be derived in a much stricter fashion, has been given by Böttcher (1952) for the dielectric case. Landauer (1952, 1978) considered dielectric properties of N-phase materials and derived the following relation (here with ε replaced by σ);

$$\sum_{i=1}^{N} \frac{\sigma_e - \sigma_i}{\sigma_i + 2\sigma_e} f_i = 0. \tag{13.59}$$

For a two-phase material ($N = 2$) we have

$$\frac{\sigma_1 - \sigma_e}{\sigma_1 + 2\sigma_e} f_1 = \frac{\sigma_e - \sigma_2}{\sigma_2 + 2\sigma_e} f_2. \tag{13.60}$$

The latter expression appears first to have been given by Bruggeman

(1935) in his treatment of dielectric properties. After a rearrangement of terms, one finds that (13.59), for $N = 2$, is identical to (13.58). When the relation for σ_e is cast in the symmetric form (13.60), it is obvious that f_1 does not have to be small. If $f_1 \ll 1$, we can approximate σ_e in the denominator by the matrix conductivity σ_2. Then (13.58) is consistent with (13.30), to lowest order in f_1. In the two-dimensional case (e.g., conduction perpendicular to fibers in a matrix), eq. (13.60) retains its form only if $2\sigma_e$ is replaced by σ_e in the two denominators (Bruggeman 1935). The effective medium theory (EMT) is also known as the self-consistent method (SCM). It is mathematically equivalent to the coherent potential approximation (CPA) used to obtain the electron band structure in alloys.

6.3. Elastic properties

Budiansky (1965, 1970) and Hill (1965b) independently derived effective medium results for the elastic properties. Several others have taken similar approaches; see Laws (1980) for a review of the field. One has

$$K_e = K_2 - f_1(K_2 - K_1)\frac{3K_e + 4G_e}{3K_1 + 4G_e}, \tag{13.61}$$

$$G_e = G_2 - f_1(G_2 - G_1)\frac{5(3K_e + 4G_e)G_e}{G_e(9K_e + 8G_e) + 6G_1(K_e + 2G_e)}. \tag{13.62}$$

Thus K_e and G_e are coupled and one must find their value by numerical iteration. When Poisson's ratio obeys $\nu_1 = \nu_2 = 0.2$, it follows that $3K_1 = 4G_1$ etc, and the relations (13.61) and (13.62) decouple (Budiansky 1965). In this special case,

$$K_e = K_2 - f_1(K_2 - K_1)\frac{2K_e}{K_1 + K_e}, \tag{13.63}$$

$$G_e = G_2 - f_1(G_2 - G_1)\frac{2G_e}{G_e + G_1}. \tag{13.64}$$

When $f_1 \ll 1$, so that $K_e \approx K_2$ and $G_e \approx G_2$, the relations (13.61) and (13.62) are consistent with (13.32) and (13.33). One may show that K_e and G_e in the effective medium theory fall inside the Hashin–Shtrikman bounds, approaching the lower bound for f_2 near 0 and the upper bound for f_2 near 1 (when $K_2 > K_1$, $G_2 > G_1$). When $G_1 = G_2$, EMT agrees with the exact result (13.57).

Example: Cement with voids. Beaudoin and Feldman (1975) and Feldman and Beaudoin (1977) measured Young's modulus E for porous autoclaved cement. They obtained for the pure cement $E_2 = 36$ GPa and $\nu_2 = 0.2$. The pores can be considered as inclusions with $K_1 = G_1 = 0$. Poisson's ratio ν_1 is undefined, but it is no restriction to put $\nu_1 = 0.2$. Then the EMT equations for K_e and G_e decouple giving $K_e = K_2(1-2f_1)$, $G_e = G_2(1-2f_1)$ and, by (3.10),

$$E_e = E_2(1-2f_1). \tag{13.65}$$

The same expression for E_e is obtained from eq. (13.32) and (13.33), i.e. for a dilute suspension of spheres, if we take $K_1 = G_1 = 0$ and note that $3K_2 = 4G_2$ when $\nu_2 = 0.2$. Since one of the phases (the pores) is extreme in its properties, we cannot expect the EMT or the dilute suspension model to be adequate when f_1 is not very small. However, fig. 13.8 shows that eq. (13.65) gives a qualitative account of E_e to quite high pore concentrations. Nielsen (1982) has discussed the elastic properties of porous cement in some detail (see also Hansen 1958, 1960).

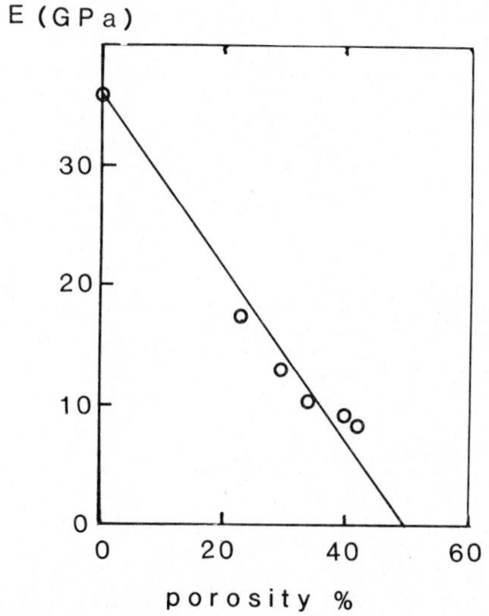

Fig. 13.8. Young's modulus E of porous cement, as a function of the vol.% porosity. The straight line is the effective-medium result. Circles are measured values (Beaudoin and Feldman 1975, Feldman and Beaudoin 1977).

6.4. Thermal expansion

Since there is an exact result for the linear expansion coefficient α, (13.21) expressed in terms of the elastic properties of the composite, there is no separate effective medium theory for α. If one uses the EMT result for K_e, the expression for α is very complicated, even in the special case $\nu_1 = \nu_2 = 0.2$ for which K_e and G_e decouple.

7. Percolation

Consider a dilute suspension of conducting spheres in an insulating matrix. There is no path which lies entirely within the conducting phase and goes through the specimen. Hence the composite is an insulator. We now increase the volume fraction of the conducting phase. Eventually one reaches a critical concentration, known as the percolation threshold f_c, at which a current can pass through the sample. It is obvious that f_c depends strongly on the geometry of the grains of the conducting phase. Examples may be constructed, for which f_c is anywhere in the interval 0 to 1. Figure 13.9 shows, in two dimensions, the idea of percolation.

The percolation phenomenon has been extensively studied, in particular because of its so-called critical behaviour when f is near f_c.

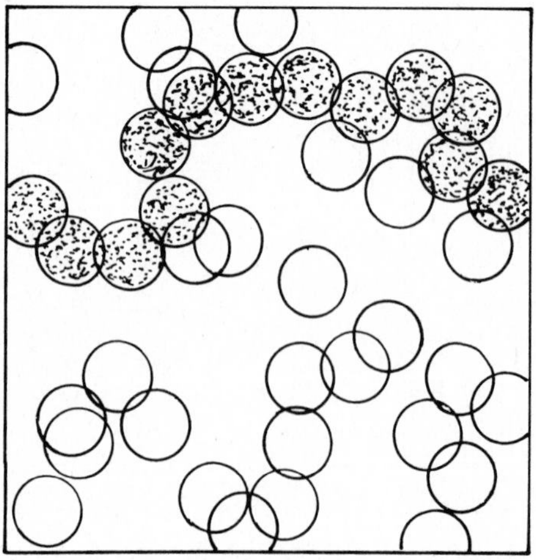

Fig. 13.9. Illustrating the idea of percolation. At a certain surface coverage by discs, there is an uninterrupted path through discs which connects two sides of the material.

Near f_c, the conductivity varies as $\sigma \sim \sigma_0(f-f_c)^\alpha$ where α is a "critical index". Much of the work in this field deals with resistor networks, along the lines reviewed by Kirkpatrick (1973). The critical behaviour is a universal feature which does not depend crucially on the kind of system studied. However, discrete networks seem to give a poor description of continuous materials (Söderberg and Grimvall 1986).

The progress in the field of percolation is so rapid that we shall only mention a few recent studies of various types. Benguigui (1984) investigated the critical index α for both conduction and elastic properties of a sheet with randomly punched holes. Kantor and Webman (1984) considered theoretically the elastic properties of percolating systems. Suen et al. (1979) made a numerical study of a regular stacking of conducting spheres in a matrix. Percolation in an anisotropic composite with elongated inclusions (carbon black in polyvinyl chloride) was studied by Balberg et al. (1983).

CHAPTER 14

ANISOTROPIC MATERIALS

1. Introduction

Fibrous or lamellar composites are typical examples of materials with anisotropic physical properties. More generally, we shall be interested in composites in which the properties of each separate phase are isotropic but the geometrical distribution of the phases is anisotropic. We shall also treat polycrystalline one-phase materials for which the property of interest is anisotropic in a single crystallite. The two cases are partly related. Consider the microstructure in fig. 14.1. Each grain in the material consists of two phases A and B which form a lamellar structure. If A and B have properly chosen and isotropic conductivities, the grains are characterised by a certain anisotropic conductivity with components $\sigma_\perp$ and $\sigma_{||}$, perpendicular and parallel to the lamellae. Thus the entire system can be viewed either as a one-phase polycrystalline material with

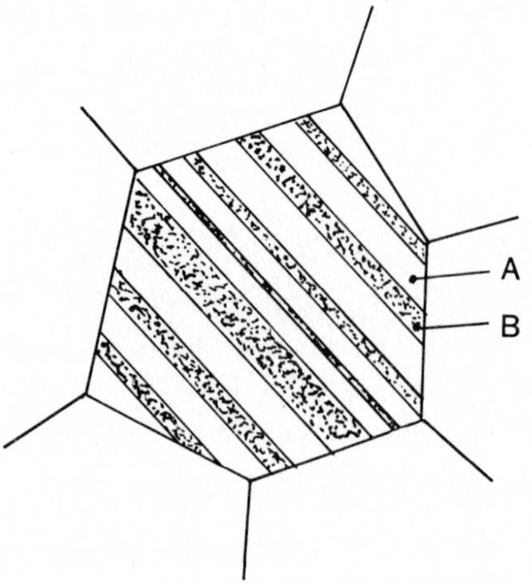

Fig. 14.1. A two-phase material may be viewed as a one-phase material with anisotropic properties (schematic illustration).

an anisotropic conductivity in each crystallite or as a two-phase material with an isotropic conductivity in each phase. A very important case is that of a polycrystalline material in which the grains are randomly oriented so that the specimen is isotropic on a large scale. One may call this a quasi-isotropic material. Some authors use the word aelotropy for the anisotropy of a single crystal and reserve the use of anisotropy to systems composed of many crystallites. We shall not make this distinction.

Much of our discussion parallels that of the preceding chapter. In particular, we are interested in the effective-conductivity tensor components $(\sigma_e)_{ij}$ of an overall anisotropic material, the effective conductivity σ_e of an isotropic polycrystalline sample and analogous parameters describing elastic and thermal expansion properties. There are bounds to such quantities, e.g. of the Hashin–Shtrikman type, and there are effective medium theories.

2. Averages for a single crystal

Let $A(\theta, \phi)$ be a physical property which is anisotropic. The angles θ and ϕ are measured relative to some crystallographic direction $[hkl]$. One can expand $A(\theta, \phi)$ in spherical harmonics $Y_{lm}(\theta, \phi)$:

$$A(\theta, \phi) = \sum_{l,m} a_{lm} Y_{lm}. \tag{14.1}$$

The average of A, over all points (θ, ϕ) on a unit sphere, is

$$\bar{A} = \int A(\theta, \phi) \frac{\mathrm{d}\Omega}{4\pi} = \sum_{l,m} a_{lm} \int Y_{lm} \frac{\mathrm{d}\Omega}{4\pi} = a_{00}, \tag{14.2}$$

where $\mathrm{d}\Omega = \sin\theta \, \mathrm{d}\theta \, \mathrm{d}\phi$. In a lattice with cubic structure it would be natural to expand $A(\theta, \phi)$ not in Y_{lm} but in a set of functions which all have cubic symmery. The Kubic harmonics K_m, introduced by von der Lage and Bethe (1947) form such a set. The first few of them are listed below. The direction in the lattice is specified by the Cartesian coordinates x, y and z which lie on a sphere with unit radius, $x^2 + y^2 + z^2 = 1$.

$$K_0 = 1, \tag{14.3}$$

$$K_1 = 0, \tag{14.4}$$

$$K_2 = x^4 + y^4 + z^4 - 3/5, \tag{14.5}$$

$$K_3 = x^2 y^2 z^2 + (1/22)K_2 - 1/105. \tag{14.6}$$

With normalised $K_m^* = K_m \gamma_m$ ($\gamma_0 = 1, \gamma_2 = (5/4)\sqrt{21}$ etc.) we have

$$A(\theta, \phi) = \sum_m a_m K_m^*, \tag{14.7}$$

and obtain for the average of $A(\theta, \phi)$

$$A = \int A(\theta, \phi) \frac{d\Omega}{4\pi} = a_0. \tag{14.8}$$

The general expression for the coefficient a_m is

$$a_m = \int A(\theta, \phi) K_m^* \frac{d\Omega}{4\pi}. \tag{14.9}$$

In practice, $A(\theta, \phi)$ is often unknown or difficult to obtain for a dense enough set of points (θ, ϕ). For instance, one may know $A(\theta, \phi)$ in only the three principal crystallographic directions [100], [110] and [111] of a cubic lattice. That information determines the three coefficients a_0, a_2 and a_3 in the sum (14.7) (a_1 is absent since $K_1 = 0$). One obtains

$$a_0 = \tfrac{1}{35}\,(10A[100] + 16A[110] + 9A[111]). \tag{14.10}$$

Betts et al. (1956a), Betts (1961) and others have given expressions for a_0 when $A[hkl]$ is known in more than three directions $[hkl]$ in a cubic lattice. An extension to hexagonal, tetragonal and trigonal lattice structures is found in Betts et al. (1956b).

Example: Anisotropic Young's modulus. In eq. (3.26) we gave an expression for Young's modulus in a single crystal of cubic symmetry; $\{E[hkl]\}^{-1} = s_{11} - sN^4$, where $s = 2s_{11} - 2s_{12} - s_{44}$ and N^4 contains the direction cosines of $[hkl]$. The direction cosines correspond to our x, y and z, and hence $N^4 = x^2y^2 + x^2z^2 + y^2z^2$. Using the identity $1 = (x^2 + y^2 + z^2)^2 = x^4 + y^4 + z^4 + 2N^4$, we have $N^4 = 1/5 - K_2/2$. Then,

$$\{E[hkl]\}^{-1} = s_{11} - s(\tfrac{1}{5} - \tfrac{1}{2}K_2). \tag{14.11}$$

The expansion coefficient a_0 is $s_{11} - s/5$. By (14.8) a_0 is the average of

E^{-1} over all directions. We can also obtain a_0 from (14.10), with $\{E[hkl]\}^{-1}$ taking the place of $A[hkl]$. One easily finds that $a_0 = s_{11} - s/5$. Thus, the approximate formula (14.10) gives the exact average in this case. However, if we express $E[hkl]$ (and not $1/E$) in Kubic harmonics K_m, we obtain an infinite series. Since that series contains terms higher than K_3, and (14.10) is fitted only to K_m with $m \ll 3$, (14.10) does not give the exact average.

3. Quasi-isotropic polycrystalline one-phase materials. Conductivity properties

Let the diagonal elements of the conductivity matrix of a single crystal be σ_a, σ_b and σ_c. A polycrystalline quasi-isotropic specimen of this material has an (isotropic) effective conductivity σ_e. One can show (Molyneux 1970) that σ_e is bounded by

$$3\left(\frac{1}{\sigma_a}+\frac{1}{\sigma_b}+\frac{1}{\sigma_c}\right)^{-1} \leqq \sigma_e \leqq \frac{1}{3}(\sigma_a+\sigma_b+\sigma_c). \tag{14.12}$$

Often, the right- and left-hand sides are referred to as the Voigt (1910) and Reuss (1929) bounds, σ_V and σ_R, respectively. There have been several attempts to establish more narrow bounds (Hashin and Shtrikman 1963b, Bolotin and Moskalenko 1967, Molyneux 1970, Dederichs and Zeller 1973, Hori 1973a,b, Schulgasser 1976b). It was shown by Schulgasser (1977) that one can construct a statistically isotropic material with the conductivity $(\sigma_a+\sigma_b+\sigma_c)/3$. Hence the upper (Voigt) bound in (14.12) is the best possible. The problem of the best lower bound seems not to be well understood. Schulgasser (1977) showed that the conductivity may be as low as $(\sigma_a\sigma_b\sigma_c)^{1/3}$, and he speculates that this may be the best lower bound.

Crystals of hexagonal or tetragonal symmetry have $\sigma_a = \sigma_b = \sigma_\perp$ and $\sigma_c = \sigma_\parallel$. Then, Bolotin and Moskalenko (1967) have derived an effective medium result. The effective conductivity σ_e is the solution to the equation

$$4(\sigma_e/\sigma_c)^3 - (1/5)(\sigma_e/\sigma_c)[4(\sigma_a/\sigma_c)^2 + 12(\sigma_a/\sigma_c) - 1] - (\sigma_a/\sigma_c)^2 = 0. \tag{14.13}$$

Example: Effective conductivity in polycrystalline hexagonal materials. Consider a material of hexagonal lattice symmetry and let $\sigma_\perp/\sigma_\parallel = 1.2$

(anisotropies are often smaller; cf. fig. 11.4). Then, σ_V: $(\sigma_\perp^2 \sigma_\parallel)^{1/3}$: σ_R = 1.007 : 1.004 : 1. Thus the bounds to σ_e often lie very close together.

4. Quasi-isotropic polycrystalline one-phase materials. Elastic properties

4.1. Cubic lattice structures

We first consider a statistically isotropic polycrystalline one-phase material in which the crystallites have a cubic lattice structure. The elastic properties of the individual crystallites are described by three constants, c_{11}, c_{12} and c_{44}. Our task is to find the effective bulk modulus K_e and shear modulus G_e. Voigt (1910) made the assumption that the strain is uniform throughout the sample. The bulk modulus then becomes

$$K_V = (c_{11} + 2c_{12})/3. \tag{14.14}$$

Similarly, Reuss (1929) assumed a uniform stress in the sample and obtained the bulk modulus

$$K_R = (c_{11} + 2c_{12})/3, \tag{14.15}$$

i.e. $K_R = K_V$ for cubic lattices. For the shear modulus G, the result is

$$G_V = (c_{11} - c_{12} + 3c_{44})/5, \tag{14.16}$$

$$G_R = 5(c_{11} - c_{12})c_{44}/[4c_{44} + 3(c_{11} - c_{12})]. \tag{14.17}$$

Hill (1952) has shown that the assumptions of Voigt and Reuss lead to upper and lower bounds on K and G, for any crystal structure. Thus,

$$K_R \leqq K_e \leqq K_V, \tag{14.18}$$

$$G_R \leqq G_e \leqq G_V. \tag{14.19}$$

From the general relation (3.10) between E, G and K,

$$\frac{1}{E} = \frac{1}{3G} + \frac{1}{9K}, \tag{14.20}$$

it is obvious that the smallest possible E, denoted E_R, is obtained with G_R and K_R on the right of (14.20). Similarly, the largest value, E_V is obtained with G_V and K_V on the right. Then

$$E_R \leqq E_e \leqq E_V. \tag{14.21}$$

Example: Reuss' expression for E. If we solve for E from (14.20), with G_R and K_R given by (14.16) and (14.14), and use (3.17)–(3.19), we get

$$\frac{1}{E_R} = s_{11} - \frac{1}{5}(2s_{11} - 2s_{12} - s_{44}). \tag{14.22}$$

This is recognised as the average of $\{E[hkl]\}^{-1}$ over all directions $[hkl]$; see the example on p. 286. The agreement is not accidental. In Reuss' approach one assumes that the stress σ is uniform in the polycrystalline sample. The effective strain is $\varepsilon_e = \sigma/E_e$. Now ε_e is the resulting strain from crystallites with a random crystallographic orientation. We have ε_e as an average of $\varepsilon = \sigma/E[hkl]$. Hence, $\varepsilon_e = \sigma/E_R$. A similar argument leads to E_V.

Hashin and Shtrikman (1962b) derived upper and lower bounds to K and G using the same variational method as for their bounds in multiphase materials. It seems not to be known whether the bounds to G are the best possible, i.e. if there are quasi-isotropic materials for which they are attained. The bounds to the bulk modulus must coincide in cubic lattices, since then $K_R = K_V$. Hence,

$$K_u = K_\ell = (c_{11} + 2c_{12})/3. \tag{14.23}$$

The bounds to the shear modulus are algebraically complicated (Hashin and Shtrikman 1962b). For elastically isotropic crystallites, i.e. when $c_{11} - c_{12} = 2c_{44}$, they simplify to

$$G_u = G_\ell = G_V = G_R = c_{44}. \tag{14.24}$$

Hershey (1954), Kröner (1958, 1967), Kneer (1965), Zeller and Dederichs (1973) and others have also studied the elastic properties of quasi-isotropic aggregates of crystals. Experimental values of the elastic coefficients of single crystals and the Voigt, Reuss and Hashin–Shtrikman bounds derived from them, have been tabulated by Simmons and Wang (1971) for essentially all systems with data available at that time.

4.2. Non-cubic lattices

Since the Voigt approach assumes a uniform strain and the Reuss approach a uniform stress, it is natural to use the elastic stiffnesses c_{ij} in the former and the elastic compliances s_{ij} in the latter case. The general expressions are (Schreiber et al. 1973)

$$K_V = \tfrac{1}{9}(c_{11}+c_{22}+c_{33})+\tfrac{2}{9}(c_{12}+c_{23}+c_{13}), \tag{14.25}$$

$$\frac{1}{K_R} = (s_{11}+s_{22}+s_{33})+2(s_{12}+s_{13}+s_{23}), \tag{14.26}$$

$$G_V = \tfrac{1}{15}(c_{11}+c_{22}+c_{33})-\tfrac{1}{15}(c_{12}+c_{13}+c_{23})+\tfrac{1}{5}(c_{44}+c_{55}+c_{66}), \tag{14.27}$$

$$\frac{1}{G_R} = \tfrac{4}{15}(s_{11}+s_{22}+s_{33})-\tfrac{4}{15}(s_{12}+s_{13}+s_{23})+\tfrac{3}{15}(s_{44}+s_{55}+s_{66}). \tag{14.28}$$

We check that these relations contain as special cases the results (14.14)–(14.17) for cubic crystals. Take $c_{11} = c_{22} = c_{33}$; $c_{12} = c_{13} = c_{23}$; $c_{44} = c_{55} = c_{66}$, with analogous equations for s_{ij}. The results for K_V and G_V follow immediately. Further, $1/K_R = 3(s_{11}+2s_{12})$ and $1/G_R = (4s_{11}-4s_{12}+3s_{44})/5$. With the relations (3.17)–(3.19) between c_{ij} and s_{ij} we recover (14.15) and (14.17).

Crystals belonging to one of the seven classes of hexagonal lattices have five independent elastic constants; c_{11}, c_{12}, c_{13}, c_{33} and c_{44}. The Voigt and Reuss bounds, when expressed in c_{ij}, take the form (Watt and Peselnick 1980, Meister and Peselnick 1966)

$$K_V = \tfrac{1}{9}[2(c_{11}+c_{12})+c_{33}+4c_{13}], \tag{14.29}$$

$$K_R = C^2/M, \tag{14.30}$$

$$G_V = \frac{1}{30}[12c_{66}+12c_{44}+M], \tag{14.31}$$

$$G_R = \frac{5}{2}\left[\frac{c_{44}c_{66}C^2}{(c_{44}+c_{66})C^2+3K_V c_{44}c_{66}}\right], \tag{14.32}$$

where $c_{66} = (c_{11}-c_{12})/2$ and the auxiliary moduli C and M are defined by

$$C^2 = (c_{11}+c_{12})c_{33}-2c_{13}^2, \tag{14.33}$$

$$M = c_{11}+c_{12}+2c_{33}-4c_{13}. \tag{14.34}$$

The analogous relations for the Voigt and Reuss bounds in trigonal and tetragonal lattices are listed by Meister and Peselnick (1966).

The Hashin–Shtrikman type bounds to K and G for hexagonal and trigonal (Peselnick and Meister 1965, Watt and Peselnick 1980), tetragonal (Meister and Peselnick 1966, Watt and Peselnick 1980), orthorhombic (Watt 1979) and monoclinic lattices (Watt 1980) are algebraically more complicated than (14.29)–(14.32). It appears not to be known if these bounds are the best possible.

4.3. The Voigt–Reuss–Hill estimation

The Voigt and Reuss estimations are easy to calculate. Hill (1952), having shown that they are rigorous bounds, suggested that one takes their arithmetic average as an estimation of the elastic properties. Usually, one refers to these averages as the Voigt–Reuss–Hill values, a term introduced by Chung (1963);

$$K_{\text{VRH}} = (K_{\text{V}}+K_{\text{R}})/2, \tag{14.35}$$

$$G_{\text{VRH}} = (G_{\text{V}}+G_{\text{R}})/2. \tag{14.36}$$

Note that E_{VRH}, derived from K_{VRH} and G_{VRH}, is not the same as $(E_{\text{V}}+E_{\text{R}})/2$.

There is no particular reason why the arithmetic (VRH) mean should be preferred. Kumazawa (1969) used the geometric mean $(K_{\text{V}}K_{\text{R}})^{1/2}$ and Shukla and Padial (1973) used the harmonic mean, $2K_{\text{V}}K_{\text{R}}/(K_{\text{V}}+K_{\text{R}})$. They will all give good estimations, if the material is only weakly anisotropic. We can define a measure of elastic anisotropy, A_{H}, by

$$A_{\text{H}} = \frac{G_{\text{V}}-G_{\text{R}}}{G_{\text{V}}+G_{\text{R}}}. \tag{14.37}$$

This measure has been applied by Chung and Buessem (1967) to single crystals of cubic lattice symmetry. The relation between A_{H} and Zener's measure of anisotropy, A_Z (3.23), in cubic crystals is

$$A_H = \frac{(A_Z - 1)^2}{1 + A_Z^2 + (19/3)A_Z} \approx 0.12(A_Z - 1)^2, \tag{14.38}$$

where the last expression refers to almost isotropic materials; $A_Z \approx 1$.

The approach discussed above is relevant for the elastic properties measured in short-time (e.g. acoustic) experiments. In long-time (e.g., static compression) experiments, the polycrystalline material may have time to relax so that it is characterised by the Reuss condition of uniform stress. Then K_e is given by K_R (Thomsen 1972).

In analogy to the relation between c_{ij} and the engineering elastic constants, one may derive relations between the single-crystal third-order elastic coefficients and the corresponding third-order quantities for the engineering elastic constants (Chang 1967); see also Chung (1967).

Example: Voigt–Reuss–Hill and Hashin–Shtrikman estimations of K_e and G_e. Table 14.1 gives Voigt–Reuss–Hill and Hashin–Shtrikman type estimations of the elastic properties of some polycrystals. The Voigt–Reuss–Hill (VRH) estimation of K is here defined by $(K_V + K_R)/2 \pm (K_V - K_R)/2$ and the Hashin–Shtrikman type (HS) by

Table 14.1
VRH and HS estimations of K_e and G_e (in GPa)

	Bulk modulus		Shear modulus	
	VRH	HS	VRH	HS
Al (fcc)	78.2[a]		25.9 ± 0.2	26.0
α-Fe (bcc)	168.7[a]		80.4 ± 7.0	80.7 ± 1.2
Si (diamond)	98.6[a]		67.1 ± 1.7	67.1 ± 0.2
NaCl (cubic)	24.7[a]		14.6 ± 0.2	14.6 ± 0.1
Zn (hcp)	68.3 ± 6.8	69.7 ± 2.2	39.5 ± 5.4	39.7 ± 1.8
Mg (hcp)	36.9	36.9	19.3 ± 0.1	19.3
Ice (hexagonal)	8.14	8.14	3.67 ± 0.02	3.67
Graphite (hexagonal)	161.2 ± 125.1	120.2 ± 84.0	109.1 ± 108.5	73.5 ± 72.7
β-Sn (tetragonal)	60.6	60.6	17.8 ± 2.1	18.2 ± 0.6
SiO_2 (trigonal)	37.7 ± 0.2	37.7	44.4 ± 3.4	44.2 ± 0.7
Al_2O_3 (trigonal)	251.1 ± 0.2	251.1	163.3 ± 2.7	163.5 ± 0.3
α-U (orthorhombic)	113.1 ± 1.7	112.9 ± 0.3	84.4 ± 3.7	84.2 ± 0.6
$CaSO_4$ (orthorhombic)	54.9 ± 2.7	54.1 ± 1.0	29.3 ± 6.2	29.6 ± 2.4

[a] K is isotropic for cubic solids.

$(K_u + K_\ell)/2 \pm (K_u - K_\ell)/2$, with an analogous definition for the shear modulus. No uncertainty interval is given when the upper and lower bounds coincide to the number of digits given. The entries (in GPa) are based on information from Simmons (1967) for cubic solids, from Peselnick and Meister (1965) for Mg and ice, from Watt and Peselnick (1980) for Zn, graphite, Al_2O_3 and SiO_2 and Sn, and from Watt (1969) for U and $CaSO_4$. See Watt and Peselnick (1980) for further examples. One notes that although both the VRH and HS bounds are narrow in many cases, they are of no use in a material such as graphite which is very anisotropic. In table 14.1, the VRH approximation lies inside the Hashin–Shtrikman bounds. This is a rule which has a few exceptions (Watt et al. 1976, Watt and Peselnick 1980).

5. Thermal expansion

Very little seems to be known theoretically about the thermal expansion of non-cubic polycrystals.

6.1. Transport properties

A material with aligned fibres embedded in a matrix, or more generally a two-phase material in which the phases have aligned cylindrical forms, may be transversely isotropic (in a statistical sense). Then it is described by two transport parameters, σ_L and σ_T (for "longitudinal" and "transverse"). For σ_L we have

$$\sigma_L = f_1\sigma_1 + f_2\sigma_2. \tag{14.39}$$

For the transverse conductivity bounds are given by the two-dimensional Hashin–Shtrikman formula, i.e. with 3 replaced by 2 in eqs. (13.17) and (13.18). When the structure fulfills the requirements of a symmetric cell material the results in ch. 13 §5.1 are applicable.

6.2. Elastic properties

A fibre composite which is transversely isotropic has the same elastic symmetry as a single crystal of a hexagonal lattice. Thus we need five independent elastic parameters which we may choose to be the transverse bulk modulus K_T, the transverse and longitudinal shear moduli, G_T and G_L, and the transverse and longitudinal Young's moduli, E_T and E_L. Other elastic quantities, e.g. the transverse and longitudinal Poisson ratios, can be expressed in terms of these five elastic parameters. In complete analogy to the approaches in this and the previous chapter, one can find Hashin–Shtrikman type bounds (Hill 1964, Hashin 1965) and effective-medium results (Hill 1965a). The field has been briefly reviewed by Laws (1980) and Hashin (1983). There is an exact result, the "coated cylinder" analogue of the "coated sphere" aggregate (ch. 13 §5.2). In this special case one has (Hashin and Rosen 1964, Hashin 1983)

$$E_L = f_1 E_1 + f_2 E_2 + \frac{4 f_1 f_2 (\nu_1 - \nu_2)^2}{f_1/k_2 + f_2/k_1 + 1/G_1}, \tag{14.40}$$

where $k_i = K_i + G_i/3$, and phase 2 is the coated fibre. If $\nu_1 = \nu_2$, we get the familiar result $E_L = f_1 E_1 + f_2 E_2$, which is usually a good approximation; E_L in (14.40), with G_1 or G_2 in the denominator, coincides with the two Hashin–Shtrikman bounds.

The Voigt and Reuss bounds to some elastic coefficients c_{ij} may coincide for certain aggregates with texture. Then c_{ij} is of course exactly given (Kröner and Wawra 1978).

6.3. Thermal expansion

There are two coefficients of thermal expansion, α_T and α_L. If the fibre and matrix materials are isotropic there is an exact solution (Levin 1967, Rosen and Hashin 1970):

$$\alpha_T = \alpha_1 + \frac{\alpha_2 - \alpha_1}{1/K_2 - 1/K_1}\left[\frac{3(1-2\nu_L)}{E_L} - \frac{1}{K_1}\right], \tag{14.41}$$

$$\alpha_L = \alpha_1 + \frac{\alpha_2 - \alpha_1}{1/K_2 - 1/K_1}\left[\frac{3}{2k_T} - \frac{3(1-2\nu_L)\nu_L}{E_L} - \frac{1}{K_1}\right]. \tag{14.42}$$

CHAPTER 15

CORRELATION AND ESTIMATION OF THERMOPHYSICAL PARAMETERS

1. Introduction

Most of the theories developed in this book centre around important parameters such as properly defined Debye temperatures $\Theta_D(n)$ and Grüneisen parameters $\gamma(n)$. In simplified versions of the theories, one may use a common Debye temperature Θ_D and Grüneisen parameter γ_G. That leads to close mathematical relationships between many thermophysical properties. For instance, the heat capacity, the thermal expansion coefficient and the quantity ρ/T (electrical resistivity divided by temperature) all have approximately the same temperature dependence. The empirical Lindemann formula allows us to replace Θ_D by a scaled melting temperature T_m. Another link to correlations is the electronic density of states at the Fermi level, $N(E_F)$.

The purpose of this chapter is to discuss correlations. Many relations are quoted in an approximate version, without mentioning the anomalous behaviour that some materials may show and which is discussed in the previous chapters. We shall also pass comment on some properties not discussed elsewhere in the book, for instance diffusion and the magnetic susceptibility.

2. Relations between the heat capacity, expansion coeffieicnt and electrical resistivity

Let Θ_D be some kind of Debye temperature, which separates high- and low-temperature phenomena. The inequality $T > \Theta_D$ means that we are in the high-temperature region. In practice, and depending on the desired accuracy, it may be sufficient with, say, $T > 0.8\Theta_D$. In the simplest models, some quantities tend asymptotically to a constant, here labelled by the index 0, when $T > \Theta_D$. For instance, the heat capacity $C_V \to (C_V)_0 = 3k_B$ per atom, the cubic expansion coefficient (due to lattice vibrations) $\beta \to \beta_0$, and $\rho(T)/T \to \rho_0$, where ρ is the (phonon-limited) electrical resistivity. It is convenient to normalise C_V, β and ρ/T so that their high-temperature asymptotic value is 1 (dimensionless). The normalised quantities are denoted $\tilde{C}$, $\tilde{\beta}$ and $\tilde{\rho}$. Then, within the simple models,

$$C_V = 3Nk_B\tilde{C}(T), \tag{15.1}$$

$$\beta(T) = \beta_0\tilde{\beta}(T), \tag{15.2}$$

$$\rho(T)/T = \rho_0\tilde{\rho}(T). \tag{15.3}$$

In an Einstein model for the lattice vibrations,

$$\tilde{C}(T) = \tilde{\beta}(T) = \tilde{\rho}(T) = \tilde{C}_E(T), \tag{15.4}$$

where

$$\tilde{C}_E(T) = \left(\frac{\Theta_E}{T}\right)^2 \frac{\exp(\Theta_E/T)}{[\exp(\Theta_E/T)-1]^2}. \tag{15.5}$$

Figure 15.1 shows the Einstein heat capacity $C_E(T/\Theta_E)$, the measured $C_p(T)$ (Hultgren et al. 1973a), $\beta(T)$ (Touloukian et al. 1975) and $\rho(T)/T$ (Bass 1982) for aluminium. The quantities are given in arbitrary units so that they coincide at T = 100 K.

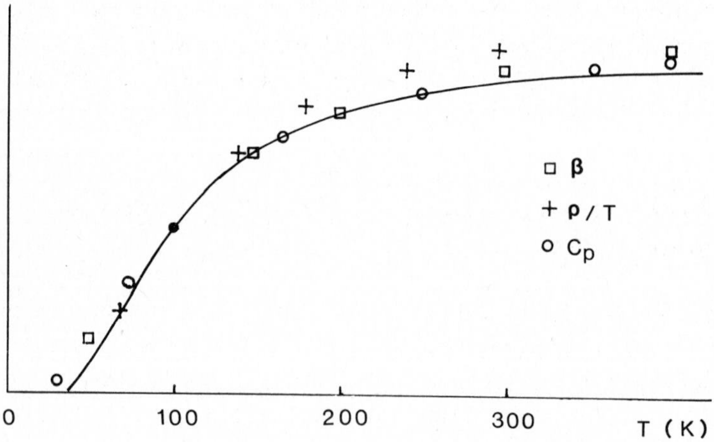

Fig. 15.1. The Einstein heat capacity (solid line) and the measured $C_p(T)$, $\beta(T)$ and $\rho(T)/T$ of Al, as a function of T and normalised to coincide at 100 K.

3. Empirical rules related to the melting temperature

Phonon frequencies vary as (force constant/atomic mass)$^{1/2}$. The Debye temperature is therefore correlated to the interatomic forces. Such forces are also related to defect energies. The potential energy of a harmonic oscillator, displaced a distance Δx from the equilibrium position, is $f(\Delta x)^2/2$, where f is a force constant. In analogy to this we expect that defect energies in a lattice can be written

$$E_\mathrm{d} = k_\mathrm{d} M \left(\frac{k_\mathrm{B}\Theta_\mathrm{D}}{\hbar}\right)^2 a^2, \tag{15.6}$$

where M is an (average) atomic mass, $M(k_\mathrm{B}\Theta_\mathrm{D}/\hbar)^2$ is a typical interatomic force constant, $2a$ is a length of the order of the interatomic distance (e.g., $(4\pi/3)a^3 = \Omega_\mathrm{a}$) and k_d is a dimensionless constant characteristic of the defect d. With reference to Lindemann's melting criterion, we have $M(k_\mathrm{B}\Theta_\mathrm{D})^2 \sim \hbar^2 k_\mathrm{B} T_\mathrm{m}/a^2$. This leads to another correlation,

$$E_\mathrm{d} = k'_\mathrm{d} k_\mathrm{B} T_\mathrm{m}. \tag{15.7}$$

Grimvall and Sjödin (1974) have discussed empirical relations like (15.6) and (15.7) in some detail. See also table 1.1.

Diffusion is an important thermophysical phenomenon which has not been dealt with in this book. One reason is that the microscopic aspects of diffusion are still very poorly understood, both the static, i.e. defect configurational (e.g., Bar-Yam and Joannopoulos 1984, Car et al. 1984) and dynamical (e.g. Rezayi and Suhl 1980, Schober and Stoneham 1982) behaviour. On the other hand, there are well-known empirical results. For instance, the activation energy Q_diff for self-diffusion in the elements is surprisingly well correlated to T_m; fig. 15.2.

In eq. (15.2) we gave the temperature dependence of the expansion coefficient $\beta(T)$ but the magnitude of β_0 was not considered. A crude, but useful, rule is (Carnelley 1879, Hidnert and Sonder 1950, Gschneidner 1964)

$$\beta(T_\mathrm{m}) = k/T_\mathrm{m}. \tag{15.8}$$

The dimensionless parameter $k \approx 0.06$. More precisely, $k = 0.059 \pm 0.015$ for elements with bcc, fcc or hcp lattices. The lowest and the highest k-values among the elements are 0.009 (grey tin) and 0.15 (Pu).

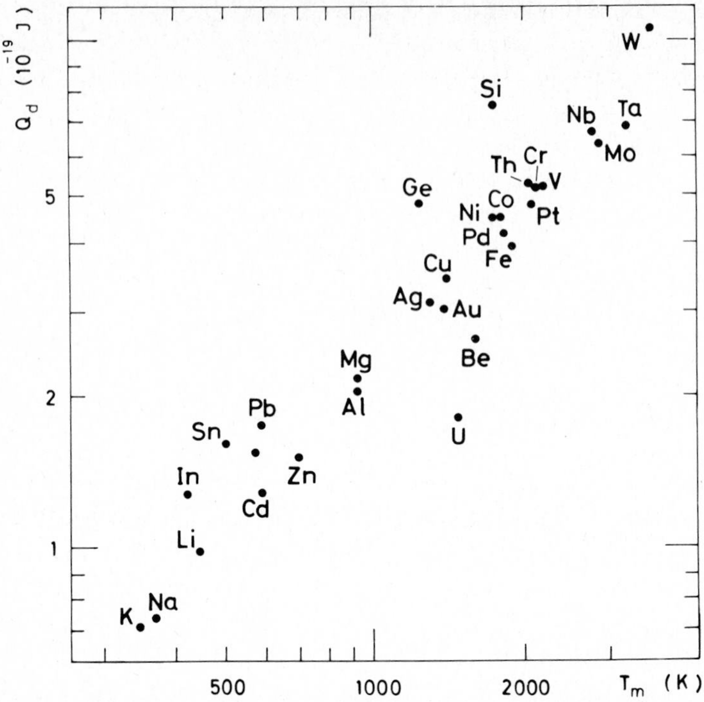

Fig. 15.2. The activation energy for self-diffusion, Q_{diff}, versus the melting temperature T_m. From Grimvall and Sjödin (1974). (Reproduced with permission from Physica Scripta.)

4. Probing electron properties near the Fermi level

According to the Sommerfeld theory, the electronic entropy and heat capacity are proportional to the electron density of states $N(E_F)$ at the Fermi level. Many other physical properties are, within simple theories, related to $N(E_F)$. For instance, the Pauli spin paramagnetism is described by the susceptibility

$$\chi_P = \frac{\partial}{\partial H}\mu_B \int_{-\infty}^{\infty} N(E)[f(E-\mu_B H)-f(E+\mu_B H)]\,dE$$

$$= 2\mu_B^2 \int_{-\infty}^{\infty} N(E)(-\partial f/\partial E)\,dE \approx 2\mu_B^2 N(E_F), \tag{15.9}$$

where $\mu_B = e\hbar/2mc$ is the Bohr magneton. The derivative with respect to H in (15.9) is taken in the weak field limit, $H \to 0$.

The susceptibility χ_L of the Landau diamagnetism is considerably more difficult to calculate, both for free-electron-like systems and transition metals. The free-electron expression is

$$\chi_L = -\frac{Ve^2k_F}{12\pi^2mc^2} = -(2/3)\mu_B^2 N_{fe}(E_F). \tag{15.10}$$

In ch. 11 we obtained an expression, $\sigma = ne^2\tau/m_b$, for the electrical conductivity. Another form of σ, which takes $v_{\boldsymbol{k}}$ to be isotropic but allows $\tau(E)$ to be energy dependent is, (11.10),

$$\sigma = (ne^2/m_b)\int \tau(E)(-\partial f/\partial E)\mathrm{d}E. \tag{15.11}$$

Then we have assumed a spherical Fermi surface with a radius $k_F = (3\pi^2 n)^{1/3}$ but allowed band effects to be approximately included through m_b.

By the Wiedemann–Franz law, the electronic part of the thermal conductivity is

$$\kappa_e = (\pi^2/3)(k_B/e)^2 T\sigma = L_0 T\sigma, \tag{15.12}$$

where L_0 is the Lorenz number. However, this relation requires that the relaxation time τ does not vary with the energy $E_{\boldsymbol{k}}$ of the electron states involved in the transport process. The analogue of (15.11) is, (12.49),

$$\kappa_e = (nk_B^2T/m_b)\int \tau(E)[(E-\mu)/k_BT]^2(-\partial f/\partial E)\mathrm{d}E. \tag{15.13}$$

We shall first let the temperature be so low, or the energy dependence of $N(E)$ so weak, that $(-\partial f/\partial \varepsilon)$ can be considered as a delta function at the Fermi level. The question now is how band effects (m_b or $N(E_F)$) and electron–phonon many-body corrections enter the physical properties mentioned above. We consider these properties in models at three levels of sophistication: (i) A strict free-electron model. (ii) A nearly-free-electron model in which the Fermi surface is assumed to be spherical, with a radius $k_F = (3\pi^2 n)^{1/3}$ and with band effects included through an isotropic effective band mass $m_b = \hbar^2 k_F/|\nabla_{\boldsymbol{k}} E_{\boldsymbol{k}}|$. (iii) Electron–phonon many-body corrections to the model (ii), when such effects should enter.

The first column in table 15.1 suggests a strong correlation between

Table 15.1

Dependence of various properties on the electron density n, band mass m_b and electron–phonon enhancement parameter λ

Property	Free-electron model	Electron band effects included through m_b	Electron–phonon mass enhancement included
C_{el}	$Amn^{1/3}$	$Am_bn^{1/3}$	$Am_b(1+\lambda)n^{1/3}$ (low T) $Am_bn^{1/3}$ (high T)
χ_P	$Bmn^{1/3}$	$Bm_bn^{1/3}$	no correction
χ_L	$-(1/3)Bmn^{1/3}$	$-(1/3)Bm(m/m_b)n^{1/3}$	no correction
σ	Cn/m	Cn/m_b	no correction
κ_e	Dn/m	Dn/m_b	no correction

$A = (\pi/3)^{2/3}(Vk_B^2T/\hbar^2)$; $B = (3/\pi^4)^{1/3}(V\mu_B^2/\hbar^2)$; $C = e^2\tau$; $D = L_0TC$.

C_{el} and χ_P. Since C_{el} and χ_P are both proportional to $N(E_F)$, this correlation may not be limited to free-electron-like systems. However, electron–phonon many-body effects in C_{el}, and contributions other than χ_P to the measured susceptibility χ, spoil the proportionality between C_{el} and χ in real metals.

It is a non-trivial matter to decide whether the band density of states, the band electron velocity $\boldsymbol{v}_{\boldsymbol{k}}$, the band mass m_b, etc. should be enhanced (Grimvall 1981). In fact, only the low-temperature heat capacity and the effective band masses measured in cyclotron resonance and de Haas–van Alphen experiments should be renormalised by a factor $1+\lambda$. In most other cases, renormalisation effects are either absent or enter in such a way as to cancel. For instance, the renormalisations of τ and m_b cancel in the static electrical conductivity, $\sigma = ne^2\tau/m_b$.

There is also an electron–electron many-body effect. Its influence on C_{el} is small in the simple metals, and it is to a large extent folded into the band structure $N(E_F)$ of transition metals. However, it is essential for the magnetic susceptibility. Within the Stoner (1938) model one can write

$$\chi_P = 2\mu_B^2 N(E_F)/[1-IN(E_F)], \tag{15.14}$$

where I is the intra-atomic Coulomb interaction between the electrons. When $IN(E_F) > 1$, χ_P is negative which implies that the system is magnetically ordered.

We get from eq. (B.3) that the electronic heat capacity contains an average of $\langle 1/\boldsymbol{v}_{\boldsymbol{k}}\rangle$ over the Fermi surface, while the expression (11.15) for the electrical conductivity contains a product of $\langle 1/v_k \rangle$ and $\langle v_k^2 \rangle$. When the electron states at the Fermi level are described by a single and isotropic band mass, $m_b = \hbar^2 k_F / |\nabla_k E_k|$, any combination of $\langle v_k^n \rangle$ can be expressed in m_b. Often, this is an oversimplification, and one should introduce different effective band masses $m_b(n)$, in analogy to $\Theta_D(n)$ and $\gamma(n)$.

Finally, we consider the case that $(-\partial f/\partial \varepsilon)$ cannot be approximated by a delta function at the Fermi level. When the electron density of states, the electron relaxation time τ, etc. vary strongly with the energy $E-\mu$, the delta-function approximation may be inadequate and one has to state precisely what is the shape and width of the "energy window" around E_F, where the electron states are probed. It follows from (8.13) that C_{el} has contributions from a region defined by the weight function $[(E-\mu)/k_BT]^2(-\partial f/\partial E) \sim [(E-\mu)/k_BT]^2 f(1-f)$. The susceptibility χ_P of the Pauli paramagnetism, eq. (15.9), and the relaxation time τ in the electrical conductivity, eq. (11.11), have a weight function $f(1-f)$. The entropy probes the electronic states weighted by $f[\ln f]+[1-f]\ln(1-f)$, eq. (8.22); the chemical potential $\mu(T)$ has a temperature dependence determined by $f[(E-\mu)/k_BT]$ or $1-f[(E-\mu)/k_BT]$, depending on the sign of $E-\mu$, eq. (8.5); the relaxation time of the electronic contribution

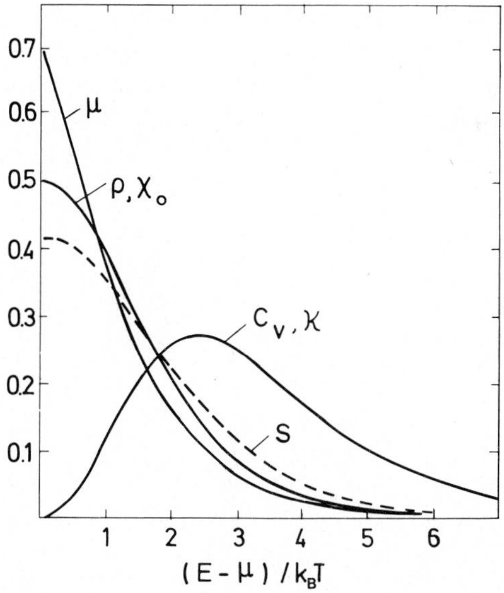

Fig. 15.3. Weight functions for Fermi surface averages of some physical parameters. The area under each curve is normalised to unity. From Grimvall (1981).

to the thermal conductivity has the same weight function as C_{el}, (15.13). Figure 15.3 summarises these cases. It is worth noting that C_e and κ_e probe the density of electron states in an energy interval of width $\sim 10k_BT$. This is much wider than the width $\sim k_BT$ often alluded to in textbooks.

APPENDIX A

The electron gas

In this appendix we recapitulate some results for the homogeneous electron gas. Let N_e electrons occupy a volume V, in which a uniform positive charge assures charge neutrality. The energy of the system per unit volume is completely determined by the number-density n of electrons,

$$n = N_e/V. \tag{A.1}$$

Often one uses the dimensionless parameter r_s instead of n, r_s being the radius of a sphere with volume V/N_e and given in units of the Bohr radius a_0 ($a_0 = \hbar^2/me^2 = 0.529 \times 10^{-10}$ [m]). Thus

$$r_s^3 = 3V/4\pi N_e a_0^3 = 3/4\pi n a_0^3. \tag{A.2}$$

In the free-electron model of a metallic element, each of the N atoms contributes Z electrons to the electron gas, Z being the valency. The atomic volume is $\Omega_a = V/N$ and hence

$$n = Z/\Omega_a. \tag{A.3}$$

The Fermi wave number k_F, the Fermi momentum $p_F = \hbar k_F$, the Fermi velocity $v_F = p_F/m$ and the Fermi energy $E_F = \hbar^2 k_F^2/2m$ can be expressed in terms of n or r_s:

$$k_F = (3\pi^2 n)^{1/3} = (me^2/\hbar^2)(9\pi/4)^{1/3} r_s^{-1}, \tag{A.4}$$

$$E_F = (\hbar^2/2m)(3\pi^2 n)^{2/3} = (me^4/2\hbar^2)(9\pi/4)^{2/3} r_s^{-2}. \tag{A.5}$$

Expressed in the dimensionless parameter r_s, one has

$$k_F = (3.63 r_s^{-1}) \times 10^{10}\,[\mathrm{m}^{-1}], \tag{A.6}$$

$$v_F = (4.20r_s^{-1}) \times 10^6 \text{ [m/s]}. \tag{A.7}$$

Energies are often given in electron volt (1[eV] = 1.602×10^{19} [J]) or in Rydberg (1[Ry] = $e^2/2a_0 = me^4/2\hbar^2$ = 13.61[eV]); 1[mRy/atom] = 0.314 [kcal/mol] = 1.31 [kJ/mol]). Then,

$$E_F = (8.03r_s^{-2}) \times 10^{-18}\text{[J]} = 50.1r_s^{-2}\text{[eV]} = 3.68r_s^{-2}\text{[Ry]}. \tag{A.8}$$

The Fermi temperature T_F is

$$T_F = E_F/k_B = (58.2r_s^{-2}) \times 10^4 \text{ [K]}. \tag{A.9}$$

The average kinetic energy per electron, $\langle E_{kin} \rangle$, is

$$\langle E_{kin} \rangle = (3/5)E_F = (3me^4/10\hbar^2)(9\pi/4)^{2/3} r_s^{-2} = 2.210r_s^{-2}\text{[Ry]}. \tag{A.10}$$

The average exchange and correlation energy $\langle E_{x,c} \rangle$ per electron cannot be given in a closed form for arbitrary r_s. The exchange energy is caused by the requirement that the total wave function is antisymmetric. As a consequence, electrons with like spins tend to stay apart. One speaks of a "Fermi hole" around an electron. This reduces the electron–electron Coulomb interaction and thus gives a negative contribution to the energy. If the electrons are represented by an antisymmetrised combination of plane waves, the exhange energy is (per electron)

$$\langle E_x \rangle = -(3e^4m/4\pi\hbar^2)(9\pi/4)^{1/3} r_s^{-1} = -0.916r_s^{-1}\text{[Ry]}. \tag{A.11}$$

The correlation energy is much more difficult to calculate. One of several approximations is due to Nozières and Pines (1966). It interpolates between the known results at high and low r_s,

$$\langle E_{x,c} \rangle = -0.916\, r_s^{-1} + 0.031 \ln r_s - 0.115\text{[Ry]}. \tag{A.12}$$

Hohenberg and Kohn (1964) and Kohn and Sham (1965) have given a general theory for the ground-state energy of an inhomogeneous electron gas, known as the Hohenberg–Kohn–Sham formalism. The total energy E is a unique functional of the electron number-density $n(\boldsymbol{r})$ given at points $\boldsymbol{r}$ in the sample, $E = E[n]$. For details see the review by Callaway and March (1984).

In ch. 1 we need some expressions for the electrostatic energy of a point ion in an electron gas. In a strict jellium model, i.e. a uniform electron gas and a uniformly smeared-out positive background, there is no electrostatic energy. Now suppose that we concentrate the positive charge in point charges Ze, surrounded by spheres with a volume Ω_a, in which the electron gas is still uniform. The negative charge in each volume element of Ω_a then interacts electrostatically with the point charge and with the negative charge in the rest of Ω_a. This electrostatic, or Coulomb, energy, expressed as an average per atom (ion plus the Z electrons inside Ω_a) is (Kittel 1963)

$$\begin{aligned}\langle E_C\rangle &= -(9e^2Z^2/10)(4\pi/3)^{1/3}\Omega_a^{-1/3}\\ &= -(9me^4/10\hbar^2)Z^{5/3}r_s^{-1} = -1.8\,Z^{5/3}r_s^{-1}\,[\mathrm{Ry}].\end{aligned} \tag{A.13}$$

This calculation of $\langle E_C\rangle$ neglects the fact that the spheres of volume Ω_a partly overlap and also leave regions in the material which are not included in the calculation. This is motivated by the simplicity of the calculations. One can also, with more effort, calculate the electrostatic energy for a lattice of positive charges embedded in an electron gas which is still assumed to be of uniform density. The result is closely related to the Madelung energy in ionic lattices, and we write for the electrostatic energy $\langle E_C\rangle$

$$\langle E_C\rangle = -\alpha_C Z^{5/3}r_s^{-1}[\mathrm{Ry}]. \tag{A.14}$$

Here Ze is the point ion charge. $\langle E_C\rangle$ from (A.13) corresponds to $\alpha_C = 1.8$ in (A.14).

APPENDIX B

The electron density of states

Electron and phonon states in a periodic lattice are labelled by a wave vector $\boldsymbol{k}$ (or $\boldsymbol{q}$). The prescription for turning a sum over all $\boldsymbol{k}$ (or $\boldsymbol{q}$) into an integral is

$$\sum_{\boldsymbol{k}} (\ldots) = \frac{V}{(2\pi)^3} \int (\ldots) \mathrm{d}^3\boldsymbol{k}, \tag{B.1}$$

where V is the volume of the crystal and $(\cdots)$ is an arbitrary function of $\boldsymbol{k}$. Let $E(\boldsymbol{k})$ (or $E_{\boldsymbol{k}}$) be the energy of an electron. It is often convenient to write the integral in (B.1) with $E(\boldsymbol{k})$ as the integration variable. For an isotropic system

$$\sum_{\boldsymbol{k}} (\ldots) = \int (\ldots) N(E) \mathrm{d}E. \tag{B.2}$$

$N(E)$ is the electron density of states,

$$N(E) = \frac{V}{(2\pi)^3} \int \mathrm{d}^3\boldsymbol{k}\, \delta(E_{\boldsymbol{k}} - E) = \frac{V}{(2\pi)^3} \int_{S_E} \frac{\mathrm{d}S}{|\nabla_{\boldsymbol{k}} E(\boldsymbol{k})|}. \tag{B.3}$$

The last integral is over that surface S_E in $\boldsymbol{k}$-space for which $E(\boldsymbol{k}) = E$. Our $N(E)$ refers to one spin direction. Other authors let $N(E)$ refer to both "spin up" and "spin down". The difference will appear as a factor of 2 in our expressions when we sum over all electron states. In an anisotropic system, we can define a directional density of states function $N(E; \boldsymbol{k})$ as

$$N(E; \boldsymbol{k}) = \frac{V}{(2\pi)^3} \frac{4\pi k^3}{\boldsymbol{k} \cdot \nabla_{\boldsymbol{k}} E(\boldsymbol{k})}. \tag{B.4}$$

Then,

$$\sum_{\boldsymbol{k}} (\ldots) = \int \mathrm{d}E \int \frac{\mathrm{d}\Omega_{\boldsymbol{k}}}{4\pi} (\ldots) N(E;\boldsymbol{k}), \tag{B.5}$$

where $\mathrm{d}\Omega_{\boldsymbol{k}}$ is a solid angle. For free electrons, $|\nabla_{\boldsymbol{k}} E(\boldsymbol{k})| = \hbar^2 k/m$ and $\mathrm{d}S = k^2 \mathrm{d}\Omega_{\boldsymbol{k}}$. Equation (B.3) gives

$$N(E) = \frac{V}{2\pi^2} \frac{mk}{\hbar^2}, \tag{B.6}$$

or

$$N(E) = \frac{V}{4\pi^2} \left(\frac{2m}{\hbar^2}\right)^{3/2} \sqrt{E}. \tag{B.7}$$

Often, the density of states is given per atom (and spin),

$$N_{\mathrm{a}}(E) = N(E)/N. \tag{B.8}$$

At the Fermi level,

$$N_{\mathrm{a}}(E) = \frac{Z}{2\pi^2} \frac{mk_{\mathrm{F}}}{n\hbar^2} = \frac{3Z}{4} \frac{1}{E_{\mathrm{F}}}. \tag{B.9}$$

For any functional form of $E(\boldsymbol{k})$ we can define an effective electronic band mass $m_{\mathrm{b}}(\boldsymbol{k})$ by

$$\hat{\boldsymbol{k}} \cdot \nabla_{\boldsymbol{k}} E(\boldsymbol{k}) = \frac{\hbar^2 |\boldsymbol{k}|}{m_{\mathrm{b}}(\boldsymbol{k})}, \tag{B.10}$$

where $\hat{\boldsymbol{k}} = \boldsymbol{k}/|\boldsymbol{k}|$ is a unit vector along $\boldsymbol{k}$. If $N_{\mathrm{fe}}(E)$ is the density of states for free electrons we have

$$N(E;\boldsymbol{k}) = \frac{m_{\mathrm{b}}(\boldsymbol{k})}{m} N_{\mathrm{fe}}(E). \tag{B.11}$$

The average band mass m_{b} at the Fermi level obeys the relations

$$N(E_{\mathrm{F}}) = \frac{V}{(2\pi)^3} \int_{S_{\mathrm{F}}} \frac{m_{\mathrm{b}}(\boldsymbol{k})|\boldsymbol{k}|}{\hbar^2} \mathrm{d}\Omega = \frac{m_{\mathrm{b}}}{m} N_{\mathrm{fe}}(E_{\mathrm{F}}). \tag{B.12}$$

APPENDIX C

The dynamical matrix

Atomic force constants. We expand the potential energy Φ of a lattice in the small displacements $u_\alpha(\kappa l)$, referring to the κth atom in the lth cell being displaced by an amount u in the α-direction ($\alpha = x, y, z$):

$$\Phi = \Phi_0 + \sum_{\kappa l \alpha} \Phi_\alpha(\kappa l) u_\alpha(\kappa l)$$

$$+ (1/2) \sum_{\kappa l \alpha} \sum_{\kappa' l' \beta} \Phi_{\alpha\beta}(\kappa l; \kappa' l') u_\alpha(\kappa l) u_\beta(\kappa' l') + \ldots, \tag{C.1}$$

where the dots denote higher-order terms. Further,

$$\Phi_\alpha(\kappa l) = \frac{\partial \Phi}{\partial u_\alpha(\kappa l)}, \tag{C.2}$$

$$\Phi_{\alpha\beta}(\kappa l; \kappa' l') = \frac{\partial^2 \Phi}{\partial u_\alpha(\kappa l) \partial u_\beta(\kappa' l')}. \tag{C.3}$$

The derivatives are evaluated with the atoms at their equilibrium positions. The quantities $\Phi_{\alpha\beta}(\kappa l; \kappa' l')$ are called atomic force constants. The physical meaning of $\Phi_{\alpha\beta}(\kappa l; \kappa' l')$ is the negative force in the α-direction on the atom (κl) that arises when the atom $(\kappa' l')$ is displaced a unit distance in the β-direction while all other atoms are held fixed at their equilibrium positions.

The dynamical matrix. From $\Phi_{\alpha\beta}(\kappa l; \kappa' l')$ we define the elements $D_{\alpha\beta}(\kappa\kappa'; \boldsymbol{q})$ of the dynamical matrix. The periodicity and symmetry of a lattice imposes several conditions on $\Phi_{\alpha\beta}$. For instance, $\Phi_{\alpha\beta}$ only depends on $l - l'$. Then,

$$D_{\alpha\beta}(\kappa\kappa'; \boldsymbol{q}) = (M_\kappa M_{\kappa'})^{-1/2} \exp[i\boldsymbol{q} \cdot (\boldsymbol{R}(\kappa) - \boldsymbol{R}(\kappa'))]$$

$$\times \sum_l \Phi_{\alpha\beta}(\kappa l; \kappa' 0) \exp[i\boldsymbol{q} \cdot \boldsymbol{R}(l)]. \tag{C.4}$$

$\boldsymbol{R}(\kappa l)$, $\boldsymbol{R}(\kappa)$ and $\boldsymbol{R}(l)$ are position vectors in the lattice, with $\boldsymbol{R}(\kappa l) = \boldsymbol{R}(l)+\boldsymbol{R}(\kappa)$. If there are r atoms in a primitive cell, κ and κ' run from 1 to r. Further, α, $\beta = x, y, z$. Thus we have a $3r \times 3r$ matrix D with elements $D_{\alpha\beta}(\kappa\kappa'; \boldsymbol{q})$ which are functions of the wave vector $\boldsymbol{q}$ lying in the first Brillouin zone of the reciprocal lattice.

Let M be a diagonal $3n \times 3n$ matrix with diagonal elements M_1, M_1, $M_1, \ldots, M_n, M_n, M_n$ where M_i is the mass of the atom with $\kappa = i$. Then we can write for the dynamical matrix D:

$$D = M^{-1/2} D_0 M^{-1/2}, \tag{C.5}$$

where D_0 is the force constant part of the dynamical matrix which does not contain any atomic masses. One can show that D and D_0 are Hermitian matrices, i.e. $D_{\alpha\beta}(\kappa\kappa'; \boldsymbol{q}) = D^*_{\beta\alpha}(\kappa'\kappa; \boldsymbol{q})$.

Example: D for an fcc lattice. As an illustration, consider an fcc monatomic lattice with only nearest-neighbour central interactions. In spite of this drastic simplification, the evaluation of the dynamical matrix is algebraically tedious. In our case there is only one atom per primitive cell, so the index κ is irrelevant and will be dropped. Suppose that the interaction potential is $\phi(r) = (F/2)(r-r_0)^2$, where F has the dimension of a force constant. We have $\phi'(r) = F(r-r_0) = 0$ and $\phi''(r) = F$ at the atomic equilibrium positions $r = r_0$. Then, $\Phi_{\alpha\beta}(l, l') = -x_\alpha x_\beta F/r_0^2$ and $\Phi_{\alpha\beta}(l, l) = \sum x_\alpha x_\beta F/r_0^2$ where the sum is over all $l \neq l'$. Since we are restricted to nearest-neighbour interactions, we take $\boldsymbol{R}(l') = (0, 0, 0)$ and $\boldsymbol{R}(l) = (\pm a/2, \pm a/2, 0)$, $(\pm a/2, 0, \pm a/2)$ or $(0, \pm a/2, \pm a/2)$ where a is the side length of the unit cell in real space. (This cell contains four atoms and is not the primitive cell with one atom that is labelled by l.) (From (C.4) we have

$$D_{\alpha\beta}(q) = M^{-1} \sum_l \Phi_{\alpha\beta}(l; 0) \exp[\mathrm{i}\boldsymbol{q} \cdot \boldsymbol{R}(l)]. \tag{C.6}$$

The right-hand side of (C.6) contains terms like $\exp[(\mathrm{i}a/2)(q_x+q_y)] + \exp[(-\mathrm{i}a/2)(q_x+q_y)] = 2\cos[(a/2)(q_x+q_y)]$. After a considerable amount of algebra we have

$$D_{ii} = (4F/M)\{2-\cos(aq_i/2)[\cos(aq_j/2)+\cos(aq_k/2)]\}, \tag{C.7}$$

$$D_{ij} = 0, \tag{C.8}$$

where i, j, k are any of x, y, z with $i \neq j \neq k$ and $\boldsymbol{q} = (q_x, q_y, q_z)$. One may remark that the bcc lattice is dynamically unstable if there are only nearest-neighbour interactions.

Secular equation for the vibrational frequencies. The phonon eigenfrequencies squared, ω^2, are obtained as solutions of

$$|D_{\alpha\beta}(\kappa\kappa';\boldsymbol{q})-\omega^2 I_{\alpha\beta}(\kappa\kappa')| = 0, \tag{C.9}$$

where || denotes a determinant and $I_{\alpha\beta}(\kappa\kappa')$ is the $3r\times 3r$ unit matrix. For each $\boldsymbol{q}$, there are $3r$ solutions which we label $\omega(\boldsymbol{q},\lambda)$.

The expansion of the potential energy in terms of the force constant matrix elements $\Phi_{\alpha\beta}$ does not require a periodic lattice, but is also applicable in a solid with lattice defects. As long as we work within the harmonic approximation there are still $3N$ eigenfrequencies, where N is the number of atoms. (More precisely there are $3N-6$ eigenfrequencies since we should exclude a translation or a rotation of the entire solid. This unimportant point is neglected.) The frequencies are solutions to

$$|M^{-1/2}\Phi M^{-1/2}-\omega^2 I| = 0. \tag{C.10}$$

M, Φ and I are $3N\times 3N$ matrices. M is diagonal, with elements $M_1, M_1, M_1, \ldots, M_N, M_N, M_N$, I is a unit matrix and Φ has the elements $\Phi_{\alpha\beta}(\kappa l;\kappa' l')$.

Some theorems on dynamical matrices. If $f(x)$ is a polynomial in x, the following relation holds for the trace (Tr) of the dynamical matrix D and its eigenvalues:

$$\sum_{\boldsymbol{q}} \operatorname{Tr} f(D) = \sum_{\boldsymbol{q}\lambda} f(\omega^2(\boldsymbol{q},\lambda)). \tag{C.11}$$

As a special case, let $f(x) = x$. Then

$$\sum_{\lambda} \omega^2(\boldsymbol{q},\lambda) = \operatorname{Tr} D(\boldsymbol{q}). \tag{C.12}$$

More generally, when $f(x) = x^n$,

$$\sum_{\lambda} \omega^{2n}(\boldsymbol{q},\lambda) = \operatorname{Tr} D^n(\boldsymbol{q}). \tag{C.13}$$

The theorem holds also in the real-space formulation of the eigenvalue problem:

$$\sum_{\text{all modes}} \omega^2 = \mathrm{Tr}\,(M^{-1/2}\Phi M^{-1/2}). \tag{C.14}$$

This is of particular importance in a non-periodic lattice (e.g., a crystal with lattice defects or an alloy).

REFERENCES

Abeles, B., 1963, Phys. Rev. **131,** 1906.

Ackerman, D.A. and A.C. Anderson, 1982, Phys. Rev. Lett. **49,** 1176.

Alers, G.A. and J.R. Neighbours, 1957, J. Appl. Phys. **28,** 1514.

Allen, P.B., 1980a, in: Superconductivity in d- and f-band Metals, H. Suhl and M.B. Maple, eds. (Academic Press, New York) p. 291.

Allen, P.B., 1980b, in: Physics of Transition Metals, P. Rhodes, ed. (Inst. Phys. Conf. Ser. No. 55) p. 425.

Allen, P.B. and M.L. Cohen, 1969, Phys. Rev. **187,** 525.

Allen, R.E., G.P. Alldredge and F.W. de Wette, 1970, Phys. Rev. **B 2,** 2570.

Allen, R.E., G.P. Alldredge and F.W. de Wette, 1971, Phys. Rev. **B 4,** 1661.

American Institute of Physics Handbook, 1972, D.E. Gray, ed. (McGraw-Hill, New York).

Anand, L., 1982, Scripta Met. **16,** 173.

Anderson, A.C., 1983, in: Thermal Conductivity, Vol. 16, D.C. Larsen, ed. (Plenum, N.Y.) p. 3.

Anderson, O.L., 1963, J. Phys. Chem. Solids **24,** 909.

Anderson, O.L., 1965, in: Physical Acoustics, vol. III B, W.P. Mason, ed. (Academic Press, New York) p. 43.

Anderson, O.L., 1966a, Phys. Rev. **144,** 553.

Anderson, O.L., 1966b, J. Phys. Chem. Solids **27,** 547.

Anderson, P.W., B.I. Halperin and C.M. Varma, 1972, Phil. Mag. **25,** 1.

Ashcroft, N.W. and D.C. Langreth, 1967, Phys. Rev. **155,** 682.

Ashcroft, N.W. and N.D. Mermin, 1976, Solid State Physics (Holt, Rinehart and Winston, New York).

Ashcroft, N.W. and J.W. Wilkins, 1965, Phys. Lett. **14,** 285.

Bäckström, G., 1985, High Temp. – High Press. **17,** 185.

Baier, W. and H. Köhler, 1986, to be published.

Bailey, A.C. and B. Yates, 1967, Proc. Phys. Soc. **91,** 390.

Bailey, A.C. and B. Yates, 1970, J. Appl. Phys. **41,** 5088.

Balberg, I., N. Binenbaum and S. Bozowski, 1983, Solid State Commun. **47,** 989.

Balian, R., M. Kléman and J.-P. Poirier eds., 1981, Physics of Defects, Les Houches Session XXXV (North-Holland, Amsterdam).

Baltes, H.P. and E.R. Hilf, 1973, Solid State Commun. **12,** 369.

Barkman, J.H., R.L. Anderson and T.E. Brackett, 1965, J. Chem. Phys. **42,** 1112.

Barron, T.H.K., 1957, Ann. Phys. (N.Y.) **1,** 77.

Barron, T.H.K., 1965, in: Lattice Dynamics, R.F. Wallis, ed. (Pergamon, Oxford) p. 247.

Barron, T.H.K. and M.L. Klein, 1962, Phys. Rev. **127,** 1997.

Barron, T.H.K. and M.L. Klein, 1974: in Dynamical Properties of Solids, Vol. 1, G.K. Horton and A.A. Maradudin, eds. (North-Holland, Amsterdam) p. 391.

Barron, T.H.K. and J.A. Morrison, 1960, Proc. Roy. Soc. A **256,** 427.

Barron, T.H.K. and R.W. Munn, 1967, Phil. Mag. **15,** 85.

Barron, T.H.K. and R.W. Munn, 1968, J. Phys. C **1,** 1.

Barron, T.H.K., W.T. Berg and J.A. Morrison, 1957, Proc. Roy. Soc. A **242,** 478.

Barron, T.H.K., W.T. Berg and J.A. Morrison, 1959, Proc. Roy. Soc. A **250,** 70.

Barron, T.H.K., A.J. Leadbetter and J.A. Morrison, 1964, Proc. Roy. Soc. A **279,** 62 [Corrigenda A **289,** 440 (1966)].

Barron, T.H.K., A.J. Leadbetter, J.A. Morrison and L.S. Salter, 1966, Acta Cryst. **20,** 125.

Barron, T.H.K., J.G. Collins and G.K. White, 1980, Adv. Phys. **29,** 609.

Barsch, G.R., 1967, Phys. Stat. Sol. **19,** 129.

Barsch, G.R. and Z.P. Chang, 1967, Phys. Stat. Sol. **19,** 139.

Bar-Yam, Y. and J.D. Joannopoulos, 1984, Phys. Rev. Lett. **52,** 1129.

Bass, J., 1972, Adv. Phys. **21,** 431.

Bass, J., 1982, in: Landolt-Börnstein New Series, Vol. III/15a, K.-H. Hellwege and J.L. Olsen, eds. (Springer, Berlin).

Batallan, F., I. Rosenman and C.B. Sommers, 1975, Phys. Rev. B **11,** 545.

Beaman, D.R., R.W. Balluffi and R.O. Simmons, 1964, Phys. Rev. **134,** A532.

Beaudoin, J.J. and R.F. Feldman, 1975, Mag. Concr. Res. **5,** 103.

Beck, H., P.F. Meier and A. Thellung, 1974, Phys. Stat. Sol. (a) **24,** 11.

Benedek, G., M. Miura, W. Kress and H. Bilz, 1984, Phys. Rev. Lett. **52,** 1907.

Benguigui, L., 1984, Phys. Rev. Lett. **53,** 2028.

Beni, G. and P.M. Platzman, 1976, Phys. Rev. B **14,** 1514.

Beran, M., 1965, Nuovo Cim. Suppl. **3,** 448.

Beran, M. and J. Molyneux, 1963, Nuovo Cim. **30,** 1406.

Berg, W.T. and J.A. Morrison, 1957, Proc. Roy. Soc. A **242,** 467.

Bergman, D.J., 1978, Phys. Reports (Phys. Lett. C) **43,** 377.

Bergman, D.J., 1980, Phys. Rev. Lett. **44,** 1285.

Bergman, D.J., 1982, Ann. Phys. **138,** 78.

Berman, R., 1976, Thermal Conduction in Solids (Clarendon Press, Oxford).

Berryman, J.G., 1980, J. Acoust. Soc. Am. **68,** 1820.

Betts, D.D., 1961, Can. J. Phys. **39,** 233.

Betts, D.D., A.B. Bhatia and M. Wyman, 1956a, Phys. Rev. **104,** 37.

Betts, D.D., A.B. Bhatia and G.K. Horton, 1956b, Phys. Rev. **104,** 43.

Bevk, J., 1973, Phil. Mag. **28,** 1379.

Bevk, J., T.B. Massalski and U. Mizutani, 1977, Phys. Rev. B **16,** 3456.

Bilz, H. and W. Kress, 1979, Phonon Dispersion Relations in Insulators (Springer, Berlin).

Blatt, F.J., 1968, Physics of Electronic Conduction in Solids (McGraw-Hill, New York).

Bloch, F., 1928, Z. Physik **52,** 555.

Bloch, F., 1930, Z. Physik **59,** 208.

Boas, W. and J.K. Mackenzie, 1950, Progr. Met. Phys. **2,** 90.

Boehler, R., I.C. Getting and G.C. Kennedy, 1977, J. Phys. Chem. Solids **38,** 233.

Bohm, D. and T. Staver, 1951, Phys. Rev. **84,** 836.

Böhmer, W. and P. Rabe, 1979, J. Phys. C **12,** 2465.

Bolotin, V.V. and V.N. Moskalenko, 1967, J. Appl. Mech. Tech. Phys. **8,** 3.

Borelius, G., 1934, Ann. Physik **20,** 57.

Born, M., 1939, J. Chem. Phys. **7,** 591.

Born, M., 1942, Rep. Progr. Phys. **9,** 294.

Born, M. and K. Huang, 1954, Dynamical Theory of Crystal Lattices (Oxford Univ. Press).
Born, M. and J.E. Mayer, 1932, Z. Physik **75,** 1.
Bottani, C.E., P.M. Ossi and F. Rossitto, 1978, J. Phys. F **8,** 1671.
Böttcher, C.J.F., 1952, Theory of Electric Polarization (Elsevier, Amsterdam).
Brace, W.F., C.H. Scholz and P.N. La Mori, 1969, J. Geophys. Res. **74,** 2089.
Brewer, L., 1967, Acta Met. **15,** 553.
Brooks, C.R. and R.E. Bingham, 1968, J. Phys. Chem. Solids **29,** 1553.
Bross, H., A. Seeger and R. Haberkorn, 1963, Phys. Stat. Sol. **3,** 1126.
Brout, R. and W. Visscher, 1962, Phys. Rev. Lett. **9,** 54.
Brown, Jr., W.F., 1955, J. Chem. Phys. **23,** 1514.
Brudnoy, D.M., 1976, J. Phys. Chem. Solids **37,** 1109.
Brüesch, P., 1982, Phonons: Theory and Experiments I (Springer, Berlin).
Bruggeman, D.A.G., 1935, Ann. Physik **24,** 636.
Brugger, K., 1964, Phys. Rev. **133,** A1611.
Brugger, K., 1965a, J. Appl. Phys. **36,** 759.
Brugger, K., 1965b, J. Appl. Phys. **36,** 768.
Brugger, K. and T.C. Fritz, 1967, Phys. Rev. **157,** 524.
Buckingham, M.J., 1951, Nature **168,** 281.
Buckingham, M.J. and M.R. Schafroth, 1954, Proc. Phys. Soc. A **67,** 828.
Budiansky, B., 1965, J. Mech. Phys. Solids **13,** 223.
Budiansky, B., 1970, J. Compos. Mater. **4,** 286.
Burke, J., 1972, J. Less-Common Metals **28,** 441.
Burnell, D.M., J. Zasadzinski, R.J. Noer, E.L. Wolf and G.B. Arnold, 1982, Solid State Commun. **41,** 637.
Burton, J.J., 1970, J. Chem. Phys. **52,** 345.
Butt, N.M., G. Heger and B.T.M. Willis, 1985, Acta Cryst. **B41,** 374.
Buyers, W.J.L. and R.A. Cowley, 1969, Phys. Rev. **180,** 755.
Caglioti, G., 1982, in: Mechanical and Thermal Behavior of Metallic Materials, Proc. Int. School of Physics "Enrico Fermi" Course 82, G. Caglioti and A.F. Milone, eds. (North-Holland, Amsterdam) p. 1.
Callaway, J., 1963, Nuovo Cimento **29,** 883.
Callaway, J. and N.H. March, 1984, in: Solid State Physics, Vol. 38, H. Ehrenreich and D. Turnbull, eds. (Academic Press, New York) p. 135.
Cape, J.A., G.W. Lehman, W.V. Johnston and R.E. DeWames, 1966, Phys. Rev. Lett. **16,** 892.
Caplin, A.D. and L.K. Nicholson, 1978, J. Phys. F **8,** 51.
Car, R., P.J. Kelly, A. Oshiyama and S.T. Pantelides, 1984, Phys. Rev. Lett. **52,** 1814.
Carr, W.J., Jr., 1961, Phys. Rev. **122,** 1437.
Carnelley, T., 1879, Ber. Deutsche Chem. Ges. **12,** 439.
Carruthers, P., 1961, Rev. Mod. Phys. **33,** 92.
Catlow, C.R.A., J. Corish, P.W.M. Jacobs and A.B. Lidiard, 1981, J. Phys. C **14,** L121.
Chadwick, G.A. and D.A. Smith, 1976, Grain Boundary Structure and Properties (Academic Press, London).
Chan, W.C., 1978, J. Phys. F **8,** 859.
Chang, R., 1967, Appl. Phys. Lett. **11,** 305.
Chen, T.S., G.P. Alldredge, F.W. de Wette and R.E. Allen, 1971, J. Chem. Phys. **55,** 3121.
Cho, S-A., 1982, J. Phys. F **12,** 1069.
Chou, M.Y., P.K. Lam and M.L. Cohen, 1983, Phys. Rev. B **28,** 4179.

Christensen, R.M., 1979, Mechanics of Composite Materials (Wiley, New York).
Christoffel, E.B., 1877, Ann. Mat. Pura Appl. **8,** 193.
Christy, R.W. and A.W. Lawson, 1951, J. Chem. Phys. **19,** 517.
Chung, D.H., 1963, Phil. Mag. **8,** 833.
Chung, D.H., 1967, J. Appl. Phys. **38,** 5104.
Chung, D.H. and W.R. Buessem, 1967, J. Appl. Phys. **38,** 2010.
Chung, D.Y., J.M. Farley and G.A. Saunders, 1975, Phys. Stat. Sol. (a) **29,** K43.
Cibuzar, G., A. Hikata and C. Elbaum, 1984, Phys. Rev. Lett. **53,** 356.
Cimberle, M.R., G. Bobel and C. Rizzuto, 1974, Adv. Phys. **23,** 639.
Cohen, M.L., 1985, in: Proceedings Int. School of Physics "Enrico Fermi", LXXXIX Course (North-Holland, Amsterdam) p. 16.
Cohen, M.L., 1986, Proc. Second Int. Conf. Science of Hard Materials, to be published.
Cole, H.S.D. and R.E. Turner, 1967, Phys. Rev. Lett. **19,** 501.
Collins, J.G. and G.K. White, 1964, Progr. Low Temp. Phys. **4,** 450.
Cook, M.A. and L.A. Rogers, 1963, J. Appl. Phys. **34,** 2330.
Copley, J.R.D., 1973, Can. J. Phys. **51,** 2564.
Cowley, E.R. and R.A. Cowley, 1965, Proc. Roy. Soc. A **287,** 259.
Cowley, E.R. and R.A. Cowley, 1966, Proc. Roy. Soc. A **292,** 209.
Cowley, R.A., 1963, Adv. Phys. **12,** 421.
Cowley, R.A., 1968, Rep. Progr. Phys. **31,** 123.
Cowley, R.A., 1970, Rev. Int. Hautes Tempér. et Réfract. **7,** 202.
Cracknell, A.P., 1984, in: Landolt-Börnstein New Series, Vol. III/13c, K.-H. Hellwege and J.L. Olsen, eds. (Springer, Berlin).
Daams, J.M., B. Mitrović and J.P. Carbotte, 1981, Phys. Rev. Lett. **46,** 65.
Date, E.H.F. and K.W. Andrews, 1969, J. Phys. D **2,** 1373.
Dato, P. and H. Köhler, 1984, J. Phys. C **17,** 3711.
Dawber, P.G. and R.J. Elliott, 1963, Proc. Roy. Soc. A **273,** 222.
Debye, P., 1912, Ann. Physik **39,** 789.
Debye, P., 1914, Ann. Physik **43,** 49.
Dederichs, P.H. and R. Zeller, 1973, Z. Physik **259,** 103.
Dederichs, P.H. and R. Zeller, 1980, Point Defects in Metals II (Springer, Berlin).
Devreese, J.T., V.E. Van Doren and P.E. Van Camp, eds. 1983, Ab Initio Calculation of Phonon Spectra (Plenum Press, New York).
Dewey, J.M., 1947, J. Appl. Phys. **18,** 578.
Dietsche, W., H. Kinder, J. Mattes and H. Wühl, 1980, Phys. Rev. Lett. **45,** 1332.
Dietsche, W., G.A. Northrop and J.P. Wolfe, 1981, Phys. Rev. Lett. **47,** 660.
Dobrzynski, L., 1969, Ann. Phys. (Paris) **4,** 637.
Dobrzynski, L. and J. Friedel, 1968, Surf. Sci. **12,** 469.
Dobrzynski, L. and G. Leman, 1969, J. Physique **30,** 116.
Domb, C., 1981, in: Perspectives in Statistical Physics, H.J. Raveche, ed. (North-Holland, Amsterdam) p. 173.
Domb, C. and L. Salter, 1952, Phil. Mag. **43,** 1083.
Doniach, S. and S. Engelsberg, 1966, Phys. Rev. Lett. **17,** 750.
Donohue, J., 1974, The Structures of the Elements (Wiley, New York).
Dugdale, J.S., 1977, The Electrical Properties of Metals and Alloys (Arnold, London).
Dugdale, J.S. and D. Gugan, 1962, Proc. Roy. Soc. A **270,** 186.
Dugdale, J.S. and A. Myers, 1985, in: Landolt-Börnstein New Series, Vol. III/15b, K.-H. Hellwege and J.L. Olsen, eds. (Springer, Berlin).

Dulong, P.L. and A.T. Petit, 1819, Ann. Chim. [2] **7,** 113, 225, 337.
Dupuis, M., R. Mazo and L. Onsager, 1960, J. Chem. Phys. **33,** 1452.
Durand, M.A., 1936, Phys. Rev. **50,** 449.
Dykhne, A.M., 1970, Zh. Eksp. Teor, Fiz. **59,** 110 [Sov. Phys. – JETP **32,** 63 (1971)].
Eckhardt, D. and W. Wasserbäch, 1978, Phil. Mag. A **37,** 621.
Ecsedy, D.J. and P.G. Klemens, 1977, Phys. Rev. B **15,** 5957.
Einstein, A., 1907, Ann. Physik **22,** 180.
Éliashberg, G.M., 1962, Zh. Eksp. Teor. Fiz. **43,** 1005 [Sov. Phys. – JETP **16,** 780 (1963)].
Elliott, R.J. and D.W. Taylor, 1964, Proc. Phys. Soc. **83,** 189.
Enderby, J.E. and N.H. March, 1966, Proc. Phys. Soc. **88,** 717.
Engel, N., 1967, Acta Met. **15,** 565.
Engquist, H.-L. and G. Grimvall, 1980, Phys. Rev. B **21,** 2072.
Epstein, S.G. and O.N. Carlson, 1965, Acta Met. **13,** 487.
Eros, S. and C.S. Smith, 1961, Acta Met. **9,** 14.
Eshelby, J.D., 1949, Proc. Roy. Soc. A **7,** 396.
Eshelby, J.D., 1957, Proc. Roy. Soc. A **241,** 376.
Eshelby, J.D., 1975, in: The Physics of Metals 2. Defects, P.B. Hirsch, ed. (Cambridge Univ. Press, Cambridge) p. 1.
Every, G.A., 1979, Phys. Rev. Lett. **42,** 1065.
Every, G.A., 1980, Phys. Rev. B **22,** 1746.
Ewing, R.H., 1971, Acta Met. **19,** 1359.
Ewing, R.H. and B. Chalmers, 1972, Surf. Sci. **31,** 161.
Farnell, G.W., 1961, Can. J. Phys. **39,** 65.
Faulkner, J.S., 1982, Progr. Materials Sci. **27,** 1.
Fedorov, F.I., 1963, Kristallographiya **8,** 213.
Fedorov, F.I., 1968, Theory of Elastic Waves in Crystals (Plenum, New York).
Feldman, J.L., 1964, Proc. Phys. Soc. **84,** 361.
Feldman, R.F. and J.J. Beaudoin, 1977, Mag. Concr. Res. **7,** 19.
Felice, R.A., J. Trivisonno and D.E. Schuele, 1977, Phys. Rev. B **16,** 5173.
Finnis, M.W. and M. Sachdev, 1976, J. Phys. F **6,** 965.
Finnis, M.W., K.L. Kear and D.G. Pettifor, 1984, Phys. Rev. Lett. **52,** 291.
Fisher, E.S. and D. Dever, 1970, Acta Met. **18,** 265.
Fisk, Z. and G.W. Webb, 1976, Phys. Rev. Lett. **36,** 1084.
Fletcher, G.C., 1978, Physica **93** B, 149.
Fletcher, G.C. and M. Yahaya, 1979, J. Phys. F **9,** 1529.
Flinn, P.A., G.M. McManus and J.A. Rayne, 1960, J. Phys. Chem. Solids **15,** 189.
Flubacher, P., A.J. Leadbetter and J.A. Morrison, 1959, Phil. Mag. **4,** 273.
Foiles, C.L., 1985, in: Landolt-Börnstein New Series III/15b, K.-H. Hellwege and J.L. Olsen, eds. (Springer, Berlin).
Fricke, H., 1953, J. Phys. Chem. **57,** 934.
Friedel, J., 1953, Phil. Mag. **44,** 444.
Friedel, J., 1969, in: The Physics of Metals I. Electrons, J.M. Ziman ed. (Cambridge Univ. Press, Cambridge) p. 340.
Friedel, J., 1974, J. Phys. Lett. (France) **35,** L59.
Friedel, J., 1980, in: Physics of Modern Materials, Vol. I (IAEA, Vienna) p. 163.
Friedel, J., 1982, Phil. Mag. A **45,** 271.
Froyen, S. and M.L. Cohen, 1984, Phys. Rev. B **29,** 3770.

Fuchs, K., 1935, Proc. Roy. Soc. A **151,** 585.
Fuchs, K., 1938, Cambridge Phil. Soc. **34,** 100.
Fung, Y.C., 1965, Foundations of Solid Mechanics (Prentice-Hall, Englewood Cliffs).
Gachon, J.C., M. Notin, C. Cunat, J. Hertz, J.C. Parlebas, G. Moraitis, B. Stupfel and F. Gautier, 1980, Acta Met. **28,** 489.
Garland, C.W. and G. Jura, 1954, J. Chem. Phys. **22,** 1108.
Gazis, D.C., R. Herman and R.F. Wallis, 1960, Phys. Rev. **119,** 533.
Geballe, T.H. and G.W. Hull, 1958, Phys. Rev. **110**, 773.
Gelatt, C.D., Jr., H. Ehrenreich and R.E. Watson, 1977, Phys. Rev. B **15,** 1613.
Gerlich, D., 1969, J. Phys. Chem. Solids, **30,** 1638.
Girifalco, L.A., 1967, Scripta Met. **1,** 5.
Gladstone, G., M.A. Jensen and J.R. Schrieffer, 1969, in: Superconductivity, Vol. 2, R.D. Parks, ed. (Dekker, New York) p. 665.
Glyde, H.R., J.P. Hansen and M.L. Klein, 1977, Phys. Rev. B **16,** 3476.
Golding, B., B.G. Bagley and F.S.L. Hsu, 1972, Phys. Rev. Lett. **29,** 68.
Gösele, U., W. Frank and A. Seeger, 1983, Solid State Commun. **45,** 31.
Graebner, J.E. and L.C. Allen, 1983, Phys. Rev. Lett. **51,** 1566.
Granato, A., 1958, Phys. Rev. **111,** 740.
Greegor, R.B. and F.W. Lytle, 1979, Phys. Rev. B **20,** 4902.
Grimvall, G., 1969a, phys. stat. sol. **32,** 383.
Grimvall, G., 1969b, Phys. kondens. Materie **9,** 283.
Grimvall, G., 1976, Physica Scripta **14,** 63.
Grimvall, G., 1981, The Electron–Phonon Interaction in Metals (North-Holland, Amsterdam).
Grimvall, G., 1983, Int. J. Thermophys. **4,** 363.
Grimvall, G., 1984a, Physica **127** B, 165.
Grimvall, G., 1984b, J. Phys. C **17,** 3545.
Grimvall, G. and I. Ebbsjö, 1975, Physica Scripta **12,** 168.
Grimvall, G. and J. Rosén, 1983, Int. J. Thermophys. **4,** 139.
Grimvall, G. and S. Sjödin, 1974, Physica Scripta **10,** 340.
Grosse, C. and J.-L. Greffe, 1979, J. Chimie Physique **76,** 305.
Grüneisen, E., 1913, Berichte Deutsche physikal. Ges. **15,** 186.
Grüneisen, E. and E.S. Goens, 1924, Z. Physik **29,** 141.
Grüneisen, E., 1933, Ann. Physik **16,** 530.
Gschneidner, K.A., Jr., 1964, in: Solid State Physics, Vol. 16, F. Seitz and D. Turnbull, eds. (Academic Press, New York) p. 275.
Gschneidner, K.A., Jr., and K. Ikeda, 1983, J. Magn. Magnetic Materials **31–34,** 265.
Gschneidner, K.A., Jr., and G.H. Vineyard, 1962, J. Appl. Phys. **33,** 3444.
Guillot, M., F. Tchéou, A. Marchand, P. Feldmann and R. Lagnier, 1981, Z. Physik B **44,** 53.
Gunton, D.J. and G.A. Saunders, 1972, J. Mater. Sci. **7,** 1061.
Hafner, J., 1983, Phys. Rev. B **27,** 678.
Hahn, H. and W. Ludwig, 1961, Z. Physik **161,** 404.
Hale, D.K., 1976, J. Mat. Sci. **11,** 2105.
Hansen, J.-P., 1970, Phys. Rev. A **2,** 221.
Hansen, T.C., 1958, Swedish Res. Inst. Cement Concrete, Royal Inst. Techn. Stockholm Medd. **33,** 1.

Hansen, T.C., 1960, Swedish Res. Inst. Cement Concrete, Royal Inst. Techn. Stockholm Handl. **31,** 1.

Harding, J.H. and A.M. Stoneham, 1981, Phil. Mag. B **43,** 705.

Harris, A.B. and H. Meyer, 1962, Phys. Rev. **127,** 101.

Harrison, W.A., 1966, Pseudopotentials in the Theory of Metals (Benjamin, New York).

Harrison, W.A., 1980, Electronic Structure and the Properties of Solids (Freeman and Co., San Francisco).

Hasegawa, H. and D.G. Pettifor, 1983, Phys. Rev. Lett. **50,** 130.

Hashin, Z., 1959, in: Nonhomogeneity in Elasticity and Plasticity, W. Olszak, ed. (Pergamon Press, New York) p. 463.

Hashin, Z., 1962, ASME J. Appl. Mech. **29,** 143.

Hashin, Z., 1965, J. Mech. Phys. Solids **13,** 119.

Hashin, Z., 1983, ASME J. Appl. Mech. **50,** 481.

Hashin, Z. and B.W. Rosen, 1964, ASME J. Appl. Mech. **31,** 223.

Hashin, Z. and S. Shtrikman, 1962a, J. Appl. Phys. **33,** 3125.

Hashin, Z. and S. Shtrikman, 1962b, J. Mech. Phys. Solids **10,** 343.

Hashin, Z. and S. Shtrikman, 1963a, J. Mech. Phys. Solids **11,** 127.

Hashin, Z. and S. Shtrikman, 1963b, Phys. Rev. **130,** 129.

Hasson, G.C., J.-Y. Boos, I. Herbeuval, M. Biscondi and C. Goux, 1972, Surf. Sci. **31,** 115.

Haussühl, S., 1973, Solid State Commun. **13,** 147.

Hearmon, R.F.S., 1946, Rev. Mod. Phys. **18,** 409.

Hearmon, R.F.S., 1961, An Introduction to Applied Anisotropic Elasticity (Oxford Univ. Press, Oxford).

Hearmon, R.F.S., 1979, in: Landolt-Börnstein New Series III/11, K.-H. Hellwege and A.M. Hellwege, eds. (Springer, Berlin).

Hearmon, R.F.S., 1984, in: Landolt-Börnstein New Series III/18, K.-H. Hellwege and A.M. Hellwege, eds. (Springer, Berlin).

Heine, V., 1967, Phys. Rev. **153,** 673.

Henderson, B., 1972, Defects in Crystalline Solids (Edward Arnold, London).

Hensel, J.C. and R.C. Dynes, 1979, Phys. Rev. Lett. **43,** 1033.

Herring, C., 1960, J. Appl. Phys. **31,** 1939.

Hershey, A.V., 1954, J. Appl. Mech. **21,** 236.

Hidnert, P. and W. Sonder, 1950, Natl. Bur. St. (US) Circ. 486.

Hill, R., 1952, Proc. Phys. Soc. A **65,** 349.

Hill, R., 1963, J. Mech. Phys. Solids **11,** 357.

Hill, R., 1964, J. Mech. Phys. Solids **12,** 199.

Hill, R., 1965a, J. Mech. Phys. Solids **13,** 189.

Hill, R., 1965b, J. Mech. Phys. Solids **13,** 213.

Hillel, A.J., 1983, Phil. Mag. B **48,** 237.

Hillel, A.J., J.T. Edwards and P. Wilkes, 1975, Phil. Mag. **32,** 189.

Hirsch, P.B., ed., 1975, The Physics of Metals. 2. Defects (Cambridge Univ. Press).

Hohenberg, P. and W. Kohn, 1964, Phys. Rev. **136 B,** 864.

Honda, K., S. Shimizu and S. Kusakabe, 1902, Z. Physik **3,** 380.

Hoover, W.G., S.G. Gray and K.W. Johnson, 1971, J. Chem Phys. **55,** 1128.

Hori, M., 1973a, J. Math. Phys. **14,** 514.

Hori, M., 1973b, J. Math. Phys. **14,** 1942.

Horton, G.K. and A.A. Maradudin, eds., 1974, Dynamical Properties of Solids, Vol. 1 (North-Holland, Amsterdam).

Housley, R.M. and F. Hess, 1966, Phys. Rev. **146,** 517.

Howard, R.E. and A.B. Lidiard, 1964, Rep. Progr. Phys. **27,** 161.

Howson, M.A., 1984, J. Phys. F **14,** L25.

Hsiang, T.Y., J.W. Reister, H. Weinstock, G.W. Crabtree and J.J. Vuillemin, 1981, Phys. Rev. Lett. **47,** 523.

Huebener, R.P. and C.G. Homan, 1963, Phys. Rev. **129,** 1162.

Hui, J.C.K. and P.B. Allen, 1975, J. Phys. C **8,** 2923.

Huiszoon, C. and P.P.M. Groenewegen, 1972, Acta Cryst. A **28,** 170.

Hultgren, R., P.D. Desai, D.T. Hawkins, M. Gleiser, K.K. Kelley and D.D. Wagman, 1973a, Selected Values of the Thermodynamic Properties of the Elements (Am. Soc. for Metals, Metals Park, Ohio).

Hultgren, R., P.D. Desai, D.T. Hawkins, M. Gleiser and K.K. Kelley, 1973b, Selected Values of the Thermodynamic Properties of Binary Alloys (Am. Soc. for Metals, Metals Park, Ohio).

Huntington, H.B., 1958, in: Solid State Physics vol. 7, F. Seitz and D. Turnbull, eds. (Academic Press, New York) p. 213.

Huntington, H.B., G.A. Shirn and E.S. Wajda, 1955, Phys. Rev. **99,** 1085.

Hwang, J.-L., 1954, J. Chem. Phys. **22,** 154.

Hybertsen, M.S. and S.G. Louie, 1984, Solid State Commun. **51,** 451.

Ihm, J. and M.L. Cohen, 1980, Phys. Rev. **21,** 1527.

Ikeda, K. and K.A. Gschneidner, Jr., 1980, Phys. Rev. Lett. **45,** 1341.

Inoue, J. and M. Shimizu, 1976, J. Phys. Soc. Japan **41,** 1211.

Isenberg, C., 1963, Phys. Rev. **132,** 2427.

Jackson, J.L. and S.R. Coriell, 1968, J. Appl. Phys. **39,** 2349.

Jacobs, R.L., 1983, J. Phys. C **16,** 273.

JANAF Thermochemical Tables, 2nd ed., 1971 (National Bureau of Standards, Washington D.C.).

Johansson, B. and A. Rosengren, 1975, J. Phys. F **5,** L15.

Julian, C.L., 1965, Phys. Rev. **137,** A128.

Kagan, Yu. and Ya.A. Iosilevskii, 1962. Zh. Eksp. Teor. Fiz. **42,** 259. [Sov. Phys. JETP **15,** 182 (1962)].

Kagan, Yu. and Ya. Iosilevskii, 1963, Zh. Eksp. Teor. Fiz. **45,** 819. [Sov. Phys. JETP **18,** 562 (1964)].

Kantor, Y. and I. Webman, 1984, Phys. Rev. Lett. **52,** 1891.

Karlsson, Å.V., 1970, Phys. Rev. B **2,** 3332.

Kasen, M.B., 1972, Acta Met. **20,** 105.

Kaspers, W., R. Pott, D.M. Herlach and H. v. Löhneysen, 1983, Phys. Rev. Lett. **50,** 433.

Kaufmann, R. and O. Meyer, 1984, Solid State Commun. **51,** 539.

Kaveh, M. and N.F. Mott, 1982, J. Phys. C **15,** L707.

Kikuchi, M., K. Fukamichi, T. Masumoto, T. Jagielinski, K.I. Arai and N. Tsuya, 1978, phys. stat. sol. (a) **48,** 175.

Kim, D.J., 1968, Phys. Rev. **167,** 545.

Kim, Y.S. and R.G. Gordon, 1974, Phys. Rev. B **9,** 3548.

Kincaid, R.L. and D.A. Huckaby, 1976, J. Chem. Phys. **65,** 2353.

Kirkpatrick, S., 1973, Rev. Mod. Phys. **45,** 574.

Kitagawa, K., M. Ueda and H. Miyamoto, 1980, Acta Met. **28,** 1505.

Kittel, C., 1949, Phys. Rev. **75,** 972.

Kittel, C., 1963, Quantum Theory of Solids (Wiley, New York).
Kittel, C., 1976, Introduction to Solid State Physics (Wiley, New York).
Kittinger, E., J. Tichý and E. Bertagnolli, 1981, Phys. Rev. Lett. **47,** 712.
Klein, M.V., 1966, Phys. Rev. **141,** 716.
Klemens, P.G., 1955, Proc. Phys. Soc. A **68,** 1113.
Klemens, P.G., 1958, in: Solid State Physics, Vol. 7, F. Seitz and D. Turnbull, eds. (Academic Press, New York) p. 1.
Klemens, P.G., 1960, Phys. Rev. **119,** 507.
Klemens, P.G., 1969, in: Thermal Conductivity, R.P. Tye, ed. (Academic Press, London) p. 1.
Klemens, P.G., 1983a, in: Thermal Conductivity, Vol. 16, D.C. Larsen ed. (Plenum, New York) p. 15.
Klemens, P.G., 1983b, in: Thermal Conductivity, Vol. 17, J.G. Hust, ed. (Plenum, New York) p. 25.
Kleppa, O.J. and S. Watanabe, 1983, Solid State Commun. **46,** 799.
Kneer, G., 1965, phys. stat. sol. **9,** 825.
Koehler, J.S. and G. DeWit, 1959, Phys. Rev. **116,** 1121.
Kohler, M., 1938, Ann. Physik **32,** 211.
Kohler, M., 1948, Z. Physik **124,** 772.
Kohler, M., 1949, Z. Physik **125,** 679.
Kohn, W. and L.J. Sham, 1965, Phys. Rev. **140 A,** 1133.
Korshunov, V.A. and A.D. Shevchenko, 1983, Solid State Commun. **48,** 577.
Kouvel, J.S., 1956, Phys. Rev. **102,** 1489.
Kovács, I. and H. El Sayed, 1976, J. Mat. Sci. **11,** 529.
Kraftmakher, Ya., A., 1972, Fiz. Tverd. Tela **14,** 392.
Kraftmakher, Ya.A., 1978, in: Thermal Expansion, Vol. 6, I.D. Peggs, ed. (Plenum, New York) p. 155.
Krishnan, R.S., R. Srinivasan and S. Devanarayanan, 1979, Thermal Expansion of Crystals (Pergamon Press, Oxford).
Kröner, E., 1958, Z. Physik **151,** 504.
Kröner, E., 1967, J. Mech. Phys. Solids **15,** 319.
Kröner, E. and H.H. Wawra, 1978, Phil. Mag. A **38,** 433.
Krumhansl, J.A., 1965, in: Lattice Dynamics, R.F. Wallis, ed. (Pergamon Press, New York) p. 523.
Krumhansl, J.A. and J.A.D. Matthew, 1965, Phys. Rev. **140,** A 1812.
Kumazawa, M., 1969, J. Geophys. Res. **74,** 5311.
Kunc, K. and R.M. Martin, 1982, Phys. Rev. Lett. **48,** 406.
Kus, F.W. and D.W. Taylor, 1980, J. Phys. F **10,** 1495.
Kus, F.W. and D.W. Taylor, 1982, J. Phys. F **12,** 837.
Lam, P.K. and M.L. Cohen, 1983, Phys. Rev. B **27,** 5986.
Landau, L.D. and E.M. Lifshitz, 1959, Theory of Elasticity (Pergamon Press, London).
Landau, L.D. and E.M. Lifshitz, 1960, Electrodynamics of Continuous Media (Pergamon Press, London).
Landauer, R., 1952, J. Appl. Phys. **23,** 779.
Landauer, R., 1978, AIP Conf. Proc. 40, J.C. Garland and D.B. Tanner, eds. p. 2.
Lang, N.D. and H. Ehrenreich, 1968, Phys. Rev. **168,** 605.
Larkin, B.K. and S.W. Churchill, 1959, A.I.Ch.E. Journal **5,** 467.
Latimer, W.M., 1921, J. Am. Chem. Soc. **43,** 818.

Latimer, W.M., 1951, J. Am. Chem. Soc. **73,** 1480.
Laubitz, M.J. and T. Matsumura, 1972, Can. J. Phys. **50,** 196.
Laubitz, M.J., T. Matsumura and P.J. Kelly, 1976, Can. J. Phys. **54,** 92.
Laws, N., 1980, in: Physics of Modern Materials (Int. Atomic Energy Agency, Vienna) p. 465.
Lawson, N.S. and A.M. Guénault, 1982, J. Phys. F **12,** 1407.
Leadbetter, A.J., 1968, J. Phys. C **1,** 1489.
Leadbetter, A.J. and G.R. Settatree, 1969, J. Phys. C **2,** 1105.
Leadbetter, A.J., D.M.T. Newsham and G.R. Settatree, 1969, J. Phys. C **2,** 393.
Leavens, C.R., 1977, J. Phys. F **7,** 163.
Leavens, C.R. and A.H. MacDonald, 1983, Phys. Rev. B **27,** 2812.
Ledbetter, H.M. and R.L. Moment, 1976, Acta Met. **24,** 891.
Ledbetter, H.M. and R.P. Reed, 1973, J. Phys. Chem. Data **2,** 531.
Lee, T.Y.R. and R.E. Taylor, 1978, Trans. ASME **100,** 720.
Leibfried, G., 1955, in: Handbuch der Physik, Vol. VII.1, S. Flügge, ed. (Springer, Berlin) p. 104.
Leibfried, G. and N. Breuer, 1978, Point Defects in Metals I (Springer, Berlin).
Leibfried, G. and W. Ludwig, 1961, in: Solid State Physics, Vol. 12, F. Seitz and D. Turnbull, eds. (Academic Press, New York) p. 275.
Leibfried, G. and E. Schlömann, 1954, Nachr. Akad. Wiss. Göttingen Math.-Phys. Kl. 2a (4), 71.
Leighton, R.B., 1948, Rev. Mod. Phys. **20,** 165.
Leung, H.K., F.W. Kus, N. McKay and J.P. Carbotte, 1977, Phys. Rev. B **16,** 4358.
Levin, B.M., 1967, Mech. Tverd. Tela **1,** 88.
Levinson, L.M. and F.R.N. Nabarro, 1967, Acta Met. **15,** 785.
Li, Y., 1976, phys. stat. sol. (a) **38,** 171.
Lidiard, A.B., 1960, Phil. Mag. **5,** 1171.
Lifshitz, I.M., 1954, Zh. Eksp. Teor. Fiz. **26,** 551.
Lifshitz, I.M., 1956, Nuovo Cim. **3** Suppl., 716.
Lindemann, F.A., 1910, Z. Physik **11,** 609.
Ling, D.D. and C.D. Gelatt, Jr., 1980, Phys. Rev. B **22,** 557.
Logachev, Yu.A. and M.S. Yur'ev, 1972, Fiz. Tverd. Tela **14,** 3336 [Sov. Phys. – Solid State **14,** 2826 (1973)].
Logan, J. and M.F. Ashby, 1974, Acta Met. **22,** 1047.
Lomer, W.M., 1958, Vacancies and Other Point Defects in Metals and Alloys, Monograph no. 23 (Institute of Metals, London) p. 85.
Lord, A.E., Jr., 1967, J. Phys. Chem. Solids **28,** 517.
Lorenz, L., 1881, Ann. Physik **13,** 422.
Ludwig, W., 1958, J. Phys. Chem. Solids **4,** 283.
MacDonald, W.M. and A.C. Anderson, 1983, in: Thermal Conductivity, Vol. 17, J.G. Hust, ed. (Plenum, New York) p. 185.
Mahanty, J., A.A. Maradudin and G.H. Weiss, 1960, Progr. Theor. Phys. **24,** 648.
Mair, S.L., 1980, J. Phys. C **13,** 2857.
Malik, U. and L.S. Kothari, 1980, Phys. Lett. A **78,** 178.
Mannheim, P.D., 1968, Phys. Rev. **165,** 1011.
Manninen, M., R. Nieminen, P. Hautojärvi and J. Arponen, 1975, Phys. Rev. B **12,** 4012.
Manninen, M., P. Jena, R.M. Nieminen and J.K. Lee, 1981, Phys. Rev. B **24,** 7057.

Maradudin, A.A., 1974, in: Dynamical Properties of Solids, Vol. 1, G.K. Horton and A.A. Maradudin, eds. (North-Holland, Amsterdam) p. 1.
Maradudin, A.A., 1981, in: Festkörperprobleme 21 (Advances in Solid State Physics), J. Treusch, ed. (Vieweg, Brunswick) p. 25.
Maradudin, A.A. and A.E. Fein, 1962, Phys. Rev. **128,** 2589.
Maradudin, A.A. and P.A. Flinn, 1963, Phys. Rev. **129,** 2529.
Maradudin, A.A. and R.F. Wallis, 1966, Phys. Rev. **148,** 945.
Maradudin, A.A., P. Mazur, E.W. Montroll and G.H. Weiss, 1958, Rev. Mod. Phys. **30,** 175.
Maradudin, A.A., P.A. Flinn and R.A. Coldwell-Horsfall, 1961, Ann. Phys. **15,** 337.
Maradudin, A.A., E.W. Montroll and G.H. Weiss, 1963, Solid State Physics Suppl. 3, Theory of Lattice Dynamics in the Harmonic Approximation (Pergamon, New York).
March, N.H. and J.S. Rousseau, 1971, Crystal Lattice Defects **2,** 1.
Marcus, M. and C.-L. Tsai, 1984, Solid State Commun. **52,** 511.
Martin, R.M., 1985, in: Festkörperprobleme **25** (Advances in Solid State Physics), P. Grosse, ed. (Vieweg, Braunschweig) p. 3.
Masri, P. and L. Dobrzynski, 1971, J. Physique **32,** 939.
Mattheiss, L.F., L.R. Testardi and W.W. Yao, 1978, Phys. Rev. B **17,** 4640.
Matthiessen, A. and C. Vogt, 1864, Ann. Physik Chem. **122,** 19.
Mayadas, A.F., M. Shatzkes and J.F. Janak, 1969, Appl. Phys. Lett. **14,** 345.
McCombie, C.W. and J. Slater, 1964, Proc. Phys. Soc. **84,** 499.
McManus, G.M., 1963, Phys. Rev. **129,** 2004.
Meaden, G.T., 1965, Electrical Resistance of Metals (Heywood, London).
Meister, R. and L. Peselnick, 1966, J. Appl. Phys. **37,** 4121.
Men', A.A. and O.A. Sergeev, 1973, High Temp. – High Press. **5,** 19.
Mertig, I., E. Mrosan and R. Schöpke, 1982, J. Phys. F **12,** 1689.
Mertig, I., G. Pompe and E. Hegenbarth, 1984, Solid State Commun. **49,** 369.
Miller, M.N., 1969, J. Math. Phys. **10,** 1988.
Milton, G.W., 1980, Appl. Phys. Lett. **37,** 300.
Milton, G.W., 1981a, J. Appl. Phys. **52,** 5286.
Milton, G.W., 1981b, J. Appl. Phys. **52,** 5294.
Milton, G.W., 1981c, Phys. Rev. Lett. **46,** 542.
Milton, G.W. and N. Phan-Thien, 1982, Proc. Roy. Soc. A **380,** 305.
Misra, R.D., 1940, Proc. Cambridge Phil. Soc. **36,** 173.
Mitchel, W., R.S. Newrock and D.K. Wagner, 1980, Phys. Rev. Lett. **44,** 426.
Mitchell, M.A., J.R. Cullen, R. Abbundi, A. Clark and H. Savage, 1979, J. Appl. Phys. **50,** 1627.
Mizutani, U. and T.B. Massalski, 1980, J. Phys. F **10,** 1093.
Molyneux, J., 1970, J. Math. Phys. **11,** 1172.
Molyneux, J. and M.J. Beran, 1965, J. Math. Mech. **14,** 337.
Montroll, E.W., 1942, J. Chem. Phys. **10,** 218.
Montroll, E.W., 1943, J. Chem. Phys. **11,** 481.
Montroll, E.W., 1944, J. Chem. Phys. **12,** 98.
Montroll, E.W. and D.C. Peaslee, 1944, J. Chem. Phys. **12,** 98.
Mooij, J.H., 1973, phys. stat. sol. (a) **17,** 521.
Mooney, D.L. and R.G. Steg, 1969, High Temp. – High Press. **1,** 237.
Moraitis, G. and F. Gautier, 1977a, J. Phys. F **7,** 1421.
Moraitis, G. and F. Gautier, 1977b, J. Phys. F **7,** 1841.

Moriarty, J.A., D.A. Young and M. Ross, 1984, Phys. Rev. B **30,** 578.
Moruzzi, V.L., A.R. Williams and J.F. Janak, 1977, Phys. Rev. B **15,** 2854.
Moruzzi, V.L., J.F. Janak and A.R. Williams, 1978, Calculated Electronic Properties of Metals (Pergamon, Oxford).
Motakabbir, K.A. and G. Grimvall, 1981, Phys. Rev. B **23,** 523.
Mott, N.F., 1952, Phil. Mag. **43,** 1151.
Mott, N.F., 1972, Phil. Mag. **26,** 1249.
Mott, N.F. and E.W. Gurney, 1940, Electronic Processes in Ionic Crystals (Clarendon Press, Oxford).
Mountfield, K.R. and J.A. Rayne, 1984, Solid State Commun. **49,** 1055.
Mueller, F.M., A.J. Freeman, J.O. Dimmock and A.M. Furdyna, 1970, Phys. Rev. B **1,** 4617.
Munn, R.W., 1969, Adv. Phys. **18,** 515.
Münster, A., 1956, Statistische Thermodynamik (Springer, Berlin).
Murnaghan, F.D., 1944, Proc. Nat. Acad. Sci. USA **30,** 244.
Musgrave, M.J.P., 1970, Crystal Acoustics (Holden Day, San Francisco).
Nagel, S.R., G.S. Grest and A. Rahman, 1984, Phys. Rev. Lett. **53,** 368.
Nakajima, S., 1967, Progr. Theor. Phys. **38,** 23.
Nedoluha, A., 1957, Z. Physik **148,** 248.
Nelmes, R.J., 1969, Acta Cryst. A **25,** 523.
Nielsen, L.F., 1982, Mater. Sci. Eng. **52,** 39.
Nielsen, O.H. and R.M. Martin, 1983, Phys. Rev. Lett. **50,** 697.
Nishiguchi, N. and T. Sakuma, 1981, Solid State Commun. **38,** 1073.
Nordheim, L., 1931, Ann. Physik **9,** 607.
Novotny, V. and P.P.M. Meincke, 1973, Phys. Rev. B **8,** 4186.
Nozières, P. and D. Pines, 1958, Phys. Rev. **111,** 442.
Nye, J.F., 1957, Physical Properties of Crystals (Oxford Univ. Press, Oxford).
Ohashi, Y.H. and K. Ohashi, 1980, Phil. Mag. A **42,** 741.
Ohta, Y. and M. Shimizu, 1982, J. Phys. F **12,** L255.
Osamura, K., N. Otsuka and Y. Murakami, 1982, Phil. Mag. B **45,** 583.
Osborn, J.A., 1945, Phys. Rev. **67,** 351.
Ott, H., 1935, Ann. Physik **23,** 169.
Overton, W.C., Jr., 1971, Phys. Lett. **37A,** 287.
Panova, G. Kh. and B.N. Samoilov, 1965, Zh. Eksp. Teor. Fiz. **49,** 456 [Sov. Phys. – JETP **22,** 320 (1966)].
Parrot, J.E. and A.D. Stuckes, 1975, Thermal Conductivity of Solids (Pion, London).
Pathak, K.N. and Y.P. Varshni, 1969, Phys. Lett. **28A,** 539.
Paul, B., 1960, Trans, Met. Soc. AIME **218,** 36.
Pearson, W.P., 1972, The Crystal Chemistry and Physics of Metals and Alloys (Wiley, New York).
Peierls, R.E., 1929, Ann. Physik **3,** 1055.
Perenboom, J.A.A.J., P. Wyder and F. Meier, 1981, Phys. Rep. **78,** 173.
Peselnick, L. and R. Meister, 1965, J. Appl. Phys. **36,** 2879.
Peter, M., W. Klose, G. Adam, P. Entel and E. Kudla, 1974, Helv. Phys. Acta **47,** 807.
Petrusevich, V.A., V.M. Sergeeva and I.A. Smirnov, 1960, Fiz. Tverd. Tela **2,** 2894 [Sov. Phys. – Solid State **2,** 2573 (1961)].
Petterson, S., 1986, to be published.

Pettifor, D.G., 1972, in: Metallurgical Chemistry, O. Kubaschewski, ed. (HMSO, London) p. 191.

Pettifor, D.G., 1977, Calphad **1,** 305.

Pettifor, D.G., 1983, in: Physical Metallurgy, R.W. Cahn and P. Haasen, eds. (North-Holland, Amsterdam) p. 73.

Pettifor, D.G. and R. Podloucky, 1984, Phys. Rev. Lett. **53,** 1080.

Phan-Thien, N. and G.W. Milton, 1982, Proc. Roy. Soc. A **380,** 333.

Phillips, J.C., 1967, in: Proc. Int. School of Physics, Enrico Fermi, 37, W. Marshall, ed. (Academic Press, New York) p. 22.

Phillips, W.A., 1972, J. Low Temp. Phys. **7,** 351.

Piñango, E.S., S. Vieira and R. Villar, 1983, Solid State Commun. **48,** 143.

Pinski, F.J., P.B. Allen and W.H. Butler, 1978, Phys. Rev. Lett. **41,** 431.

Pohl, R.O., 1960, Phys. Rev. **118,** 1499.

Pokorny, M. and G. Grimvall, 1984, J. Phys. F **14,** 931.

Potzel, W., W. Adlassnig, U. Närger, Th. Obenhuber, K. Riski and G.M. Kalvius, 1984, Phys. Rev. B **30,** 4980.

Powell, R.L., H.M. Roder and W.J. Hall, 1959, Phys. Rev. **115,** 314.

Rayleigh, J.W., 1892, Phil. Mag. **34,** 481.

Rayleigh, Lord, 1894, Theory of Sound (Dover, New York). (Reprinted 1945).

Rayleigh, Lord, 1900, Scientific Papers 2 (Cambridge Univ. Press, Cambridge) p. 441.

Read, T.A., 1940, Phys. Rev. **58,** 371.

Reissland, J.A., 1973, The Physics of Phonons (Wiley, London).

Rezayi, E.H. and H. Suhl, 1980, Phys. Rev. Lett. **45,** 1115.

Ridley, N. and H. Stuart, 1970, Metal Science J. **4,** 219.

Rieder, K.H. and W. Drexel, 1975, Phys. Rev. Lett. **34,** 148.

Reuss, A., 1929, Z. Angew. Math. Mech. **9,** 55.

Rösch, F. and O. Weis, 1976a, Z. Physik B **25,** 101.

Rösch, F. and O. Weis, 1976b, Z. Physik B **25,** 115.

Rosen, B.W. and Z. Hashin, 1970, Int. J. Eng. Sci. **8,** 157.

Rosén, J. and G. Grimvall, 1983, Phys. Rev. B **27,** 7199.

Ross, R.G., P. Andersson, B. Sundqvist and G. Bäckström, 1984, Rep. Prog. Phys. **47,** 1347.

Rossiter, P.L., 1976, Phil. Mag. **33,** 1015.

Roufosse, M. and P.G. Klemens, 1973, Phys. Rev. B **7,** 5379.

Rowe, J.M., J.J. Rush, N.J. Chesser, K.H. Michel and J. Naudts, 1978, Phys. Rev. Lett. **40,** 455.

Sahni, V.C. and P.W.M. Jacobs, 1982, Phil. Mag. **46 A,** 817.

Sahu, D. and S.D. Mahanti, 1983, Solid State Commun. **47,** 207.

Salter, L., 1956, Proc. Roy. Soc. A **233,** 418.

Samara, G.A., L.C. Walters and D.A. Northrop, 1967, J. Phys. Chem. Solids **28,** 1875.

Schapery, R.A., 1968, J. Composite Materials **2,** 380.

Schober, H.R. and P. H. Dederichs, 1981, in: Landolt-Börnstein New Series, Vol. III/13a, K.-H. Hellwege and J.L. Olsen, eds. (Springer, Berlin).

Schober, H. and A.M. Stoneham, 1982, Phys. Rev. B **26,** 1819.

Schreiber, E., O.L. Anderson and N. Soga, 1973, Elastic Constants and their Measurement (McGraw-Hill, New York).

Schulgasser, K., 1976a, J. Math. Phys. **17,** 378.

Schulgasser, K., 1976b, J. Appl. Phys. **47,** 1880.

Schulgasser, K., 1977, J. Phys. C **10,** 407.
Schwartz, J.W. and C.T. Walker, 1967a, Phys. Rev. Lett. **16,** 97.
Schwartz, J.W. and C.T. Walker, 1967b, Phys. Rev. **155,** 969.
Seeger, A., 1973, Crystal Lattice Defects **4,** 221.
Sekimoto, H., 1978, Nucl. Sci. & Eng. **68,** 351.
Sellmyer, D.J., 1978, in: Solid State Physics, Vol. 33, H. Ehrenreich, F. Seitz and D. Turnbull, eds. (Academic Press, New York) p. 83.
Sellmyer, D.J., 1981, in: Landolt-Börnstein New Series, Vol. III/13a, K.-H. Hellwege and J.L. Olsen, eds. (Springer, Berlin).
Sevillano, E., H. Meuth and J.J. Rehr, 1979, Phys. Rev. B **20,** 4908.
Shapiro, J.N., 1970, Phys. Rev. B **1,** 3982.
Shimizu, M., 1974, Phys. Lett. **50A,** 93.
Shimizu, M., 1981, Rep. Progr. Phys. **44,** 329.
Shukla, M.M. and N.T. Padial, 1973, Rev. Bras. Fis. **3,** 39.
Shukla, R.C. and E.R. Cowley, 1971, Phys. Rev. B **3,** 4055.
Shukla, R.C. and R. Taylor, 1976, J. Phys. F **6,** 531.
Shukla, R.C. and L. Wilk, 1974, Phys. Rev. B **10,** 3660.
Simmons, G., 1967, Graduate Res. Center Southern Methodist Univ. **36,** 1.
Simmons, G. and H. Wang, 1971, Single Crystal Elastic Constants and Calculated Aggregate Properties: A Handbook (MIT Press, Cambridge, Mass.).
Simmons, R.O. and R.W. Balluffi, 1960, Phys. Rev. **117,** 52.
Slack, G.A., 1957, Phys. Rev. **105,** 832.
Slack, G.A. and D.W. Oliver, 1971, Phys. Rev. B **4,** 592.
Slack, G.A., 1979, in: Solid State Physics, Vol. 34, H. Ehrenreich, F. Seitz and D. Turnbull, eds. (Academic Press, New York) p. 1.
Slater, J.C., 1940, Phys. Rev. **57,** 744.
Smith, T.F. and T.R. Finlayson, 1976, J. Phys. F **6,** 709.
So, C.B. and C.H. Woo, 1981, J. Phys. F **11,** 325.
Söderberg, M. and G. Grimvall, 1983, J. Phys. C **16,** 1085.
Söderberg, M. and G. Grimvall, 1986, J. Appl. Phys. **59,** 186.
Söderkvist, J. and G. Grimvall, 1985, Phys. Scripta, **32,** 323.
Soffer, S.B., 1967, J. Appl. Phys. **38,** 1710.
Soma, T., Y. Kitani and H.-M. Kagaya, 1984, Solid State Commun. **50,** 1007.
Sondheimer, E.H., 1950, Proc. Roy. Soc. A **203,** 75.
Sondheimer, E.H., 1952, Adv. Phys. **1,** 1.
Speich, G.R., A.J. Schwoeble and B.M. Kapadia, 1980, ASME J. Appl. Mech. **47,** 821.
Steinemann, S.G., 1978, J. Mag. Mag. Mats. **7,** 84.
Steinemann, S.G., 1979, J. Mag. Mag. Mats. **12,** 191.
Stesmans, A., 1982, Solid State Commun. **44,** 727.
Stillinger, F.H. and T.A. Weber, 1980, Phys. Rev. B **22,** 3790.
Stoner, E.C., 1938, Proc. Roy. Soc. A **165,** 372.
Stoner, E.C., 1945, Phil. Mag. **36,** 803.
Stratton, J.A., 1941, Electromagnetic Theory (McGraw-Hill, New York).
Stratton, R., 1953, Phil. Mag. **44,** 519.
Stratton, R., 1962, J. Chem. Phys. **37,** 2972.
Stripp, K.F. and J.G. Kirkwood, 1954, J. Chem. Phys. **22,** 1579.
Strössner, K., W. Henkel, H.D. Hochheimer and M. Cardona, 1983, Solid State Commun. **47,** 567.

Stroud, D. and N.W. Ashcroft, 1972, Phys. Rev. B **5,** 371.

Suck, J.-B., H. Rudin, H.-J. Güntherodt, H. Beck, J. Daubert and W. Gläser, 1980, J. Phys. C **13,** L167.

Suck, J.-B., H. Rudin, H.-J. Güntherodt and H. Beck, 1981, J. Phys. C **14,** 2305.

Suen, W.M., S.P. Wong and K. Young, 1979, J. Phys. D **12,** 1325.

Sundqvist, B., J. Neve and Ö. Rapp, 1985, Phys. Rev. B **32,** 2200.

Swalin, R.A., 1962, Thermodynamics of Solids (Wiley, New York).

Taillefer, L., G.G. Lonzarich and P. Strange, 1985, to be publ.

Takeno, S., 1963, Progr. Theor. Phys. **30,** 144.

Taylor, B., H.J. Maris and C. Elbaum, 1971, Phys. Rev. B **3,** 1462.

Thiessen, M., 1986, Int. J. Thermophys., Vol. **7,** to be published.

Thirring, H., 1913, Phys. Zeitschrift **14,** 867.

Thomsen, L., 1972, J. Geophys. Res. **77,** 315.

Thomsen, L. and O.L. Anderson, 1969, J. Geophys. Res. **77,** 981.

Thurston, R.N., 1965, Proc. IEEE **53,** 1320.

Thurston, R.N. and K. Brugger, 1964, Phys. Rev. **133,** A1604. (Erratum Phys. Rev. **135,** AB3).

Tiwari, M.D. and B.K. Agrawal, 1973a, J. Phys. F **3,** 2051.

Tiwari, M.D. and B.K. Agrawal, 1973b, Phys. Rev. B **7,** 4665.

Tiwari, M.D. and B.K. Agrawal, 1973c, Phys. Rev. B **8,** 1397.

Tiwari, M.D., G. Thummes and H.H. Mende, 1981, Phil. Mag. B **44,** 63.

Tomlinson, P.G., 1979, Phys. Rev. B **19,** 1893.

Tosi, M.P. and F.G. Fumi, 1963, Phys. Rev. **131,** 1458.

Toth, L.E., 1971, Transition Metal Carbides and Nitrides (Academic Press, New York) p. 143.

Touloukian, Y.S. and E.H. Buyco, 1970, Specific Heat. Thermophysical Properties of Matter, Vols. 4 and 5 (Plenum, New York).

Touloukian, Y.S., R.W. Powell, C.Y. Ho and P.G. Klemens, 1970, Thermal Conductivity. Thermophysical Properties of Matter, Vols. 1 and 2 (Plenum, New York).

Touloukian, Y.S., R.W. Powell, C.Y. Ho and M.C. Nicholaou, 1973, Thermal Diffusivity. Thermophysical Properties of Matter, Vol. 10 (Plenum, New York).

Touloukian, Y.S., R.K. Kirby, R.E. Taylor and P.D. Desai, 1975, Thermal Expansion. Thermophysical Properties of Matter, Vol. 12 (Plenum, New York).

Touloukian, Y.S., R.K. Kirby, R.E. Taylor and T.Y.R. Lee, 1977, Thermal Expansion. Thermophysical Properties of Matter, Vol. 13 (Plenum, New York).

Truesdell, C. and R.A. Toupin, 1960, Handbuch der Physik, Vol. 3/1, S. Flügge, ed. (Springer, Berlin) p. 226.

Turley, J. and G. Sines, 1971, J. Phys. D **4,** 264.

Vaks, V.G., S.P. Kravchuk and A.V. Trefilov, 1980, J. Phys. F **10,** 2105.

van Attekum, P.M.Th.M., P.H. Woerlee, G.C. Verkade and A.A.M. Hoeben, 1984, Phys. Rev. B **29,** 645.

van Beek, L.K.H., 1967, Progr. Dielectrics, Vol. 7, J.B. Birks, ed. p. 69.

van den Beukel, A. and S. Radelaar, 1983, Acta Met. **31,** 419.

Varley, J.H.O., 1956, Proc. Roy. Soc. **237,** 413.

Vegard, L., 1921, Z. Physik **5,** 17.

Vieira, S. and M. Hortal, 1971, J. Phys. C **4,** 1703.

Vineyard, G.H. and G.J. Dienes, 1954, Phys. Rev. **93,** 265.

Voigt, W., 1910, Lehrbuch der Kristallphysik (Leipzig, Teubner). (Reprinted 1928).
von der Lage, F.C. and H.A. Bethe, 1947, Phys. Rev. **71,** 612.
Wachtman, J.B., Jr., W.E. Tefft, D.G. Lam, Jr. and C.S. Apstein, 1961, Phys. Rev. **122,** 1754.
Walker, E. and M. Peter, 1977, J. Appl. Phys. **48,** 2820.
Wallace, D.C., 1970, in: Solid State Physics, Vol. 25, H. Ehrenreich, F. Seitz and D. Turnbull, eds. (Academic Press, New York) p. 301.
Wallace, D.C., 1972, Thermodynamics of Crystals (Wiley, New York).
Waller, I., 1923, Z. Physik **17,** 398.
Waller, I., 1925, Diss. Uppsala (Sweden).
Wallis, R.F., 1975, in: Dynamical Properties of Solids, Vol. 2, G.K. Horton and A.A. Maradudin, eds. (North-Holland, Amsterdam) p. 441.
Walpole, L.J., 1966, J. Mech. Phys. Solids **14,** 151.
Walpole, L.J., 1969, J. Mech. Phys. Solids **17,** 235.
Walpole, L.J., 1981, Adv. Appl. Mech. **21,** 169.
Warren, B.E., 1969, X-ray Diffraction (Addison-Wesley, Reading, Mass.) p. 238.
Watt, J.P., 1979, J. Appl. Phys. **50,** 6290.
Watt, J.P., 1980, J. Appl. Phys. **51,** 1520.
Watt, J.P. and L. Peselnick, 1980, J. Appl. Phys. **51,** 1525.
Watt, J.P., G.F. Davies and R.J. O'Connell, 1976, Rev. Geoph. Space Phys. **14,** 541.
Watts, R.K., 1977, Point Defects in Crystals (Wiley, New York).
Weaire, D., M.F. Ashby, J. Logan and M.J. Weins, 1971, Acta Met. **19,** 779.
Weiner, D., A. van den Beukel and P. Penning, 1975, Acta Met. **23,** 783.
Werthamer, N.R., 1969, Phys. Rev. B **1,** 572.
White, G.K., 1969, in: Thermal Conductivity, R.P. Tye, ed. (Academic Press, London) p. 69.
White, G.K., J.G. Collins, J.A. Birch, T.F. Smith and T.R. Finlayson, 1978, Inst. Phys. Conf. Ser. 39: Transition Metals, M.J.G. Lee, J.M. Perz and E. Fawcett, eds., p. 420.
White, R.M., 1970, Quantum Theory of Magnetism (McGraw-Hill, New York).
Wiedemann, G. and R. Franz, 1853, Ann. Physik **89,** 497.
Wiener, O., 1912, Abh. Math.-Physik. Kl. Königl. Sächs. Ges. Wiss. **32,** 509.
Wiesmann, H., M. Gurvitch, M. Lutz, A. Gosh, B. Schwarz, M. Strongin, P.B. Allen and J.W. Halley, 1977, Phys. Rev. Lett. **38,** 782.
Willis, B.T.M. and A.W. Pryor, 1975, Thermal Vibrations in Crystallography (Cambridge Univ. Press, Cambridge).
Willis, J.R., 1981, Adv. Appl. Mech. **21,** 1.
Wilson, A.H., 1954, The Theory of Metals (Cambridge Univ. Press, Cambridge).
Wohlfarth, E.P., 1974, phys. stat. sol. (a) **25,** 285.
Wolfe, J.P., 1980, Physics Today **33** (December) 44.
Wollenberger, H.J., 1982, in: Physical Metallurgy, R.W. Cahn and P. Haasen, eds. (North-Holland, Amsterdam) p. 1139.
Wu, T.T., 1966, Int. J. Solids Struct. **3,** 1.
Yamada, Y., N. Hamaya, J.D. Axe and S.M. Shapiro, 1984, Phys. Rev. Lett. **53,** 1665.
Yamashita, J. and S. Asano, 1974, Progr. Theor. Phys. **51,** 317.
Yamashita, J., S. Wakoh and S. Asano, 1975, J. Phys. Soc. Jpn **39,** 344.
Yates, B., 1972, Thermal Expansion (Plenum Press, London).
Yates, B., M.J. Overy and O. Pirgon, 1975, Phil. Mag. **32,** 847.
Yin, M.T. and M.L. Cohen, 1982, Phys. Rev. B **26,** 5668.

Yin, M.T. and M.L. Cohen, 1984, Phys. Rev. B **29,** 6996.
Young, D.A. and B.J. Alder, 1974, J. Chem. Phys. **60,** 1254.
Yussouff, M. and J. Mahanty, 1966, Proc. Phys. Soc. **87,** 689.
Yussouff, M. and J. Mahanty, 1967, Proc. Phys. Soc. **90,** 519.
Zeller, R. and P.H. Dederichs, 1973, phys. stat. sol. (b) **55,** 831.
Zeller, R.C. and R.O. Pohl, 1971, Phys. Rev. B **4,** 2029.
Zen, E-an., 1956, Amer. Min. **41,** 523.
Zener, C., 1948, Elasticity and Anelasticity of Metals (Univ. Chicago Press, Chicago).
Zener, C. and S. Bilinski, 1936, Phys. Rev. **50,** 101.
Ziman, J.M., 1960, Electrons and Phonons (Clarendon, Oxford).
Ziman, J.M., 1970, Models of Disorder (Cambridge Univ. Press, Cambridge).
Ziman, J.M., 1979, Models of Disorder (Cambridge Univ. Press, Cambridge).

AUTHOR INDEX *

* This list also includes the authors referred to by 'et al.' in the text.

SUBJECT INDEX

MATERIALS INDEX *

* Single data points in graphs are not included in this list.